METEOROLOGICAL MONOGRAPHS

VOLUME 31 | DECEMBER 2003 | NUMBER 53

A HALF CENTURY OF PROGRESS IN METEOROLOGY: A TRIBUTE TO RICHARD REED

Edited by

Richard H. Johnson
Robert A. Houze Jr.

American Meteorological Society
45 Beacon Street, Boston, Massachusetts 02108

ISBN 1-878220-58-6
ISSN 0065-9401

Published by the American Meteorological Society
45 Beacon St., Boston, MA 02108

Printed in the United States of America
by Sheridan Books, Inc.

TABLE OF CONTENTS

PREFACE

World War II saw a rapid growth in meteorology, marked by substantial increases in the workforce, expansion of meteorology in academia, and the emergence of new subdisciplines. As subfields developed, leaders arose, but only a few made profound contributions across many of those fields. Richard J. Reed is one of those few. On 15 January 2002 a symposium was held at the American Meteorological Society's (AMS) 82d Annual Meeting in Orlando, Florida, to pay tribute to the extraordinary scientific career of Richard J. Reed. The major theme of the symposium was the broad scope and significant impact of a half century of Reed's research, spanning virtually the entire field of meteorology (from the boundary layer through the troposphere into the stratosphere and mesosphere, from the Tropics to the Arctic, from the scale of turbulence to the mesoscale to the synoptic scale, from fog to deep convection, and from tropical easterly waves to polar lows to explosive cyclogenesis). Reed's unique insight in the interpretation of observations has led to fundamental discoveries that have shaped the course of the science in many of these areas.

To paint the background for the papers in this volume, Richard Reed himself starts the monograph off with a brief autobiography. His words provide a window into his life as a scientist and help us understand how he has contributed so significantly to a half century of progress in meteorology. In looking back over Reed's achievements of the past 50 years, one has to wonder how he was able to repeatedly strike out onto new frontiers and immediately contribute authoritative works in each new area. The following excerpt from his autobiography sheds light on his unique love of the science and his contributions that grew out of that devotion.

> From the start I loved meteorology with a deep-seated passion and experienced aesthetic feelings toward meteorological phenomena that motivated and sustained my future endeavors. These emotional attachments lasted throughout my career. Without them it is unthinkable that I could have achieved what I did. My choice of specific research topics was determined or inspired by a number of circumstances, ranging from healthy curiosity, to obligations imposed by others, to opportunities proffered by outside parties, to the occasional desire to strike out in new directions, to contact with directly experienced weather events, and, in my most original work, to a series of improbable events. In a word, there was no common thread except for an insatiable desire to observe and understand atmospheric phenomena.

At the Reed Symposium in Orlando, 12 invited talks were presented in four sessions covering many of Reed's key contributions to the atmospheric sciences: fronts, frontogenesis, and the tropopause (Lance Bosart, Brian Hoskins, Melvyn Shapiro); the stratosphere (James Holton, Taroh Matsuno, Richard Lindzen); tropical studies, numerical prediction, and Reed's leadership role in the sciences (Robert Burpee, Anthony Hollingsworth, John Perry); and polar lows, extratropical cyclones, and mesoscale flows (Erik Rasmussen, Ying-Hwa Kuo, and Clifford Mass). Eight of these talks have been expanded into the papers that now appear in this monograph. In addition to putting Reed's contributions into a broad context, these papers contain valuable reviews of the subject areas as well as a number of new research findings.

Several common themes regarding the nature of Reed's research emerge from the symposium and the papers appearing in this monograph. One is the unusual care and thoroughness with which Reed treats and analyzes observations. His attention to detail and ability to synthesize diverse observations placed in a theoretical context has enabled him to lucidly expose underlying mechanisms of a wide range of physical processes. The care and patience with which he conducts research was exemplified in the symposium talk by Holton, who pointed out that in Reed's discovery of the quasi-biennial oscillation (QBO), he delayed publication of this groundbreaking result until late arrival of data from Nairobi, Kenya, in order to confirm the global extent of the oscillation.

Also evident, even from a casual reading of Reed's papers, is an unusual clarity in exposition, which has no doubt contributed to the significant and lasting impact of his research. At the symposium, Lindzen noted that as a theoretician he was drawn to Reed's observational studies because of the way in which he was able to distill key results and present them in a clear, concise manner. A number of Reed's pioneering studies has opened up new approaches and avenues for research. As explained by Bosart, Reed's early work on upper-level fronts, which introduced potential vorticity to examine intrusions of stratospheric air into the troposphere, has been extended and expanded in subsequent decades, even to this day where it remains the principal tool of upper-level front/tropopause analysis. While sophisticated models have been developed in recent years to describe the process of tropopause folding in greater detail than before, Shapiro emphasized that the basic findings still confrm the earliest studies of this phenomenon by Reed in the 1950s.

With seemingly boundless energy and contagious enthusiasm, Reed has tackled many problems across the broad spectrum of the atmospheric sciences, and has inspired several generations of students and colleagues. It is fitting that we pay tribute to his many achievements with a symposium and special monograph on his research achievements. Many people beyond the authors and symposium lecturers have contributed substantially to the success of both of these endeavors and they should be acknowledged: Gail Cordova and Rick Taft (Colorado State University); Mike Wallace, Candace Gudmundsen, and Kathryn Stout (University of Washington); Fred Sanders (Marblehead, Massachusetts); Rick Anthes (UCAR); and the AMS staff (Joyce Annese, Claudia Gorski, Bryan Hanssen, Ken Heideman, Ron McPherson, Gretchen Needham, Jennifer Rosen, Yale Schiffman, and Keith Seitter).

Richard H. Johnson and Robert A. Houze Jr.
Editors

CONTRIBUTORS

Lance Bosart
Department of Earth and Atmospheric Sciences
University at Albany
State University of New York
ES 227, 1400 Washington Ave.
Albany, NY 12222
E-mail: bosart@atmos.albany.edu

Robert W. Burpee
Cooperative Institute of Marine and Atmospheric Studies
4600 Rickenbacker Causeway
Miami, FL 33149

Anthony Hollingsworth
European Centre for Medium-Range
Weather Forecasts
Shinfield Park, Reading RG2 9AX
United Kingdom
E-mail: dia@ecmwf.int

James Holton
Department of Atmospheric Science
University of Washington
Box 351640
Seattle, WA 98195
E-mail: holton@atmos.washington.edu

Brian Hoskins
Department of Meteorology
Earley Gate
P.O. Box 243, Reading RG6 6BB
United Kingdom
E-mail: b.j.hoskins@reading.ac.uk

Richard Lindzen
Program in Atmospheres, Oceans and Climate, 54-1720
Massachusetts Institute of Technology
Cambridge, MA 02139
E-mail: rlindzen@mit.edu

John Perry
6205 Talley Ho Lane
Alexandria, VA 22307
E-mail: johnperry@cox.net

Erik Rasmussen
Hoejbakkestraede 2
DK-2620 Albertslund
Denmark
E-mail: erik.rasmussen@mobilixnet.dk

Adrian Simmons
European Centre for Medium-Range Weather Forecasts
Shinfield Park, Reading RG2 9AX
United Kingdom
E-mail: Adrian.Simmons@ecmwf.int

Pedro Viterbo
European Centre for Medium-Range Weather Forecasts
Shinfield Park, Reading RG2 9AX
United Kingdom
E-mail: Pedro.Viterbo@ecmwf.int

OTHER PRESENTERS

Y.-H. Kuo and R. Reed

Ying-Hwa Kuo and Melvyn A. Shapiro
National Center for Atmospheric Research, Boulder, Colorado
"Modeling and Prediction of Explosive
Marine Cyclogenesis"

Clifford F. Mass
University of Washington, Seattle, Washington
"Mesoscale Circulations"

Taroh Matsuno
Frontier Research System for Global Change, Tokyo, Japan
"Personal Retrospective of the Development of Thoughts
on the Mechanisms of Sudden Stratospheric Warmings"

Melvyn A. Shapiro
NOAA/Environmental Technology Laboratory, Boulder, Colorado
"The Development of Concepts of the Tropopause and the Exchange of Trace Constituents between the Stratosphere and Troposphere"

Chapter 1

A Short Account of My Education, Career Choice, and Research Motivation

RICHARD J. REED

I graduated from Braintree (Massachusetts) High School in June 1940 in the period between the Great Depression — the effects of which were still being felt — and U.S. entry in World War II. Though I was a good student with well-rounded abilities and no strong leaning toward any particular subject or career choice, it was assumed that I would join two mathematically and scientifically talented classmates in attending the Massachusetts Institute of Technology (MIT). Certainly math and science had been two of my favorite subjects, and I would have welcomed the opportunity to go to a school like MIT or another elite school such as Harvard. However, the tuition at these institutions was out of reach for my family, even when supplemented by a scholarship awarded at my graduation. Consequently it was decided that I would enroll in the fall at a less expensive school, Boston College, where I would satisfy my love for numbers by majoring in accounting. Actually it was the depression mentality, not the fondness for numbers, which shaped my choice of major. Decent jobs were scarce in those days, and I was reminded by family and friends that certified public accountants (CPAs) were well paid and reliably employed.

My scholarship having run out at the end of the freshman year and having no other source of funds for the next year's tuition, I took a job in a factory with the purpose of earning enough money to return to college the following year. Then fate intervened. On 7 December 1941 the Japanese attacked Pearl Harbor, and war was declared. Imbued with the near-universal love of country that marked the public mood in those genuinely patriotic times, I enlisted in the U.S. Navy at the bottom rank — apprentice seaman — and in January 1942 reported to Newport, Rhode Island, for boot training. At the finish of the training I was invited, as were all recruits, to select from a long list of specialties (e.g., boatswain, yeoman, etc.) the three that I found most attractive. Among the three I selected was aerology (the naval term for meteorology), probably motivated by a long, but ill-defined, interest in weather. I had never read on the subject or been exposed to it in high school or college, and I had never maintained a weather station as was commonly done by young weather enthusiasts, though I did keep a daily record of the local wind direction taken from the wind vane atop a nearby church. In any event the authorities decided that aerology was to be my specialty and assigned me to an on-the-job training program at the nearby naval air station at Quonset Point, Rhode Island, rather than to the customary training school at Lakehurst, New Jersey, which was already full to capacity.

On the practical side I was instructed in map plotting and weather observation by enlisted men stationed there, the foremost of whom was Albert Lewis, a colorful character from Arkansas with limited academic background but high native intelligence. Lewis rose to the rank of chief petty officer while I was still billeted at Quonset Point and subsequently became a commissioned officer, attaining the rank of lieutenant commander by war's end. Such were the unusual opportunities that existed for advancement during the war years when merit rather than formal credentials and time served in rank was the basis for promotion. Lewis was an excellent weather analyst and forecaster who either by direct instruction or by example helped me acquire skills in weather analysis and forecasting. Although analysis and forecasting were mainly the duties of the officers, enlisted men shared these duties increasingly as they rose in rank. Lewis's map analyses were a thing of beauty that had a profound effect on my aesthetic appreciation of meteorology. The emotional charge I have always felt in analyzing frontal cyclones stemmed from his early example.

Freshly minted young ensigns Max Edelstein and Alvin Morris, the latter to become a longtime friend after the war, were assigned the job of teaching the trainees the elements of meteorology. To aid their instruction they suggested that we read a popular — and deservedly so — elementary textbook by Blair (1937). I have never forgotten this experience. Once started on the book I could not put it down, staying up that night until I had finished reading it and feeling at the conclusion that I had thoroughly absorbed the material despite my relatively weak scientific background. If there ever was a case of love at first sight for a scientific subject, I experienced it that day (and night). There are those who view unusual ability in math and physics as the key to scientific success and its manifestation in a particular subject as largely a matter of accident. I have never subscribed to this view. The aesthetic feelings aroused in me by weather patterns and the fascination I felt for weather phenomena as physically evolving entities have always seemed to me inborn facets of my being. I cannot picture any other field of study having had the same emotional effect.

Given the chance to progress at a pace determined by merit alone, I advanced three grades in a little over a year to the rank of aerographer mate first class. At that stage, undoubtedly at Al Lewis's instigation, I was nominated to take the examination for a fleet appointment to the U.S. Naval Academy. After a brief period of preparation at a navy facility in Norfolk, Virginia, I took the exam and passed with flying colors whereupon I went Annapolis with the expectation of being admitted. But then dame fortune smiled on me as never before, as anyone who has trouble visualizing me as a career naval officer can appreciate. Despite having better than 20–20 vision, I was found to have astigmatism, which at that time, but assuredly not now, was grounds for rejection. As a consolation prize, I was sent under the navy's V-12 program to Dartmouth College for one year (three semesters) to acquire further credits in standard academic subjects (math, physics, English, history, etc.) at the conclusion of which I was slated to go to midshipman school and be commissioned an officer in the U.S. Navy Reserve. But again lady luck intervened. Shortly before the year ended an announcement was posted inviting trainees to become candidates for a lone position that was available for an additional year (three semesters) of study in the meteorology program at the California Institute of Technology (Cal Tech). I jumped at the chance, and in view of my background was not surprised when I was selected. Thus my formal academic training in meteorology began.

From my earliest days as an enlisted man, I had heard of the notorious Irving P. Krick who headed the meteorological unit at Cal Tech, a special division of the Aeronautical Engineering Department. Now I was to be trained in his preserve—a somewhat disconcerting thought! Fortunately Krick was on duty as a colonel in the air force so that I was not directly exposed to his influence, and his cohorts in the department taught, with acceptable competence, the less demanding analysis and forecasting courses, subjects in which I was already well versed. The more demanding meteorology courses, physical and dynamical, were taught by regular members of the Cal Tech faculty—the eminent geophysicist Beno Gutenberg and the brilliant aeronautical engineer Homer Joe Stewart—as were supporting math courses. So despite the Krick influence I received a sound undergraduate education in meteorology.

Having completed eight semesters at the college level, two at Boston College and three each at Dartmouth and Cal Tech, I was awarded a bachelor of science degree in meteorology from Cal Tech in June 1945. From there I was sent to midshipman school at the University of Notre Dame in South Bend, Indiana, and commissioned an ensign in November of that year. Meanwhile the war with Japan had ended (that in Europe had concluded while I was still at Cal Tech) so that after commissioning, rather than being sent to the Pacific theater as originally planned, I was assigned to air station duty in Rhode Island, within hailing distance of my home near Boston, Massachusetts. After being released from active duty in June 1946, I enrolled at MIT, entering the Meteorology Department in September 1946 and graduating with an doctor of science (Sc.D.) degree in June 1949. The relatively short period for attaining the Sc.D. was the norm for the department in those days, when demonstration of the ability to perform original research, rather than the accomplishment of noteworthy research, was the goal.

The department at MIT had been founded by Carl-Gustav Rossby, the foremost meteorologist of his day, and subsequently was headed by Sverre Petterssen, the leading synoptic meteorologist of the era. By the time I arrived, Henry Houghton, a man valued for his wisdom and leadership ability as well as his scientific acumen, was in charge. Houghton taught physical meteorology, Bernand Haurwitz and Victor Starr dynamic meteorology, James Austin synoptic meteorology, and Hurd Willett long-range forecasting and climate change. Their courses were always well prepared and were uniformly excellent in content. I was indeed fortunate to have the benefit of instruction by such knowledgeable and dedicated teachers. Though synoptic meteorology was always my primary interest, I wrote my Sc.D. thesis under Willett on the seemingly exotic subject "The Effect of Atmospheric Motions on Ozone Distribution and Variations," motivated largely by what I understood as an unwritten rule that mainline synoptic meteorology was not a suitable area for an MIT doctoral dissertation. The rule, if ever such existed, was broken a few years later by my early collaborator Fred Sanders. On the other hand, I may have been drawn to Willett by his interest in solar–weather relationships and the notion, posed at one time by Haurwitz (then no longer at MIT), that the source of such a relationship, if any, might reside in the ozone layer. Certainly neither at that time nor at any later time was I sympathetic to the idea of a direct solar effect on weather. I merely wished to understand the already well-documented ozone–weather relationship and for this purpose sought out as thesis advisor the faculty member most closely associated with the problem.

After completing the Sc.D., I was offered a position on a project, sponsored by the Office Naval Research (ONR), known as the Pressure Change Project, and held the position for the next five years. The project was led primarily by Austin and had as its goal studying the mechanism of atmospheric pressure change in cyclones and anticyclones. Austin subscribed to a thermal mechanism that had little appeal to me; I preferred instead a more dynamic approach based on, admittedly, ill-defined ideas related to vorticity. This divergence of approach was not a source of conflict, since earlier Austin had made extensive use of vorticity in his Sc.D. thesis, written under Pettersen. The truth is that Austin, to his credit, made little effort to influence my thinking or control my activities, allowing me to follow my interests in whatever direction they carried me. Besides preparing project reports of minor importance, I published a couple of articles in the *Journal of Meteorology* on atmospheric ozone (the initial one representing my first publication in a refereed journal) and a jointly authored paper on atmospheric cooling by melting snow, which was inspired by a late season, surprise snowstorm in Boston. During this period I read critically the works of Bjerknes and Palmén, became familiar with, and intrigued by, the seemingly odd ideas of E. Kleinschmidt Jr., and was drawn to Charney's work on baroclinic instability, which seemed to me to provide a more realistic explanation of cyclone development than the prevailing Norwegian model. I also taught my first class, basically a synoptic laboratory.

By far the most important papers that I published during the postdoctoral period were a pair on upper-tropospheric frontogenesis and the related topic of tropopause folding or stratospheric intrusions. The first of these was coauthored with Fred Sanders, then working on his dissertation. Fred and I were kindred spirits who fed each other's discontent with what we regarded as the nearly blind acceptance by many meteorologists of the model of cyclone development advocated by the Norwegian School. It seemed obvious to us — and we were not alone in this view — that fronts often strengthened during cyclogenesis rather than providing sharp preexisting thermal discontinuities on which cyclones formed. Also we objected to the view, suggested by cross-section analyses of the Norwegian School, that fronts extended with more or less equal strength throughout the troposphere rather than, at least for truly strong and unambiguous fronts, being mainly confined to restricted regions of the atmosphere. These regions were the lower troposphere, particularly the near-surface layers, and in some instances the layer near the tropopause. Fred deserves the main credit for proposing this dichotomy, which we felt not only better described nature but equally importantly encouraged more open thinking on atmospheric structure. Following the completion of our joint paper Fred decided to concentrate on low-level fronts, as he did in his thesis (and in the subsequently published version of the thesis), and suggested that I continue treating upper-tropospheric fronts, as I did in my paper on tropopause folding. The papers on upper-level fronts made extensive use of potential vorticity in distinguishing between tropospheric and stratospheric air and in tracing air trajectories. I would be remiss if I failed to acknowledge the part played by Victor Starr in stimulating my interest in potential vorticity and in calling my attention to its role as a tracer. Victor also encouraged skepticism regarding accepted ideas, a necessary attitude for the young scientist intent upon breaking fresh ground.

During the five-year postdoctoral stint I stayed alert for a teaching opportunity at a reputable university. Some offers were received that seemed worthy of consideration, but in each case Houghton advised me against accepting them. Thus when an offer came for a position at the University of Washington, at that time a little-known player in the meteorological firmament, he surprised me by urging that I give the offer serious consideration. Trusting his judgment, I looked further into the proposition and decided to make the move to Seattle. I owe Houghton much credit for his sage advise on important issues and am deeply indebted to Austin for both his direct and behind-the-scenes efforts to advance my career. Like my father and Al Lewis before him I count him as one of my main boosters — those who served not so much as mentors but as confidence builders and promoters of my interests. At Washington, I was to meet another great booster, Phil Church, the chairman of the Department of Meteorology and Climatology.

In accepting the position at the University of Washington, I had agreed to work on an arctic project being negotiated by Phil with the Air Force Cambridge Research Center (AFCRC) of the Geophysical Research Directorate (GRD). In the next several years a number of reports, M.S. theses, and published papers on arctic weather analysis and forecasting resulted from this effort, only one of which represented a substantial contribution. An "Arctic Forecast Guide" for naval forecasters, prepared while I was a consultant to the Navy Weather Research Facility at Norfolk, Virginia, was a by-product of this work. I pursued a number of other interests besides arctic meteorology during those early years at Washington. With Ed Danielsen, like Fred Sanders a soul mate who shared my reservations about the Norwegian model, I published a further paper on upper-tropospheric fronts. Nominally Ed was my first doctoral student, but he was so independent in nature that I always felt that he viewed his dissertation — much of it conceived before my arrival at Washington and later carried out separately from our joint research — as falling in a special unsupervised category.

I also published a couple of papers on graphical prediction, which at that time was competitive with numerical prediction as a method of preparing prognostic surface charts. The latter soon surpassed the graphical method, not to my surprise since I always regarded graphical prediction more as a teaching tool for synoptic labs than as a long-term approach to forecasting. Nonetheless, during my forthcoming sabbatical at the National Meteorological Center (NMC) in Suitland, Maryland, the equations and procedures involved in the graphical method were programmed and the output of the numerical version used for several years as background for manually prepared surface prognoses. Eventually, in 1966, the more sophisticated mainline numerical surface prognoses reached the stage where they could be used directly and the graphically based prognoses slowly faded into oblivion.

One contribution published during those early years was written purely for fun. This was a paper entitled "Flying Saucers over Mt. Rainier," inspired by a picture in a Seattle newspaper of a saucer-shaped cloud formation downwind of that majestic mountain. Informed by my colleague Bob Fleagle that the original sighting of flying saucers was made by the pilot of a light plane flying in the vicinity of Mount Rainier, I did an analysis of the later situation and proposed on the basis of the investigation that the flyer in the original sighting likely mistook lenticular clouds for vehicles from outer space. Needless to say, this proposal had no impact on the true believers. In fact it outraged some. I have always been proud of this offbeat paper and disappointed that it has not received greater notice in the literature on the subject. I freely admit to having a passionate dislike of UFOs, sasquatches, Loch Ness monsters, and other products of deluded minds.

My second major paper of this early period — the paper in which I first described the phenomenon that came to be known as the quasi-biennial oscillation (QBO) — had its origin in an improbable series of events, which I will now relate. Some will say that I was only the codiscoverer of this amazing phenomenon, and I will not contest them. As with all discoveries bits and pieces of the final picture emerged over a period of time, and two British meteorologists, Veryard and Ebdon, published a paper similar to mine in the same month. But when I made the discovery, I felt — and justifiably so — that I had found and described in its essential form a bizarre phenomenon not recognized as such in previous work. Indeed much of the preceding literature was unknown to me at the time. For reasons that are now obscure, I was invited to present a review paper on the stratospheric circulation at the annual meeting of the American Meteorological Society (AMS) held in Boston in January 1960. Perhaps the invitation stemmed in part from my early work on atmospheric ozone and in part from my participation in summer seminars held by McGill University at Stanstead College in Quebec, Canada. In the beginning, the seminars were devoted to arctic meteorology, which was then my main field of research. But subsequently they took on an increasingly stratospheric complexion that, according to my recollection, stemmed from an interest on the part of some of the participants, including myself, in the stratospheric sudden warming phenomenon, a rare event that was seen most strikingly in the arctic stratosphere. Whatever the reason for my invitation to speak at Boston, I felt a need to be brought up to date on the latest work being done on stratospheric meteorology and decided, accordingly, to attend an AMS-sponsored meeting being held on that subject at Minneapolis, Minnesota, in September 1959. Conveniently, as it turned out, the organizer of the meeting, Arthur Belmont, requested all speakers to bring multiple preprints of their contributions for distribution to the participants — a procedure that I encountered there for the first time — and I made a point of collecting a copy of each. My own participation was confined to chairing one of the sessions.

The following November my colleague, Joost Businger, who had accompanied me on the trip to Minneapolis, visited my office one morning and asked what I was doing. I replied that I was working on the Boston paper. "What do you plan to say about the strange winds at Christmas Island reported by Colonel Frank McCreary?" he inquired. I responded that I was so busy watching the time clock, as chair of the session at which McCreary spoke, that I had not paid much attention to what he said; but, no matter, I had a copy of the preprint of his paper on the shelf nearby. After Businger left, I perused the paper and was intrigued by figures showing the alternation between easterly and westerly winds at different levels in the equatorial stratosphere during the two years of nuclear testing at Christmas Island, and in the seeming downward progression of the changes. This behavior was contrary to the accepted picture at that time of persistent Krakatoa easterlies between 25 and 30 km, as first seen from the drift of the dust from the volcanic eruption of Krakatoa in 1883, and essentially steady Berson westerlies underneath, as first observed in wind soundings taken in Africa by the German meteorologist von Berson. At that time my office shelves were laden with booklets containing mean-monthly sounding data for U.S. stations, including those operated in the Pacific by U.S. personnel. I feverishly pulled these from the shelves and roughly tabulated the stratospheric wind data for nine stations in the equatorial Pacific. It turned out that at one of these stations, Canton Island, there were enough high reaching soundings to establish that a pattern of downward-propagating easterly and westerly winds occurred for at least two cycles and that all stations in the sample shared the pattern. Intuitively (or wishfully) I assumed that the cycle was a regular feature of the equatorial stratosphere and so reported it in Boston.

The discovery came too late to announce in the abstract of my paper published in the *Bulletin of Meteorology*, but is on record in a preprint of the paper, faded copies of which still exist (Reed 1959). Formal publication of the results in the *Journal of Geophysical Research* was delayed until March 1961 (Reed et al. 1961), the same month in which Veryard and Ebdon published similar findings in the *Meteorological Magazine* (Veryard and Ebdon 1961). The delay was caused by the time that elapsed between my initial announcement and the arrival of confirming data from Nairobi, Kenya, a station on the far side of the world from the Pacific stations. I thought the oscillation must be of global extent but felt the need to establish the fact before submitting the paper for formal publication. Even so, one of the reviewers rejected this seminal paper, perhaps because it conflicted with the prevailing view on equatorial stratospheric winds. Fortunately the editor had decided to proceed with publication before the negative review was received.

Eventually AFCRL decided to end its support of arctic research and, in the person of Ed Kessler, persuaded me to turn my efforts instead to looking into synoptic applications of radar data. For this purpose they supplied the department with a vertically pointing 1.87-cm radar, and I began studying, in collaboration with Carl Kreitzberg, a Ph.D. candidate, time strips of the radar echoes observed in the Pacific storms that passed overhead.

Subsequently AFCRL loaned us an Army–Navy Ground Meteorology Device (ANGMD-I) sounding system for taking serial ascents in the storms, thereby allowing better definition of the thermal, moisture, and wind structures in the storms. In the winter and spring of 1961, extensive data were gathered in five storms that traversed the area. Carl and I began analyzing the data before I left on sabbatical, and during my absence he completed the analyses and used the findings as the basis of his doctoral dissertation. Carl was my second Ph.D. (if Ed Danielsen is counted as my first).

Upon returning from the sabbatical at NMC, I turned my attention increasingly to stratospheric meteorology, obtaining support for the effort from the National Science Foundation (NSF). The NSF administrative procedures were simpler than for the air force–sponsored research and, at least at that time, allowed principal investigators more freedom in modifying grant objectives when unexpected opportunities arose. The change to NSF support was indeed a welcome one; I am deeply grateful for the 35 years of unbroken support that followed. Topics studied during the period between my first and second sabbaticals (1962–68) were the QBO, the stratospheric sudden warming phenomenon, the transport of trace substances in the stratosphere, tidal motions in the stratosphere and mesosphere, and a variety of other features of the circulation at high levels. Some of these studies employed wind data acquired from rocket launches at a sparse network of firing ranges. The rocket data opened up a new region of the atmosphere to exploration, and I have to thank Willis Webb, the head of the meteorological unit at the White Sands Missile Range in New Mexico, for inviting me to become a consultant to the unit and providing me with the incentive to enter this new realm. An unexpected bonus of the work on tidal winds took place when Richard Lindzen joined me as a postdoc in 1964–65 and took on the problem of explaining the tidal winds theoretically. He made some progress with the theory before leaving the University of Washington for a second postdoc at the University of Oslo. While in Oslo, and subsequently at the National Center for Atmospheric Research (NCAR), he succeeded brilliantly in achieving his objective. It was the source of much satisfaction when later observational work that I carried out with a number of my students confirmed Lindzen's predictions. It should be mentioned also that Dick was the *de facto* supervisor of my fifth Ph.D. student Donald McKenzie, who wrote his dissertation on an aspect of tidal theory. My third and fourth doctoral students, Stuart Muench and John Perry, had earlier written dissertations on the stratospheric sudden warming phenomenon.

As the time for my second sabbatical neared, I was approached, probably on Bob Fleagle's recommendation, to spend the year in Washington, D.C., as the executive scientist of the U.S. Committee for the Global Atmospheric Research Program (GARP). Thus I spent the year from September 1968 to August 1969 assisting Jule Charney, the committee chairman, and Joe Smagorinsky and Vern Suomi, the vice chairmen, in drawing up the conceptual plan for U.S. participation in this large international undertaking. My motivation in accepting an administrative post was twofold: I was a firm backer of Jule Charney's idea of conducting an experiment to gather a global dataset, and I welcomed the opportunity to participate in a program that had as one of its objectives improving U.S.–Soviet relationships by bringing U.S. and Soviet scientists together in cooperative ventures. It is often forgotten in this time of more harmonious relationships between our two nations that many scientists, on both sides of the Iron Curtain, shared this idealistic motivation for participating in GARP.

Although the busy year in Washington, D.C., afforded little time for research, the GARP experience had a profound effect on my future research. As part of my duties I was involved with working groups that were convened to recommend subprograms in areas that were felt to contribute importantly to the overall objectives. Two of the subjects treated by these groups particularly caught my fancy—tropical meteorology and clear-air turbulence—and upon returning home, I decided to leave stratospheric meteorology behind and strike out in new directions. No doubt the occasional urge felt by scientists, as well as by those in other walks of life, to try something new and different played a part in the decision. The venture into clear-air turbulence—actually begun shortly before leaving on the GARP stint—was relatively short lived, resulting after my return in two papers coauthored with Ken Hardy of GRD. The venture into tropical meteorology lasted for a much longer time. A first paper on the structure and properties of synoptic-scale wave disturbances in the equatorial western Pacific drew much favorable attention, and this together with my GARP experience no doubt led to my appointment as the chairman of the panel charged with drawing up the U.S. plan for participation in the GARP Atlantic Tropical Experiment (GATE). While waiting for the experiment to begin, I did further work on the Pacific waves, aided by my student Richard Johnson. Dick subsequently became my sixth Ph.D. recipient, writing his dissertation on the role of precipitation downdrafts in cumulus–synoptic-scale interaction.

During the experiment, held in 1974 off the west coast of Africa, I was stationed for nearly three months at the headquarters in Dakar, Senegal, as an internationally appointed advisor to Joachim Kuettner, the leader of the experiment. Watching the weather unfold each day on the maps prepared on site and on the satellite imagery, I became fascinated with the easterly or African waves that crossed the region with great regularity, and upon returning home determined to make them the main object of my research. I regard the output of this research to be one of my most important contributions to meteorology, though more for its definitive nature than for its originality. Toby Carlson and particularly Bob Burpee must be credited with the seminal work on the subject. My research on the waves spanned more than a decade, ending with a pair of papers written in collaboration with Tony Hollingsworth while on sabbatical leave at the European Centre for Medium-Range Weather Forcasts (ECMWF). In studying convective activity in the waves I was struck, as were others, by the strong diurnal cycle that was present whether they were located over land or sea. This observation sparked a general interest on my part in the nature of the diurnal variation of tropical convection. Subsequently several papers were written on the diurnal cycle in the GATE area and, based on data from the First GARP Global Experiment (FGGE), in the tropical Pacific as well. Some of these papers turned up interesting regional differences in the satellite signatures of the deep convection that to my knowledge have never been explained nor for that matter widely acknowledged. In view of the current emphasis on simulating cloud effects in the general circulation models employed in global warming studies, I can only hope that these important differences will eventually be recognized and used to test the fidelity of the models. Incidentally, I must credit Bill Gray's work on tropical convection for greatly stimulating my own interest.

In the late 1970s while still working on African waves I became interested in small synoptic-scale cyclones that, risking controversy, I will refer to here as polar lows. This interest stemmed from the connection in my mind between the convectively enhanced cyclonic systems that formed off Africa and the observed small cyclones that occasionally developed during winter in cold Pacific masses poleward of the jet stream and that were seen by satellite imagery to involve the organization of deep convection. After the lapse of several years following my beginning paper on the subject, I resumed research on polar lows — whether of the Pacific variety (sometimes referred to as comma clouds) or of the so-called true type that form over high-latitude ocean bodies far removed from middle-latitude influences. The research included observational studies done with Warren Blier prior to my ECMWF sojourn, a combined observational and theoretical study carried out with Charles Duncan while I was at ECMWF, and, after my return home, studies that employed the Pennsylvania State University (Penn State)–NCAR mesoscale model (MM5) as a diagnostic tool. Blier earned his Ph.D. on one such study, becoming number seven on my list of doctoral students. Mark Albright and Jim Bresch also did yeoman work in application of the model to specific cases of polar lows.

Shortly after starting the work on cold-air cyclogenesis, I undertook studies of two events that produced damaging winds in western Washington: one connected with a vortex formation in the lee of Mount Olympus during the passage of an intense synoptic-scale cyclone and the other associated with downslope winds in the lee of the Cascades. The first of the studies was undertaken in response to a request from the Washington State Department of Transportation that I join a team of experts set up to investigate the causes of the collapse of the Hood Canal Bridge during the windstorm. The second was started of my own volition, simply out of a desire to document the conditions underlying a destructive event of a type known to occur occasionally during winter in and near the western foothills of the Cascades. This was only one of several instances, dating back to the April snowstorm in Boston, in which my research was prompted by curiosity regarding a directly experienced event.

Before embarking on the sabbatical at ECMWF I began research on a subject — explosive cyclogenesis — that was attracting much interest at the time. Sanders and Gyakum had written the seminal paper on the subject (1980) and follow-up papers (Gyakum 1983a,b, Anthes et al. 1983; Bosart 1981; Bosart and Lin 1984, Uccellini 1986; Uccellini et al. 1984, 1985) had looked in depth at two remarkable cases that had been observed in the western Atlantic. Aware of an equally remarkable case in the eastern Pacific — a region hardly noted for extreme events — I undertook, with Mark Albright's help, a synoptic study of the case. During the late stages of the study, Rick Anthes paid a visit to the Department of Atmospheric Sciences and upon seeing the results of the research carried out thus far suggested that I collaborate with his junior colleague at NCAR — Bill Kuo — in an effort to extend the investigation by modeling the storm with use of the Penn State–NCAR mesoscale model. The synoptic study was completed and the results submitted for publication prior to my departure for England, but sufficient time did not remain to begin the collaboration with Bill. This had to await my return. During the year at ECMWF I became much impressed with the ability of state-of-the-art numerical models to reproduce storm developments realistically so that I returned home not only eager to apply model diagnostics to explosive cyclogenesis, but also to polar lows and other synoptic-scale and mesoscale entities.

I have already alluded to the modeling effort on polar lows. Even more productive was the work done with Bill on explosive cyclogenesis, nearly 10 papers resulting from our joint endeavor. The first of these treated the eastern Pacific case; the others were based on cases observed in the Atlantic. Among specific items examined in the latter was the evolution of the storm structures, the airflow within the storms, the sensitivity of the developments to various physical processes — such as latent heat release in clouds and heat flux from the ocean surface, and the storm developments from the perspective of the "potential vorticity thinking" advocated by McIntyre, Hoskins, and Robertson. Mark Stoelinga collaborated with Bill and me on the most ambitious of the potential vorticity papers and subsequently, working with the same case, Mark used a highly sophisticated scheme of potential vorticity inversion to assess the role of frictional and diabatic processes in the development. For this achievement he was awarded a doctorate, becoming the eighth Ph.D. on my list of nine. In addition to the sensitivity studies carried out with Bill, I took advantage of my ECMWF connection to join with Adrian Simmons in looking into the role of latent heat release and surface fluxes in the development of a number of other major Atlantic storms.

My baseball team was completed with Jordan Power's successful attainment of his Ph.D. In his work he employed the Penn State–NCAR model to investigate three cases of mesoscale gravity wave development, evaluating the effects on the developments of such factors as background atmospheric structure, physical processes, and grid resolution. Before closing I would like to remark on two papers that I coauthored, one that appeared early in the post-ECMWF period and the other published in 2001 with Bill Kuo and associates, which, barring a miraculous recovery of my health, is likely to be my final research contribution. In the first, coauthored with Dan Keyser, we presented a generalization of the Petterssen frontogenesis function and examined its role in the forcing of vertical motion in idealized cyclones. I had developed the basic idea while teaching my graduate synoptic course and showed Dan my class notes during a visit he made to the University of Washington. He informed me that he had recently obtained a similar result and suggested that we collaborate on a paper. I agreed. But at the time I was preparing to leave on sabbatical, precluding any immediate effort on my part. Dan took over and expanded the topic in his inimitable style, producing a paper that despite the fact that I can take little credit for its final form engenders a special feeling of pride. Perhaps this is because the underlying idea was a theoretical one — at least from the standpoint of a lowly synoptic meteorologist.

The second paper dealt with a small or mesoscale cyclone in the Mediterranean Sea that bore a close resemblance to a tropical cyclone. The paper, which appeared in a recent issue of *Meteorology and Atmospheric Physics*, documented the storm development and behavior in detail and tested the ability of the NCAR model to simulate the sequence of events. A unique feature of the study, in the sense that I had not employed it previously in my work, was the use of adjoint sensitivity to alter the initial conditions at one point in the simulation and from the revised initial state to obtain an improved forecast. This procedure, in combination with a vortex implant method, resulted in a considerable improvement. The Mediterranean mesoscale cyclone bears a similarity to the tropical cyclones of low latitudes and the polar lows of high latitudes. An interesting challenge is to develop a unified theory of these phenomena. To some extent Emanuel and Rotunno have already done this.

In summary, during my early education I demonstrated the capacity for a career in science but felt no strong urge to pursue a scientific career, much less a career in meteorology. It was an accident of World War II that I was shunted into this field as an enlisted man in the navy and a matter of rare good fortune that I was given the opportunity to obtain a B.S. degree in meteorology while still in uniform. However, it was no accident that once introduced to the subject, my future course was set. From the start I loved meteorology with a deep-seated passion and experienced aesthetic feelings toward meteorological phenomena that motivated and sustained my future endeavors. These emotional attachments lasted throughout my career. Without them it is unthinkable that I could have achieved what I did. My choice of specific research topics was determined or inspired by a number of circumstances, ranging from healthy curiosity, to obligations imposed by others, to opportunities proffered by outside parties, to the occasional desire to strike out in new directions, to contact with directly experienced weather events, and, in my most original work, to a series of improbable events. In a word there was no common thread except for an insatiable desire to observe and understand atmospheric phenomena.

REFERENCES

Anthes, R. A., Y.-H. Kuo, and J. R. Gyakum, 1983: Numerical simulations of a case of explosive marine cyclogenesis. *Mon. Wea. Rev.*, **111,** 1174–1188.

Blair, T. A., 1937: *Weather Elements*. Prentice Hall, 411 pp.

Bosart, L. F., 1981: The Presidents' Day snowstorm of 18-19 February 1979: A subsynoptic-scale event. *Mon. Wea. Rev.*, **109,** 1542–1566.

——, and S. C. Lin, 1984: A diagnostic analysis of the Presidents' Day storm of February 1979. *Mon. Wea. Rev.*, **112,** 2148–2177.

Gyakum, J. R., 1983a: On the evolution of the QE II storm. I: Synoptic aspects. *Mon. Wea. Rev.*, **111,** 1137–1155.

——, 1983b: On the evolution of the QE II storm. II: Dynamic and thermodynamic structure. *Mon. Wea. Rev.*, **111,** 1156–1173.

Reed, R. J., 1959: Circulation of the stratophere. *Bull. Amer. Meteor. Soc.*, **40,** pg. 629.

——, W.J. Campbell, L.A. Rasmusssen, and D.G. Rogers, 1961: Evidence of a downward-propagating annual wind reversal in the equatorial stratosphere. *J. Geophys. Res.*, **66,** 813–818.

Sanders, F., and J. R. Gyakum, 1980: Synoptic-dynamic climatology of the "bomb." *Mon. Wea Rev.*, **108,** 1589–1606.

Uccellini, L. W., 1986: The possible influence of upstream upper-level baroclinic processes on the development of the QE II storm. *Mon. Wea. Rev.*, **114,** 1019–1027.

——, P. J. Kocin, R. A. Petersen, C. H. Wash, and K. F. Brill, 1984: The Presidents' Day Cyclone of 18-19 February 1979: Synoptic overview and analysis of the subtropical jet streak influencing the pre-cyclogenetic period. *Mon. Wea. Rev.*, **112,** 31–55.

——, D. K. Keyser, K. F. Brill, and C. H. Wash, 1985: The Presidents' Day Cyclone of 18-19 February 1979: Influence of upstream trough amplification and associated tropopause folding on rapid cyclogenesis. *Mon. Wea. Rev.*, **113,** 962–988.

Veryard, R. G., and R. A. Ebdon, 1961: Fluctuations in tropical stratospheric winds. *Meteor. Mag.*, **90,** 125–143.

FIG. 1.1. Dick Reed at age 4 or 5.

FIG. 1.2. Dick Reed (right) and brother Bob (left) on leave from U.S. Navy, about 1943.

FIG. 1.3. (Newly minted) Ensign, U.S. Navy, 1945.

FIG. 1.4. MIT graduate student (Cal Tech letterman in track on sweater).

FIG. 1.5. Research Staff members and students, Department of Meteorology, MIT, circa 1952: (top row, left to right) Dick Reed, Alan Faller, Don Friedman (behind ball); (bottom row) Ike Vanderhoven, Ed Kessler (right).

FIG. 1.6. Dick Reed lecturing, circa 1955.

Fig. 1.7. (left to right) C.-P. Chang, Vern Kousky, Dick Reed, Mike Wallace, and Jim Holton, 1969 in Reed's office, at the time of joint work on easterly waves and QBO. Photo taken by Prof. Michio Yanài of UCLA.

Fig. 1.8. Photo on Great Wall of China, 1974, AMS Delegation to China: (left to right) Dave Atlas, Betty Kellogg, Ken Spengler, Lucille Atlas, unknown, unknown, Dick Reed, Dave Johnson, Joan Reed, Will Kellogg: rest unknown.

Fig. 1.9. At Peking University, 1974, AMS Delegation to China: (left to right) Tsou Ching-meng, Dave Johnson, Chou Pei-yuan, Dick Reed.

FIG. 1.10. Dick Reed with Wendell Nuss in weather map room, about 1990.

FIG. 1.11. Dick Reed, daily coffee.

Chapter 2

Tropopause Folding, Upper-Level Frontogenesis, and Beyond

LANCE F. BOSART

Department of Earth and Atmospheric Sciences, University at Albany, State University of New York, Albany, New York

"Potential vorticity on isentropic surfaces provides a useful tool for the study of frontogenesis. During the formation of a katafront, or subsiding cold front, it is possible for the tropopause to become folded and for a thin slice of the stratospheric air to descend to the middle or even the lower troposphere."— (Reed 1955)

1. Introduction

Richard J. Reed has made highly original and substantive research contributions to the field of meteorology in a career that has spanned more than five decades. The gathering here in Orlando in January 2002 for the Reed Symposium attests to Dick's lifetime achievements and influence on the field. It is a great honor for me to have the opportunity to speak at the Reed Symposium to help highlight Dick's many important contributions and, most importantly, to show how his creative insight has helped to ignite and sustain new research inquiries and directions.

I first got to know Dick Reed while I was a graduate student at the Massachusetts Institute of Technology (MIT) back in the 1960s. Dick visited MIT on occasion to renew old friendships, interact with the faculty, and percolate new scientific ideas. I especially enjoyed listening to his spirited discussions with Frederick Sanders. Dick and Fred would "argue" about science and the weather for what seemed like hours. Fascinating stuff for a newcomer to the field because their research interests were broad, deep, and highly original. My other student memory of Dick Reed is walking into my Ph.D. thesis defense to discover him sitting there with the rest of the faculty as a "surprise" visitor. I remember thinking that "now I am doomed for sure," but it turned out that I worried needlessly because Dick spent more time arguing with Fred than paying attention to me! In the years that have followed it has been my great pleasure to count Dick Reed as a professional colleague and personal friend.

The purpose of this chapter is threefold: 1) to delineate Dick Reed's seminal contributions to the subject of upper-level fronts and frontogenesis and associated tropopause folding, 2) to show how the results of Dick Reed's research on these and related subjects have stimulated new research directions and opportunities through theoretical, observational, and numerical investigations of the structure and life cycles of fronts and cyclones, and 3) to report on an analysis of a case of intense upper-level frontogenesis that uncommonly occurred in southwesterly flow aloft during 8–10 December 1978. The focus of this paper will be on the "big picture." Emphasis will be placed on how the introduction of important scientific concepts and methodologies, and the associated findings from Dick Reed's earlier work on upper-level frontogenesis and tropopause folding, have triggered, and continue to promote, a wealth of new scientific studies on cyclone life cycles and frontal phenomena.

To appreciate the essential structure of an upper-level front, note that the background middle-troposphere thermal gradient in middle latitudes averages $1\,^{\circ}\mathrm{C}\ (100\ \mathrm{km})^{-1}$ in the cool season. A respectable upper-level front can easily have a horizontal temperature gradient of $10\,^{\circ}\mathrm{C}\ (100\ \mathrm{km})^{-1}$ with a corresponding vertical wind shear of $20–40\ \mathrm{m\ s^{-1}}$ over a 1–2-km layer. The horizontal length of an upper-level frontal zone typically ranges from 1000 to 2000 km, although Morgan and Nielsen-Gammon (1998) show an example where the frontal length exceeds 8000 km. As first shown by Reed and Sanders (1953) and Reed (1955), an upper-level front can be associated with a tropopause fold whereby a region of stratospheric air of thickness 1–2 km and width 100–200 km extrudes into the middle and lower troposphere, analogous to how liquid chocolate can be folded deep into cake batter when making marble cake. One significant aspect of upper-level fronts is that they serve as freeways for the transport of high potential vorticity (PV) stratospheric air, ozone, and other chemical constituents that normally reside in the stratosphere into the middle and lower troposphere.

Indeed, Danielsen (1964, 1968) used PV conservation along isentropic trajectories to establish compelling evidence for the downward transport of radioactive debris in tropopause folds subsequent to atomic bomb tests. In Danielsen's view, tropopause folds could be associated with lateral and vertical mixing processes in conjunction with horizontal and vertical heat and moisture fluxes as a part of typical middle-latitude cyclone life cycles. Instead of radioactive debris "safely" residing in the stratosphere for years as had been widely assumed would happen, the debris could be returned to the lower troposphere and the surface in conjunction with middle-latitude cyclogenesis relatively rapidly (a couple of weeks) after atomic bombs were detonated at remote locations. Danielsen's research, based on Reed's early ideas of upper-level frontogenesis and tropopause folding using PV conservation concepts, had the immediate impact of jolting technologically advanced nations into appreciating that even remotely released atomic chickens could still come home to roost.

Interested readers should consult a number of review papers in the literature for specific details and references on individual topics. For example, Shapiro (1981) has discussed frontogenesis and geostrophically forced secondary circulations associated with jet streaks and fronts, resulting in a review paper on the subject (Shapiro 1983). Fronts and jet streaks are examined from a theoretical perspective in Bluestein (1986) and from an observational perspective in Keyser (1986). Comprehensive reviews of the structure and dynamics of upper-level frontal zones and the tropopause appear in Keyser (1986), Keyser and Shapiro (1986), and Keyser (1999). Related articles, including review papers, on cyclone life cycles and surface fronts can be found in Browning (1990), Eliassen (1990), Holopainen (1990), Hoskins (1990), Reed (1990), Shapiro and Keyser (1990), and Uccellini (1990), all from the Erik Palmén Symposium held in Helsinki, Finland, in 1988, and in Bosart (1999), Browning (1999), Davies (1999), Farrell (1999), Holopainen (1999), Keyser (1999), Shapiro et al. (1999), Simmons (1999), and Thorpe (1999), all from the Symposium on the Life Cycles of Extratropical Cyclones held in Bergen, Norway, in 1994.

The chapter is organized as follows. Dick Reed's substantive contributions to upper-level frontogenesis and tropopause folding deduced from his use of PV conservation are highlighted in section 2. Parallel work on this subject by other scientists working with Reed and/or independently is discussed in section 3. Section 4 contains a brief review of stratospheric–tropospheric exchange processes that occur in conjunction with upper-level frontogenesis and tropopause folding. Section 5 emphasizes the dynamics of upper-level frontogenesis, tropopause folding, and associated cyclogenesis. Section 6 presents the results of the application of "PV thinking" to selected modern problems whose roots can be traced back to Dick Reed's early research. Section 7 contains a brief case study of atypical upper-level frontogenesis (7–10 December 1978) in southwesterly flow. Future research directions and conclusions are given in sections 8 and 9, respectively.

2. Early ideas on upper-level frontogenesis and tropopause folding

Bjerknes and Palmén (1937) took advantage of some of the first systematic radiosonde observations across Europe to investigate the structure of the tropopause and an associated upper-level front. In their interpretation, the upper-level front was manifest as a circumpolar boundary that marked the upward extension of the polar front and separated polar from tropical air around the hemisphere. Their cross-sectional analysis showed a modest "S shaped" tropopause (weak tropopause fold) near where the weaker portion of the upper-level front intersected the tropopause. No special dynamical signatures were attributed to upper-level fronts. No association was made between cyclogenesis and upper-level frontogenesis beyond the inference that upper-level fronts were merely manifestations of the postulated circumpolar polar front in the upper troposphere. No direct linkage was hypothesized to exist between upper-level fronts and stratospheric–tropospheric exchange processes.

Progress in understanding upper-level fronts accelerated when Reed and Sanders (1953) and Reed (1955) boldly argued that upper-level fronts represented regions of intense temperature contrast (their definition of a "frontal zone") embedded within broader baroclinic zones. The crux of this then-new thinking is contained in the introduction of Reed (1955) and is reproduced here as follows:

The theoretical meteorologist, on the other hand, has in recent years abandoned the concept of an atmosphere containing discontinuities in temperature or temperature gradient and has tended to think more and more in terms of a continuous medium characterized by broad zones of relatively strong temperature concentration. The comparative fruitfulness of this approach in problems relating to the stability of baroclinic zonal currents and the numerical prediction of pressure or height changes has raised justifiable doubts as to whether fronts play a leading role in atmospheric development. Sutcliffe (1952), in a review article which expresses opinions similar to the foregoing, has emphasized the present uncertain position of the frontal concept. However, he has also stated, and correctly, that 'for some reason, not so far as we know yet adequately explained, the baroclinic atmosphere does frequently develop towards discontinuity in temperature, often very sharp near the surface, and the field of vertical motion, in the lower troposphere at least, does tend to organize itself in gentle upgliding motions in these regions'. Since, as stressed by Palmén (1949), there are instances of distinct fronts in the upper troposphere, there appears no reason for restricting the preceding statement to the lower troposphere, Godson (1951) has suggested the term 'hyperbaroclinic zone' for the sharply bounded region of temperature concentration and 'frontal zone' for the broad baroclinic region without distinct boundaries. The writer, however, prefers to refer to baroclinic zones as such and to reserve the designation 'frontal zone' for the narrow, sharply bounded regions of intense temperature contrast. The latter terminology will be employed in this article.

Adoption of this viewpoint enabled Reed and Sanders (1953) and Reed (1955) to conduct detailed case studies of upper-level fronts and upper-level frontogenesis and to interpret their findings from a dynamical perspective. Particularly important was their ability to 1) deduce the three-dimensional circulations associated with upper-level fronts out of which came an appreciation for the role that subsidence along the warm boundary of the upper-level frontal zone played in frontogenesis via the tilting mechanism, and 2) show how this same subsidence could result in the folding of stratospheric air into the troposphere as deduced from the use of PV as a tracer of air motion.

Shown in Fig. 2.1 (Fig. 7 in Reed 1955) is an example of an upper-level front on the 500-hPa surface at 0300 UTC 15 December 1953. At this time, the upper-level front was in its most mature phase, having intensified and worked its way around from the west side to the east side of the synoptic-scale trough over the previous 24 h (not shown). Wind reports of 72, 77, and 88 m s^{-1} at Wallops Island, Virginia; Washington, D.C.; and New York, New York, respectively, attest to the intensity of the implied frontal zone on the basis of the thermal wind relationship. The inferred intensity of the frontal zone at 500 hPa is derived from Fig. 2.1 by the $-13°$C temperatures at Wallops Island, and New York, as compared to the $-27°$ and $-25°$C temperatures, respectively, at Pittsburgh, Pennsylvania, and Rome, New York, from which we estimate a peak thermal gradient of 10°C (100 km)$^{-1}$.

A cross section along the bold dash–dot line in Fig. 2.1 is presented in Fig. 2.2 (Fig. 13 of Reed 1955). As noted by Reed (1955), it shows a "thin tongue of stratospheric air slicing down into the troposphere" (bounded by the

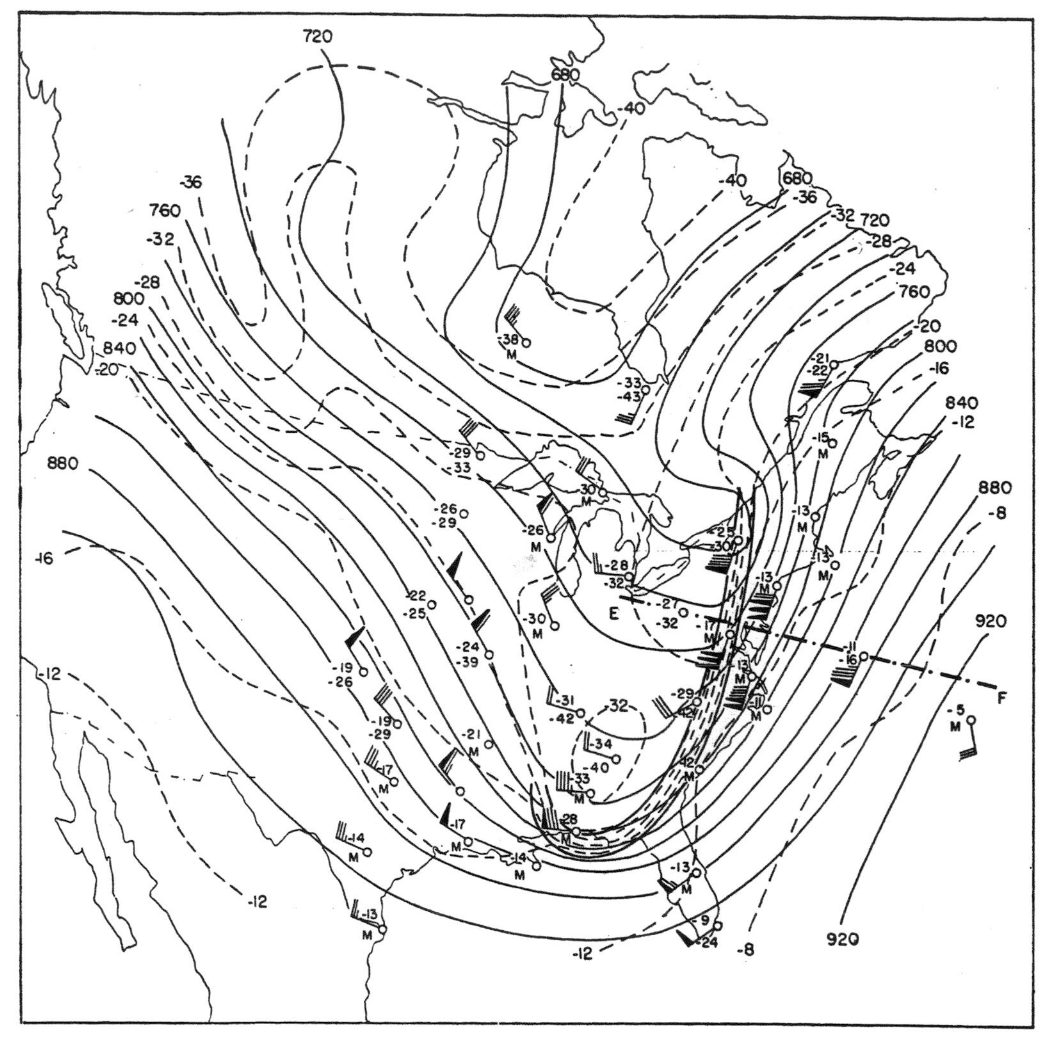

FIG. 2.1. The 500-hPa heights (solid, every 200 ft or 60 m) and isotherms (dashed, every 4°C) for 0300 UTC 15 Dec 1953. Standard station plotting model is used for winds (kt), temperature (°C), and dewpoint temperature (°C). Heavy solid lines represent boundaries of the frontal zone. Heavy dash–dot line defines the cross section shown in Fig. 2.2. Source: Fig. 7 from Reed (1955).

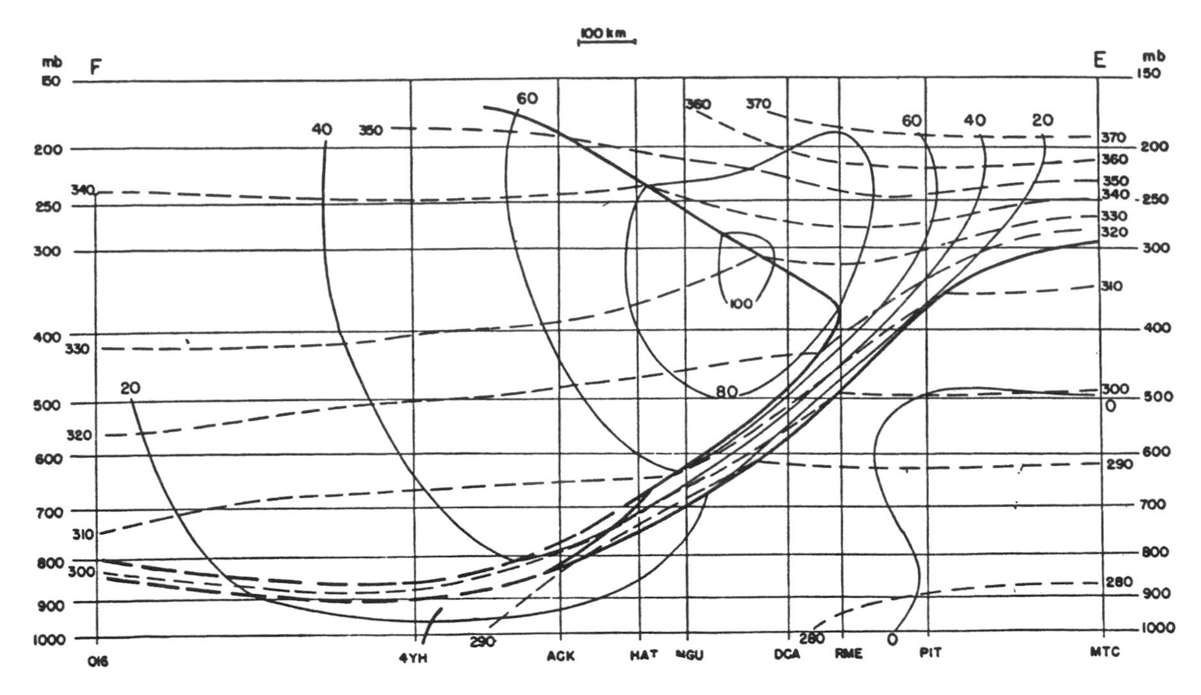

FIG. 2.2. Cross section of isotachs (solid, every 10 kt) and potential temperature (dashed, every 10 K) along line E–F depicted in Fig. 2.1. Heavy solid line denotes tropopause boundaries. Source: Fig. 13 from Reed (1955).

heavy solid lines in Fig. 2.2) where it blends with a thin, dry stable layer near 900 hPa in the warm air mass over the Atlantic Ocean. The now-familiar signature of a tropopause fold, and its associated stratospheric air, was deduced for the first time from comparing the PV distribution along selected isentropic surfaces, mindful that stratospheric PV values are an order of magnitude larger than tropospheric PV values. Reed (1955) also used successive positions of the 310-K isentropic surface in a section normal to the flow to deduce that subsidence was maximized along the warm boundary of the frontal zone (his Fig. 10, not shown). Together, these and other figures enabled Reed (1955) to conclude that "the frontal zone does not separate air of immediate polar and tropical origin. Rather, it forms within the polar airmass. Because of the strong subsidence at the warm boundary, the air at this boundary and adjacent to it on the warm side acquires temperatures characteristic of tropical air. But from the point of view of origin, it cannot be called tropical air."

Reed's (1955) paper marked the death knell to the idea that a continuous and deep polar front separating polar from tropical air encircled the Northern Hemisphere. He found that upper-level frontal zones were discontinuous, seldom extended to the surface, tended to form in northwesterly flow containing cold-air advection, and could be associated with subsidence that maximized on the warm boundary of the baroclinic zone, leading to frontogenesis via the tilting mechanism. By using PV as a tracer, Reed was able to document aspects of the life cycle of upper-level fronts and their dynamical importance to stratospheric–tropospheric exchange processes.

The findings from Reed (1955) were used by Reed and Danielsen (1959) to portray a new schematic model of the tropopause and an associated upper-level frontal zone as shown in Fig. 2.3.

Although the results of Reed and Sanders (1953) and Reed (1955) showing that the primary upper-level front-

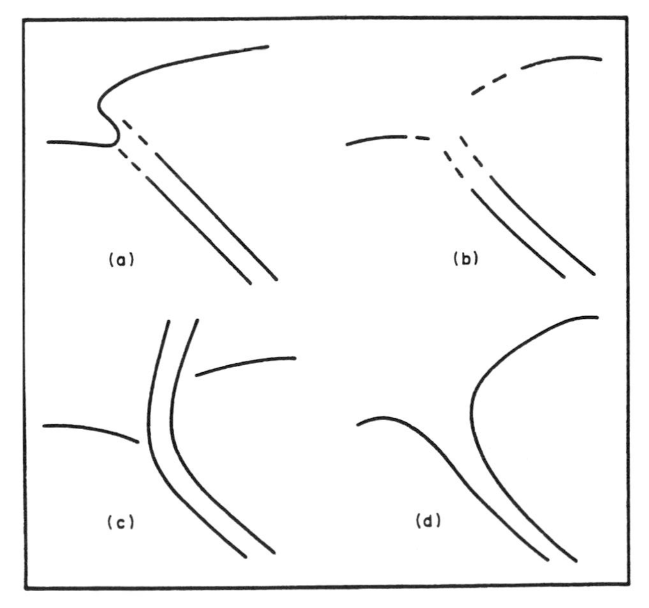

FIG. 2.3. Schematic diagrams of models used in the past for analyzing upper-level fronts and tropopauses: (a) Bjerknes and Palmén (1937), (b) Palmén and Nagler (1949), (c) Berggren (1952), and (d) Reed and Danielsen (1959). Source: Fig. 1 from Keyser and Shapiro (1986).

ogenesis mechanism was tilting associated with the strongest subsidence occurring along the warm boundary of a frontal zone were confirmed and accepted by others (e.g., Newton 1954; Staley 1960), controversy remained as to whether a preexisting front resided in the troposphere prior to the downward extrusion of ozone-rich, high-PV air [see, e.g., Newton (1958) and the footnote on the bottom of p. 257 of Palmén and Newton (1969)]. Aircraft tracer studies by, for example, Danielsen (1964, 1968) and Briggs and Roach (1963), along with further PV analyses of upper-level fronts and investigations of the tropopause (e.g., Danielsen 1959; Staley 1960), erased many of the lingering doubts as to the veracity of Reed's proposed mechanism for upper-level frontogenesis. Shown in Fig. 2.4 is a schematic illustration taken from Danielsen (1968) of a vertical secondary circulation that is conducive to upper-level frontogenesis and tropopause folding.

3. Stratospheric–tropospheric exchange and the "dry slot"

In this section the subject of stratospheric–tropospheric exchange is briefly reviewed as related to upper-level frontogenesis and the development of the "dry slot" in extratropical cyclones (a much more extensive review of this topic is found elsewhere in this volume). In a series of papers, Danielsen (1964, 1966, 1967, 1968) and Danielsen and Bleck (1967) conducted very detailed case studies of the three-dimensional airflow in extratropical cyclones. This research was motivated in part by Reed's pioneering studies of upper-level fronts and his use of PV as a tracer to help deduce the three-dimensional airflow in these cyclones. In these studies, PV again was used as a tracer along with measurements of ozone and other radioactive tracers obtained from research aircraft flights. Synthesis of the disparate observations enabled Danielsen and his coworkers to deduce the existence and structure of the dry slot in extratropical cyclones prior to its routine detection in

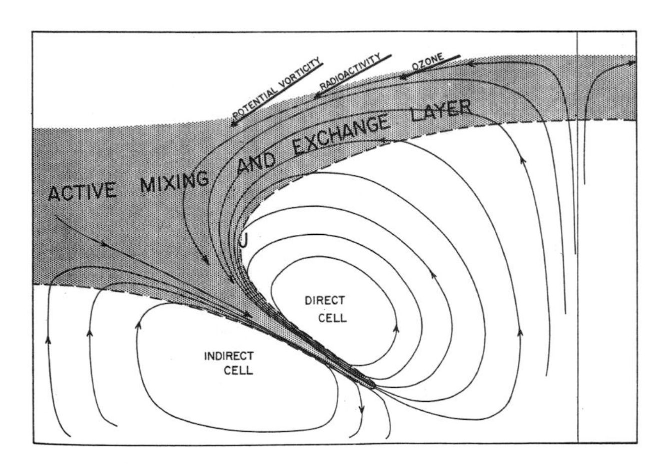

FIG. 2.4. Schematic of mean circulation relative to the tropopause, including a folded tropopause. Source: Fig. 14 from Danielsen (1968).

satellite imagery. They manually prepared isentropic trajectories (Danielsen 1961) to show that although air in the dry slot had a previous history of descent (several hundred hectopascals over a 24-h period for an average descent rate of ~ 5 cm s^{-1}), air near the tip of the dry slot could be rising. They then deduced that dry air with a previous history of subsidence could result in convective destabilization if it were to overspread a low-level warm, moist airstream as it began to rise. Carr and Millard (1985) confirmed this interpretation and Browning (1990, 1999) has provided ample evidence that the dry slot within extratropical cyclones can be a fertile breeding ground for convectively driven mesoscale circulations.

Although the use of PV as a tracer of atmospheric motions enabled Reed, Danielsen, and colleagues to adopt a Lagrangian perspective in studies of cyclogenesis and upper-level frontogenesis, it also became apparent that PV nonconservation was also important. To explore the issue of PV nonconservation, note that following the motion the time rate of change of PV can be written as [see, e.g., Eq. (2.11) in Keyser and Shapiro (1986)]:

$$\frac{dP}{dt} = -(\zeta_\theta + f)\frac{\partial \dot{\theta}}{\partial p} + \frac{\partial \theta}{\partial p}\left[\hat{\boldsymbol{k}} \cdot \left(\nabla_\theta \dot{\theta} \times \frac{\partial \vec{\boldsymbol{V}}}{\partial \theta}\right)\right] - \frac{\partial \theta}{\partial p}\left[\hat{\boldsymbol{k}} \cdot \left(\nabla_\theta \times \vec{\boldsymbol{F}}\right)\right] \quad (1)$$

In this equation $\hat{F}$ and $\dot{\theta} \equiv d\theta/dt$ are the contributions of friction and the diabatic process to PV nonconservation, θ is the potential temperature, ζ_θ is the relative vorticity, and the subscript θ indicates an operation on a θ surface. Shapiro (1976) demonstrated that (I) could be further simplified with the assumption that the vertical gradient of diabatic heating associated with clear-air turbulence (CAT) dominated the frictional contribution to PV change to yield

$$\frac{dP}{dt} = g^2(\zeta_\theta + f)\frac{\partial^2}{\partial p^2}\left(-\rho \overline{w'\theta'}\right). \quad (2)$$

In this equation, $w \equiv dz/dt$, primed quantities denote turbulent motions, ρ is the atmospheric density, the overbar refers to an ensemble average, and $\overline{w'\theta'}$ is the vertical eddy flux of potential temperature. Keyser and Shapiro (1986) noted that the vertical eddy flux of potential temperature on the right-hand side of (2) would be large and negative (downward) in regions of CAT above and below the jet core where the Richardson number is typically small (<1) and could be expected to contribute to a PV nonconservation, with increasing PV within the jet core itself. Keyser and Shapiro (1986) presented schematics (their Figs. 12 and 13 taken from Shapiro 1976) to illustrate how nonconservation of PV might originate above an upper-level frontal zone. Keyser and Shapiro (1986) argue that CAT would act to offset the required vertical spreading of isentropes at the level of maximum wind on the cyclonic shear side of a jet embedded in a strong upper-level front. Their argument rests on the idea that destabilization is resisted at the level of maximum wind by CAT processes so that vorticity on the isentropic surfaces must increase locally if PV is conserved. Likewise, as demonstrated by Browning and Watkins (1970), Browning (1971), Bosart and Garcia (1974), and Shapiro (1978), this same CAT mechanism would act to decrease the static stability within the frontal shear layers above and below the level of maximum wind. An example of the PV nonconservation process, taken from Sanders et al. (1991), is shown in Fig. 2.5. The downward movement and decrease in magnitude of PV to progressively lower potential temperatures is consistent with expectation from static stability changes associated with the CAT-induced differential diabatic heat flux.

The nonconservative behavior of PV in the circumstances described above raises the issue as to whether PV can properly be viewed as a tracer of atmospheric motions. This issue has been debated vociferously by Danielsen (1990), Haynes and McIntyre (1990), Keyser and Rotunno (1990), and McIntyre (1999). The debate goes back to Shapiro (1976) and Keyser and Shapiro (1986) who argued that the nonconservation of PV associated with frictional and diabatic processes resulted in a PV distribution in upper-level fronts and tropopause folds [see previous paragraph, Eqs. (1) and (2), and Fig. 2.5] that differed in subtle but important ways from the distribution of PV that would be expected in association with the pure advection of a passive tracer in turbulent flow. At the heart of the dispute was how to define PV: as the scalar product of the Reynolds-averaged absolute vorticity and potential temperature gradient or the Reynolds average of the scalar product of the absolute vorticity and the potential temperature gradient. Shapiro (1976) used the first interpretation whereas Danielsen (1990) used the second interpretation. As shown by Keyser and Rotunno (1990), practical considerations dictate that the PV as computed from gridded datasets is consistent with the first interpretation because the temperatures and winds themselves are already averaged quantities representative of grid volumes.

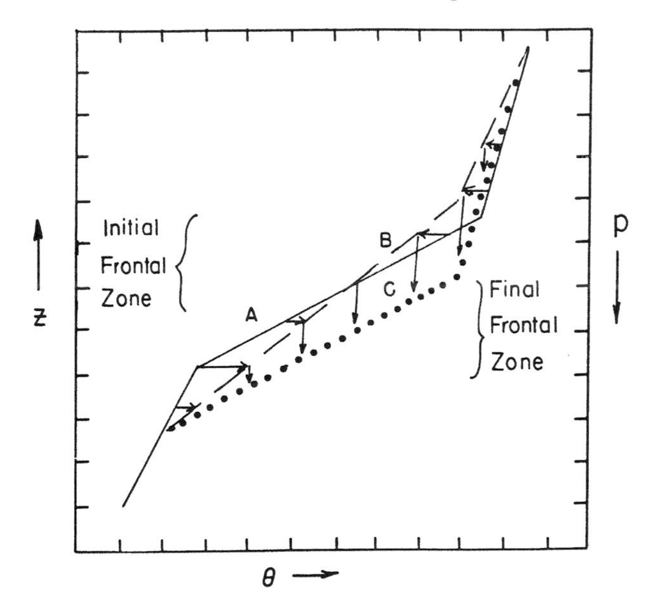

FIG. 2.5. Idealized vertical profile of potential temperature illustrating a hypothetical mechanism for migration of the frontal zone downward toward colder potential temperature. The initial profile is denoted by A. The modified profile resulting from turbulent heat transfer is the dashed line B, with horizontal arrows showing the diabatic heating or cooling. The dotted line C shows the final state, taking into account the frontal-scale subsidence, with arrows showing vertical displacement. Source: Fig. 10 from Sanders et al. (1991).

To conclude this section, note that the issue of stratospheric–tropospheric exchange as first addressed by Reed, Danielsen, and Staley (among others) is still a subject of active research and debate. The paper by Lamarque and Hess (1994) is typical of how the problem of stratospheric–tropospheric mass exchange in conjunction with tropopause folding is being attacked from a PV perspective. Current research in this area was stimulated by Wei (1987) who suggested that the problem could be better understood from a global instead of a regional or local perspective (see, e.g., Holton et al. 1995; Wirth and Egger 1999; Juckes 2000; Gettelman and Sobel 2000). These recent papers have established that the degree of tropospheric–stratospheric exchange is very sensitive to the computational method and that individual synoptic events are important to the overall process. Juckes (2000), for example, argues that air from the troposphere is effectively ingested into the stratosphere in association with downward motion (descent of the tropopause exceeds the local downward motion of the air) and that the extreme dryness of the lower stratosphere may be caused by the passage of air in this region through the very cold ($<-80°C$) temperature regions associated with synoptic-scale anticyclones of midlatitudes. To help untangle these, and other, issues, PV conservation, nonconservation, and attribution techniques should be helpful.

4. Tropopause folds and cyclogenesis

Inspection of the 1000- and 500-hPa maps for 0300 UTC 15 December 1953 in Reed (1955, his Figs. 4 and 7) reveals that a deepening surface cyclone situated over northern New York lies just downstream from the exit region of an intense upper-level frontal zone. Reed (1955) and other earlier observational studies (e.g., Staley 1960; Bosart 1970; Shapiro 1970) did not establish a direct relationship between tropopause folding and surface cyclogenesis. Bleck (1973, 1974) simulated cyclogenesis using a simple model that employed PV conservation on surfaces of constant potential temperature. In his numerical simulations there was a near-simultaneous extrusion of stratospheric air through the tropopause fold with the surface cyclogenesis. Boyle and Bosart (1986) made a similar finding on the basis of a

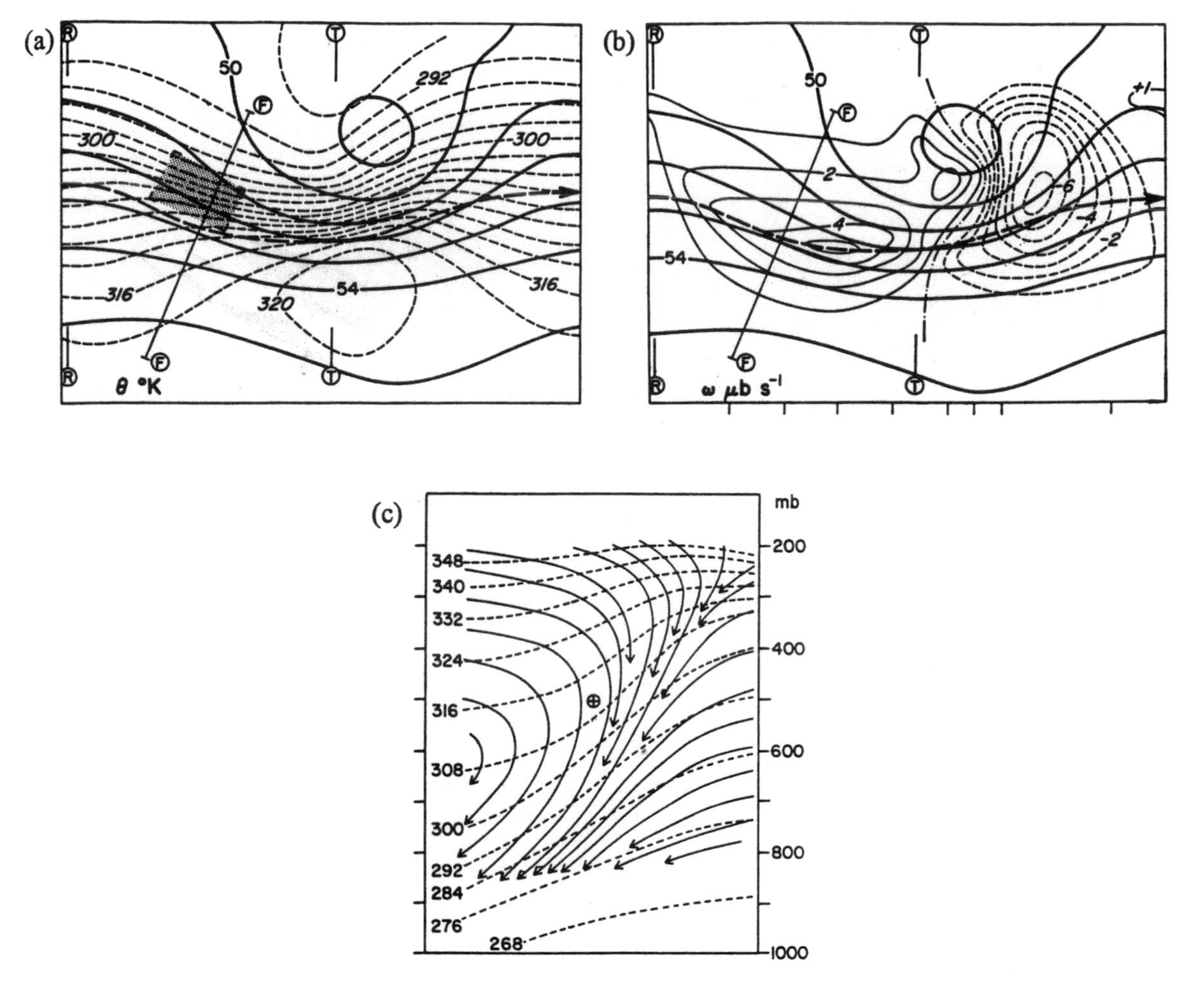

FIG. 2.6. Depiction of structure of and vertical circulations in an upper-level jet-front system derived from integration of a baroclinically unstable wave to finite amplitude in an adiabatic, β-plane, primitive-equation channel model: (a) 500-hPa geopotential height (contour interval 100 m, 50 denotes 5000 m; thick solid), potential temperature (contour interval, 2 K; dashed), and jet stream axis (thick dashed arrow); (b) as in (a) except for pressure-coordinate vertical velocity [contour interval, 1μb s^{-1}; thin solid (positive), dot–dashed (zero), dashed, (negative)] replacing potential temperature; (c) streamlines of total transverse circulation (u, ω) and isentropes (contour interval, 8 K; dashed) for the vertical section, FF, crossing the northwesterly flow inflection, shown in (a) and (b). Circled plus sign indicates location of 500-hPa jet stream (Newton and Trevisan 1984). Source: Fig. 3 from Keyser (1999).

case study of intense cyclogenesis over North America. In a case of Alpine cyclogenesis, Bleck and Mattocks (1984) and Mattocks and Bleck (1986) found that cyclogenesis to the lee of the Alps followed subsequent to a strong jet impinging on the poleward side of the mountain barrier. In all of these studies, however, there was little consideration of whether tropopause folding played an active or passive role in surface cyclogenesis.

Uccellini et al. (1985), Uccellini (1986), and Whittaker et al. (1988), in a study of the "infamous" Presidents' Day storm of 19 February 1979 (Bosart 1981; Uccellini et al. 1984), showed that an appreciable tropopause fold, manifest by a dark spot in satellite water vapor imagery representative of sinking air, preceded the rapid surface cyclogenesis phase of the storm. At issue in these (and other) papers was the extent and importance of lower- versus upper-level forcing in the cyclogenesis process. Good arguments were made for the importance of both contributions. An intense case of oceanic cyclogenesis over the North Atlantic Ocean on 9 September 1978 [the Queen Elizabeth II (QEII) storm] produced a similar debate about the relative importance of upper- versus lower-level forcing on cyclogenesis (see, e.g., Gyakum 1983a,b; 1991; Uccellini 1986). An unresolved issue from these (and other) studies is whether tropopause folding and upper-level frontogenesis, even if they precede surface cyclogenesis, is a necessary condition for cyclogenesis. Evidence from recent theoretical studies using idealized numerical models points toward a more passive role for upper-level frontogenesis and tropopause folding in cyclogenesis as will be discussed in the next section.

5. Dynamics of upper-level frontogenesis and cyclogenesis

a. Overview

Reed and Sanders (1953), Newton (1954), Reed (1955), Sanders (1955), and Staley (1960) provided convincing evidence for the importance of three-dimensional circulations to frontogenesis on the basis of the application of the Miller (1948) frontogenesis equation to individual case studies. Sawyer (1956), Eliassen (1962), Hoskins and Bretherton (1972), Keyser et al. (1992a,b), Rotunno et al. (1994), and Wandishin et al. (2000), among others, have provided a theoretical framework for understanding the growth of vertical circulations associated with fronts near the surface and in the free atmosphere [see also excellent review papers by Eliassen (1990) and Keyser (1999) for extensive discussion of this topic]. The transverse vertical circulations conducive to upper-level frontogenesis and tropopause folding illustrated schematically in Fig. 2.4 depend upon horizontal confluence, shearing deformation, and horizontal tilting for their existence and can be understood from this theoretical perspective. The earlier studies by Sawyer (1956) and Eliassen (1962) established the general applicability of inferring the behavior of vertical circulations on the basis of qualitative inferences from the Sawyer-Eliassen equation. An especially attractive way of understanding the theoretical basis for the growth of canonical vertical circulations in frontal zones has been provided most recently by Hakim and Keyser (2001) in terms of Green's functions for the Sawyer-Eliassen

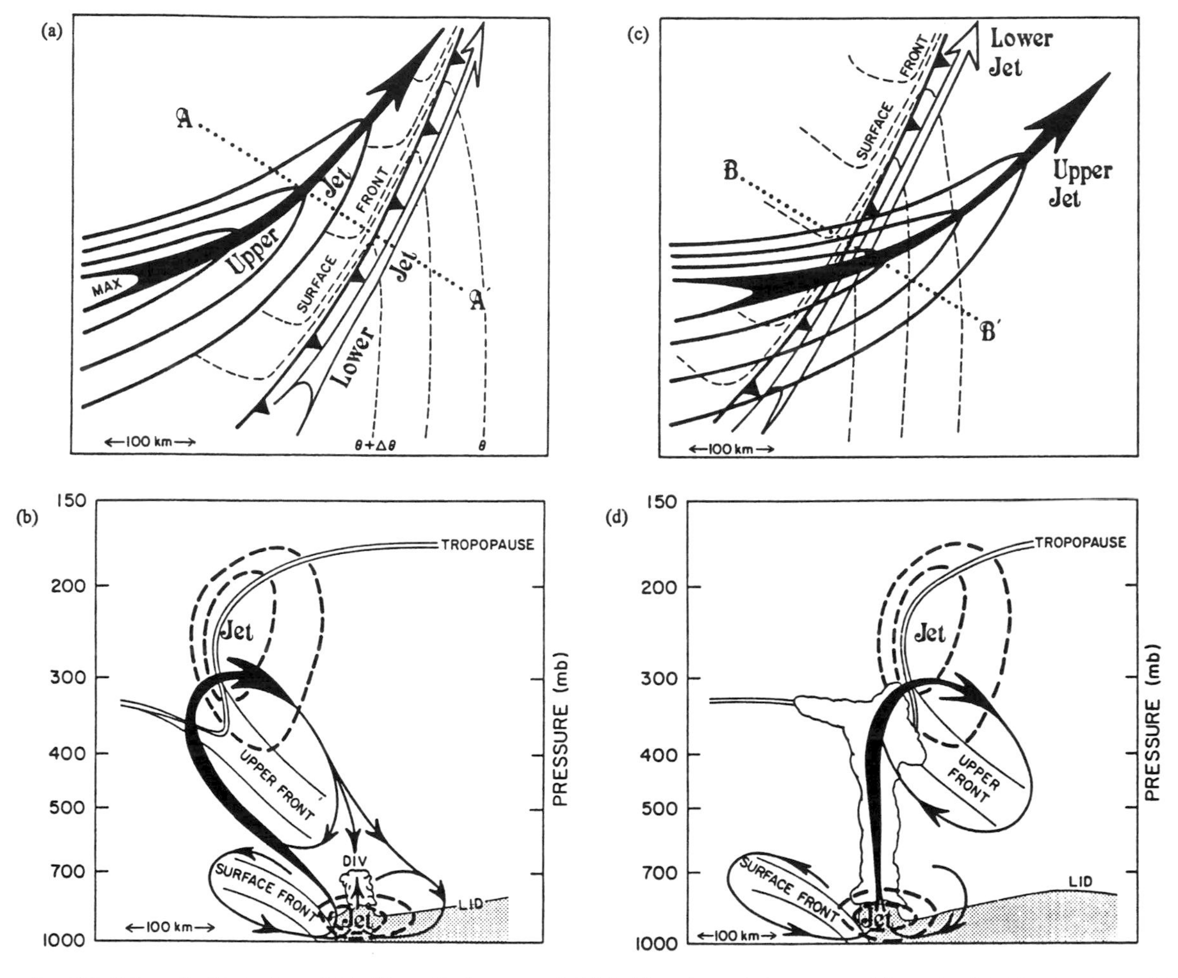

FIG. 2.7. Schematic illustration of (a), (b) vertically uncoupled (c), (d) and coupled upper- and lower-level jet-front systems: (a) horizontal map showing upper-level jet exit region (isotachs, heavy solid lines; jet axis, solid arrow) aligned along and displaced toward cold side of surface frontal zone (isentropes, dashed lines; cold front, conventional symbols) and low-level jet (jet axis, open arrow; (b) cross section along AA′ indicated in (a) depicting upper- and lower-level jets (isotachs, thick dashed lines), upper-level and surface frontal zones (bounded by thin solid lines), tropopause (double solid lines), moist boundary layer (stippled) capped by lid, and streamlines of transverse ageostrophic circulation (heavy arrows, strength of circulation proportional to width); (c) as in (a) except for upper-level jet-exit region aligned across surface frontal zone; and (d) as in (b) except for cross section along BB′ indicated in (c) (Shapiro 1982). Source: Fig. 10 from Keyser (1999).

equation. This section will provide a brief overview of the dynamics relevant to upper-level frontogenesis and cyclogenesis with emphasis on the character of the vertical circulations.

Newton and Trevisan (1984) conducted one of the earliest idealized simulations of an upper-level jet front system. With the assumption of two-dimensional, steady-state flow, they showed that associated vertical circulations were driven by horizontal confluence and horizontal shear. The structure and extent of the vertical circulation associated with the Newton and Trevisan (1984) simulation are shown in Fig. 2.6 (Fig. 3 of Keyser 1999). Comparison of Figs. 2.6a and 2.6b reveals that subsidence is maximized along the jet axis in a region of cold-air advection embedded in an area of confluence in a jet entrance region at 500 hPa. Comparison of the streamlines of the total transverse circulation and potential temperature in a cross section cutting through the northwesterly flow in Fig. 2.6c with Figs. 2.6a and 2.6b establishes that subsidence is maximized along the warm boundary of the developing baroclinic zone in the mid-troposphere. The Newton and Trevisan (1984) idealized simulation bears many similarities to the observational findings of Reed (1955).

The extent to which the transverse circulations in the upper and lower troposphere are vertically coupled or uncoupled is also of interest. Shapiro (1982) presented a schematic illustration of coupled versus uncoupled circulations, reproduced here as Fig. 2.7 from Keyser (1999, his Fig. 10). In Figs. 2.7a,b a jet exit region is approaching a low-level frontal zone. The cross section through the exit region of the jet front system in Figs. 2.7b depicts weak subsidence ahead of the front with a sloping ascent zone extending rearward behind and above the front (anafront case). In this situation the lower- and upper-level transverse circulations are uncoupled and there is little chance for deep convection to develop in the warm air ahead of the front because of subsidence. Once the upper-level jet-front system has overspread the low-level frontal zone so that the exit region of the jet front system lies in the warm air (Figs. 2.7c), deep ascent is possible in the warm air ahead of the front as the upper- and lower-level transverse circulations become vertically coupled (Figs. 2.7d). Forecasters will recognize the inherent potential for deep convection to develop in the situation illustrated in Figs. 2.7d if the air ahead of the front is warm, moist, and unstable (convective available potential energy, CAPE, is positive). The quantitative underpinnings for the Shapiro (1982) schematics shown in Figs. 2.7 were provided by Hakim and Keyser (2001). Shown in Fig. 2.8 is their semigeostrophic transverse circulation solution for an upper-level jet-front system uncoupled (Fig. 2.8a) and coupled (Fig. 2.8b) to a lower-level frontal zone. Comparison of Figs. 2.7d with Fig. 2.8a and Fig. 2.7d with Fig. 2.8b shows that when the exit region of the jet-front system lies

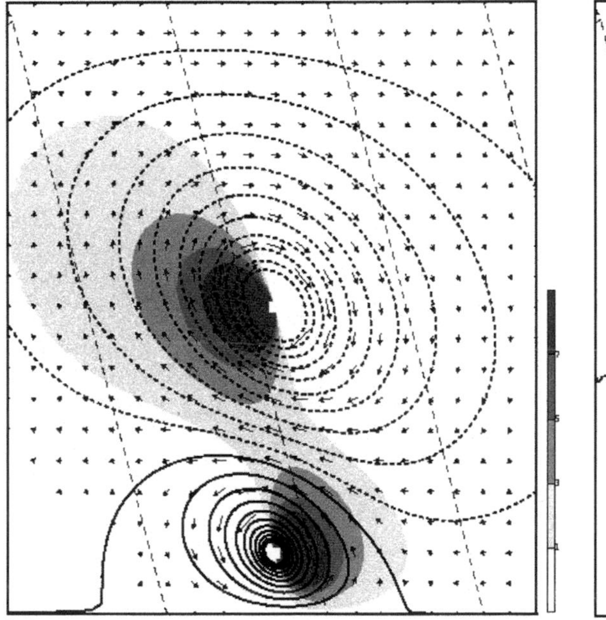

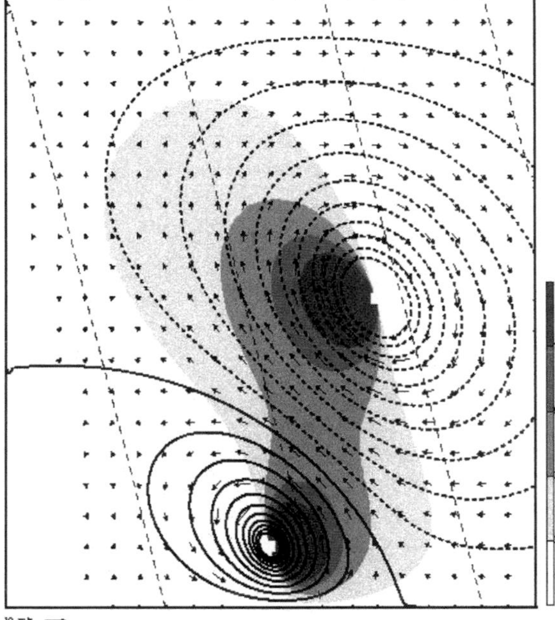

FIG. 2.8. (left) Semigeostrophic solution for an upper-level jet streak exit region overlying a surface frontal zone (refer to Figs. 2.7a,b). Values of F^2, S^2, and N^2 coincide with those for the control case (see Hakim and Keyser 2001). Streamfunction is given by thick lines (negative values dashed) every 2×10^3 m^2 s^{-1}; positive values of vertical motion ω are shaded every 2 cm s^{-1} starting at 1 cm s^{-1}, and absolute momentum is given by thin dashed lines every 30 m s^{-1}. Vectors depict the ageostrophic circulation, with ω scaled by a factor of 100 commensurate with the inverse aspect ratio of the physical dimensions of the domain, which is square in the plotting coordinates. (right) As in the left panel except for an upper-level jet streak exit region ahead of a surface frontal zone (refer to Figs. 2.7c,d). Source: Figs. 6 and 7 from Hakim and Keyser (2001).

in the warm air ahead of the front, the resulting vertical circulation is deep, in good agreement with the Shapiro (1982) schematics.

b. Rotunno et al. (1994)

Rotunno et al. (1994) used an idealized primitive equation similar to that employed by Keyser et al. (1992a) to investigate the relationship between upper-level frontogenesis and cutoff cyclone development. The evolution of potential temperature, pressure, and **Q** vectors at 6 km on days 6–9 of the simulation are shown in Fig. 2.9 (Fig. 13 of Rotunno et al. 1994). At day 6 (Fig. 2.9a) the flow at 6 km is mostly zonal with just the suggestion for weak cold advection in the northwesterly flow behind the trough. More subtle is the orientation of the **Q** vectors. They point toward warmer (colder) air where the geostrophic flow is weakly confluent (diffluent) behind the trough axis. Subsidence is maximized where the **Q** vectors are weakly divergent along a northwest–southeast-oriented line that roughly approximates where the flow changes from confluent to diffluent. Northeast (southwest) of the line delineating the maximum subsidence, the tilting term will contribute positively (negatively) to frontogenesis.

By days 7 and 8, a noticeable increase in cold advection is apparent just west of the trough axis as a cutoff cyclone develops poleward of the strengthening

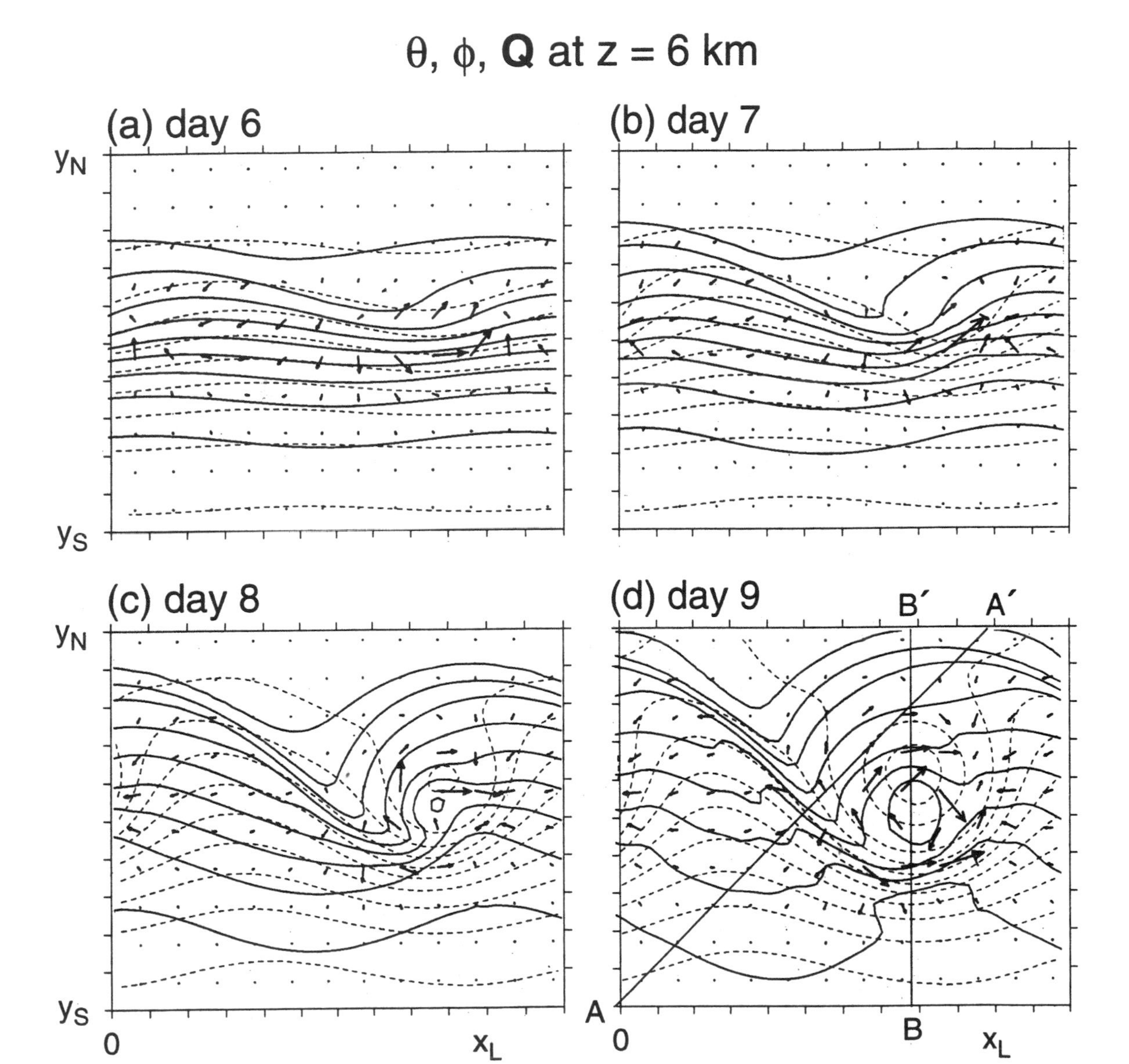

FIG. 2.9. Evolution of θ (solid lines, CI=5 K), ϕ (dashed lines, CI=1000 m^2 s^{-2}), and **Q** (vectors plotted every 4Δx; units not relevant) at $z = 6$ km for (a) day 6, (b) day 7, (c) day 8, and (d) day 9. Source: Fig. 13 from Rotunno et al. (1994).

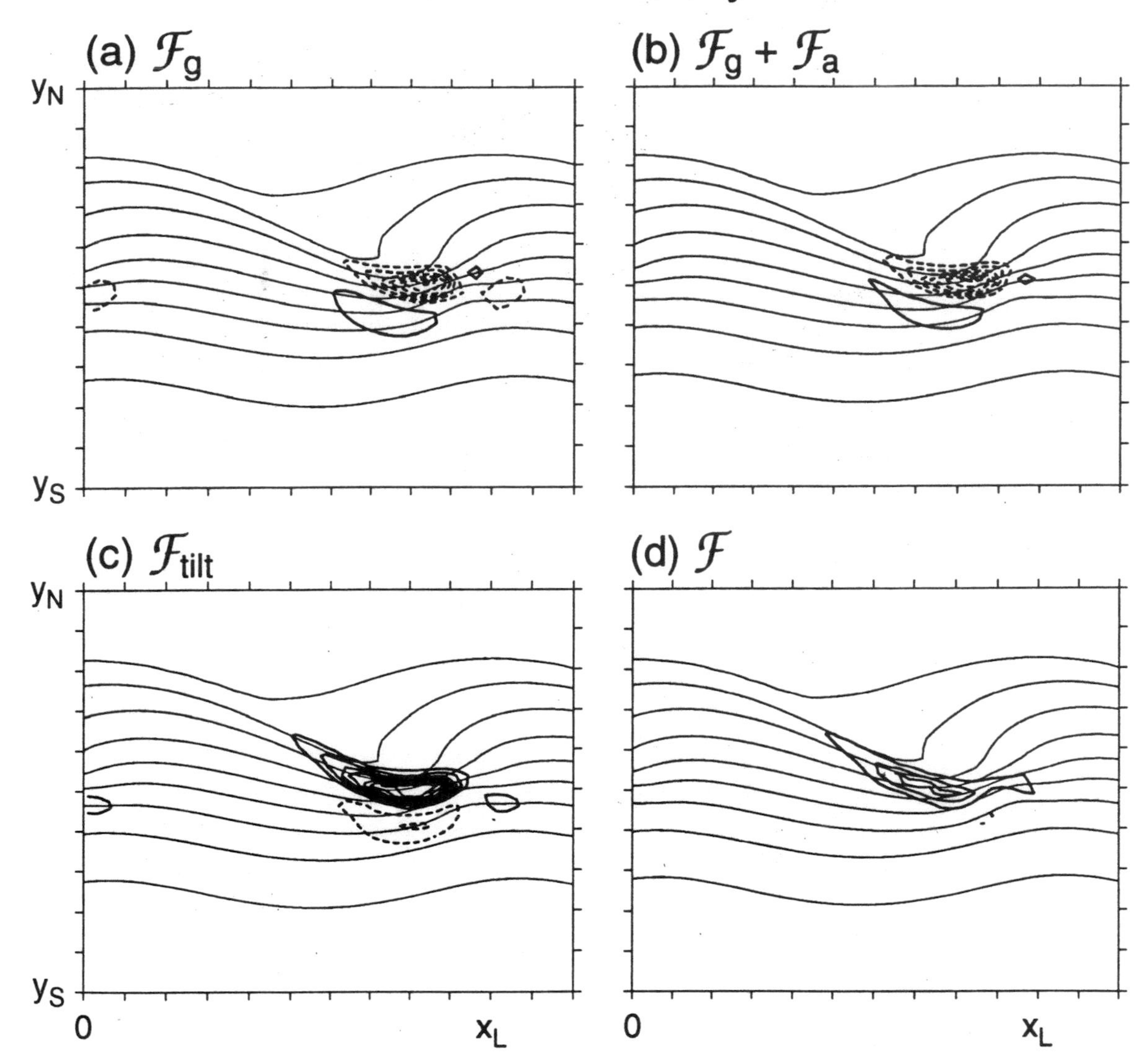

FIG. 2.10. Plot of θ (light solid lines, CI = 5 K) at day 7 at z = 6 km on the same domain as shown in Fig.2.9b displayed with (a) F_g, (b) F_g+F_a, (c) F_{tilt}, and (d) F. For all F, CI = 20 (K/100 km)^2/10^5 s. Source: Fig. 19 from Rotunno et al. (1994).

baroclinic zone (Figs. 2.9b,c). On the basis of the divergence of **Q** vectors across the baroclinic zone, subsidence is likely in the cold advection region (shown later in this chapter). Rotunno et al. (1994) argue that one way to interpret the onset of significant cold advection along the jet as the thermal gradient tightens across the front is that higher potential temperatures are brought down to the 6-km level from above by tilting-induced subsidence. To make this interpretation more explicit, displayed in Fig. 2.10 (Fig. 19 in Rotunno et al. 1994) are the geostrophic, ageostrophic, and tilting contributions to frontogenesis, and the total frontogenesis at 6 km for day 7. Comparison of Fig. 2.10c and 2.9b reveals that the positive tilting contribution to frontogenesis is maximized in the core of the baroclinic zone where the **Q** vectors are divergent. Furthermore, the **Q** vectors point toward colder (warmer) air on the cold (warm) side of the baroclinic zone, indicative of geostrophic frontolysis (frontogenesis) in Fig. 2.9b. This inference is confirmed by the geostropic frontogenesis dipole shown in Fig. 2.10a. The north-south orientation of the dipole indicates frontogenesis (frontolysis) where the flow is confluent to the south (diffluent to the north). The total frontogenesis shown in Fig. 2.10d reflects the overall dominance of the tilting contribution. Geostrophic frontogenesis, a secondary contributor, acts to weaken the tilting contribution, while the ageostrophic contribution to frontogenesis is minimal at day 7.

Shapiro (1981) and Keyser and Pecnick (1985) argued that a key element of upper-level frontogenesis was a combination of cold-air advection and horizontal confluence in northwesterly flow downstream of an upper-level ridge. In their view, subsidence associated with the cold-air advection was maximized toward the warm side of the baroclinic zone so that tilting contributed positively to frontogenesis. This so-called Shapiro effect contribution to upper-level frontogenesis is consistent with the early ideas of Reed and Sanders (1953) and Reed (1955). Rotunno et al. (1994) find that the Shapiro effect seems generally applicable in their idealized simulation of upper-level frontogenesis with the exception that the cold-air advection along the flow becomes critical only after strong midlevel subsidence develops by day 7. According to Keyser (1999), an interpretation of this discrepancy (relayed to him as a personal communication from Rotunno) "is that the Shapiro effect occurs in their model, but at a later (i.e., finite-amplitude) stage of baroclinic-wave development; their analysis is concerned with the early (i.e., linear) stage of wave evolution, where the phase lag between the midtropospheric temperatures and geopotential patterns is too small to establish along-front cold-air advection as required by the Shapiro effect."

c. Wandishin et al. (2000)

Just as Reed and Sanders (1953) and Reed (1955) demonstrated the great value of using PV as a tracer to deduce the three-dimensional motions during upper-level frontogenesis and tropopause folding, Hoskins et al. (1985) illustrated the great utility of employing PV thinking to address problems in dynamic meteorology. Similarly, Hoskins and Berrisford (1988), borrowing on an original idea by Kleinschmidt (1950), introduced "dynamic tropopause" (DT) maps, defined on the basis of a constant PV surface (e.g., 1.5×10^{-6} m^2 s^{-1} K kg^{-1}), as a useful means to elucidate important circulation features [see Wirth (2000) for a discussion of the thermal tropopause versus the DT in application to balanced flow anomalies]. A particularly valuable aspect of PV thinking has been the development of PV attribution techniques to help deduce cause and effect mechanisms during cyclogenesis (e.g., Davis and Emanuel 1991; Davis 1992; Davis et al. 1993; Davis et al. 1996; Hakim et al. 1996; Nielsen-Gammon and Lefevre 1996). Wandishin et al. (2000) applied the PV approach to study upper-level frontogenesis within a developing baroclinic wave. They note that, "Previous studies of upper-level frontogenesis have emphasized the role of the vertical circulation in driving stratospheric air down into the midtroposphere. Here, a PV-based approach is adopted that focuses on the generation of a folded tropopause."

To facilitate comparison with earlier work, Wandishin et al. (2000) apply a PV perspective to the upper-level front and developing baroclinic wave simulated by Rotunno et al. (1994) by replacing potential temperature by the height of the DT in the Miller (1948) frontogenesis equation. Their resulting "foldogenesis function" provides a way to quantify the effect of the wind on the slope of the DT and is written as

$$F = \frac{\mathrm{d}\nabla z_T}{\mathrm{d}t} = F_\mathrm{h} + F_\mathrm{w}, \tag{3}$$

where

$$F_\mathrm{h} = \hat{\boldsymbol{n}} \cdot \left[-\left(\frac{\partial u}{\partial x}\frac{\partial z_T}{\partial x} + \frac{\partial v}{\partial x}\frac{\partial z_T}{\partial y}\right)\hat{\mathrm{i}} - \left(\frac{\partial u}{\partial y}\frac{\partial z_T}{\partial x} + \frac{\partial v}{\partial y}\frac{\partial z_T}{\partial y}\right)\right], \tag{4}$$

$$F_\mathrm{w} = \nabla \mathrm{w} \cdot \hat{\mathrm{n}}. \tag{5}$$

Here, z_T is the height of the DT, the u and v wind components are measured along the DT surface, ∇z_T is evaluated along the DT, and $\hat{\mathrm{n}}$ is a horizontal unit vector perpendicular to the DT in the direction of the gradient of DT height. Note that (4) is just the standard form of Petterssen's (1956) frontogenesis equation with potential temperature replaced by z_T. A schematic of how the terms on the right-hand side of (3) contribute to "foldogenesis" is shown in Fig. 2.11 (Fig. 4 in Wandishin et al. 2000).

Descent changing to ascent in the direction of increasing DT height will steepen the DT (Fig. 2.11a). Horizontal confluence/convergence across a sloping DT

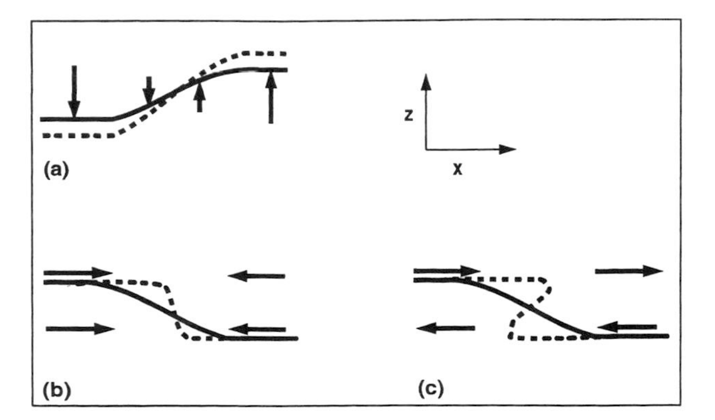

FIG. 2.11. Schematic diagrams of foldogenesis. The thick solid and dashed lines denote the tropopause at times t_o and $t_o + \Delta t$, respectively. Arrows indicate sense and magnitude of the winds in the vicinity of the tropopause. (a) Diagram illustrating differential vertical motion. If the gradient of the vertical motion is of the same sense as the height gradient, the tropopause steepens. (b) Diagram illustrating confluence/convergence. The differential advection of the tropopause causes it to steepen, but as the tropopause approaches vertical the differential advection decreases. (c) Diagram illustrating vertical shear. The winds along the tropopause are identical to the confluence/convergence case at t_o, but differential advection by the vertical shear continues after the tropopause becomes vertical to produce a tropopause fold. Source: Fig. 4 from Wandishin et al. (2000).

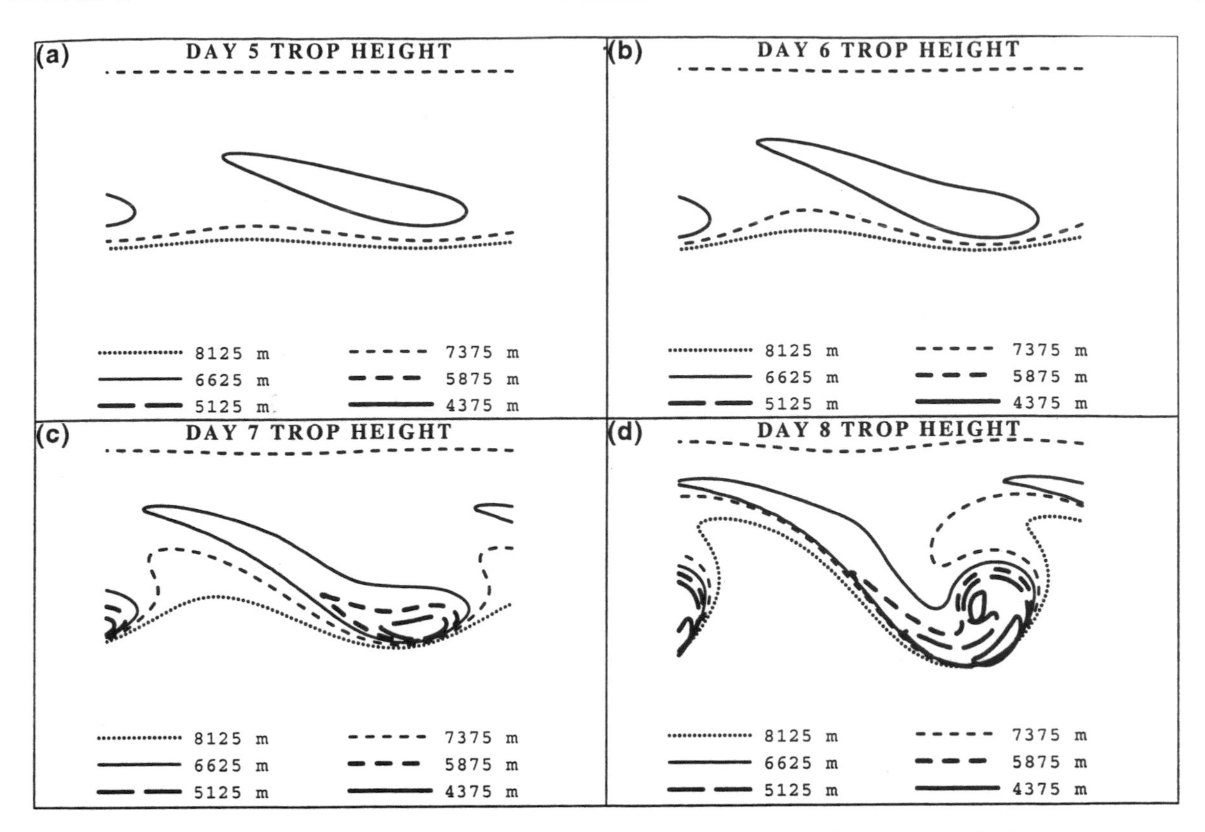

FIG. 2.12. Plot of θ (light solid lines, CI=5 K) at day 7 at z=6 km on the same domain as shown in Figs. 2.12c and 2.13c displayed with (a) F_g, (b) F_g+F_a, (c) F_{tilt}, and (d) F_s. For all F_s, CI=20 $(K/100\ km)^2/10^5$ s. Source: Fig. 19 from Rotunno et al. (1994).

will steepen the DT (Fig. 2.11b). Vertical shear in the presence of a sloping DT will steepen the DT and can result in a folded DT (Fig. 2.11c). Horizontal confluence/convergence by itself can steepen the DT until it is vertical, but it cannot produce a tropopause fold. Note also that at the point where the DT approaches the vertical, the slope becomes infinite and (3) is no longer generally applicable.

The DT is mapped using the contour superposition method (Morgan and Nielsen-Gammon 1998) for days 5–8 in Fig. 2.12 (Fig. 8 in Wandishin et al. 2000). The DT, nearly undisturbed on day 5 (Fig. 2.12a), steepens somewhat on day 6 (Fig. 2.12b), steepens strongly by day 7, becoming nearly vertical in the base of the trough (Fig. 2.12c), and then folds by day 8 (Fig. 2.12d). By days 7 and 8 the DT descends toward the ground as judged by the appearance of the 5875-m, 5125-m, and 4375-m closed height contours. The descent of the DT occurs in conjunction with upper-level frontogensis (not shown, but see Figs. 6 and 7 in Wandishin et al. 2000). Comparison of the DT structure on days 6–7 (Figs. 2.12b,c) with the 6-km fields of potential temperature, pressure, and **Q** vectors from Rotunno et al. (1994) for the same time periods (Figs. 2.9a,b) shows that the steepening of the DT to a "vertical wall" and the descent of the DT toward the ground occurs where subsidence is maximized in the region of **Q** vector divergence. Further comparison of Fig. 2.12c with the day 7 frontogenetical function at 6 km (Fig. 2.10d) and the potential temperature and perturbation height at 4125 m (Fig. 6a of Rotunno et al. 1994, not shown) suggests that the DT vertical wall lies near the southern boundary of the region of intense frontogenesis due to tilting.

Fig. 2.13 (Fig. 9 in Wandishin et al. 2000) depicts the height of the DT and the foldogenesis function for days 5–8. In general, Fig. 2.13 illustrates that foldogenesis (foldolysis) occurs upstream (downstream) of the trough axis throughout the life cycle of the baroclinic wave, that the foldogenesis function increases by nearly a factor of 100 between days 5 and 8, and that the foldogenesis mechanism is operative while the baroclinic wave is in its quasi-linear growth phase 3 days before the appearance of the tropopause fold on day 8. The reconciliation of the foldogenesis viewpoint of Wandishin et al. (2000) with the Shapiro effect (cold-air advection in confluent flow and forced subsidence) mechanism of Rotunno et al. (1994) requires consideration of how the presence of vertical wind shear across an already tilted DT can result in a further steepening of the DT (foldogenesis) in the presence of an upstream-tilted baroclinic wave. As yet, no detailed observations exist to test these different physical interpretations of upper-level frontogenesis. Progress can probably be made through additional numerical studies using existing "datasets" obtained from operational short-range prediction models and future research models. The adaptation of a PV perspective in these numerical studies is likely to pay dividends, given the general applicability of the PV approach to studies of upper-level frontogenesis and tropopause folding.

6. Applications of PV thinking

a. Overview

The application of PV thinking concepts (Hoskins et al. 1985; Hoskins 1990; McIntyre 1999) to the real atmosphere has its roots in the pioneering use of PV as a tracer by Reed, Sanders, and Danielsen (and others) to help deduce the three-dimensional structure of cyclones and upper-level fronts. The cornerstone and value of the PV perspective, whether approached observationally or theoretically, is that the relevant dynamics can be inherently connected with the observations to help facilitate a cause and effect understanding of the physical mechanisms at work in the atmosphere.

b. Cyclogenesis

Potential vorticity inversion techniques (e.g., Davis et al. 1996) have been used to diagnose the balanced dynamics of cyclones. An example from Davis et al. (1996, their Fig. 22) is shown in Fig. 2.14 for 12- and 24-h forecast simulations of the explosive western Atlantic cyclogenesis event of 4–5 January 1989. Plotted in Fig. 2.14 are cross sections of geopotential height perturbations associated with upper-level PV (Fig. 2.1a,d), interior (diabatic produced) PV (Fig. 2.14b,e), and lower-boundary potential temperature (Figs. 2.14c,f) through the surface cyclone. Height perturbations associated with upper-level PV dominate much of the troposphere and lag the surface cyclone by almost one-quarter wavelength, as is characteristic of an intensifying baroclinic system. However, the biggest contributor to surface height perturbations (>300 m at 0000 UTC 5 January 1989) is from diabatically generated interior PV. Negative height perturbations associated with a maximum in lower boundary potential temperature are weaker and mostly confined to the region ahead of the cyclone. Note also that the prominent height perturbations associated with the upper-level PV are consistent with the idealized adiabatic simulation of this case by Reed et al. (1994). As noted in Davis et al. (1996), the

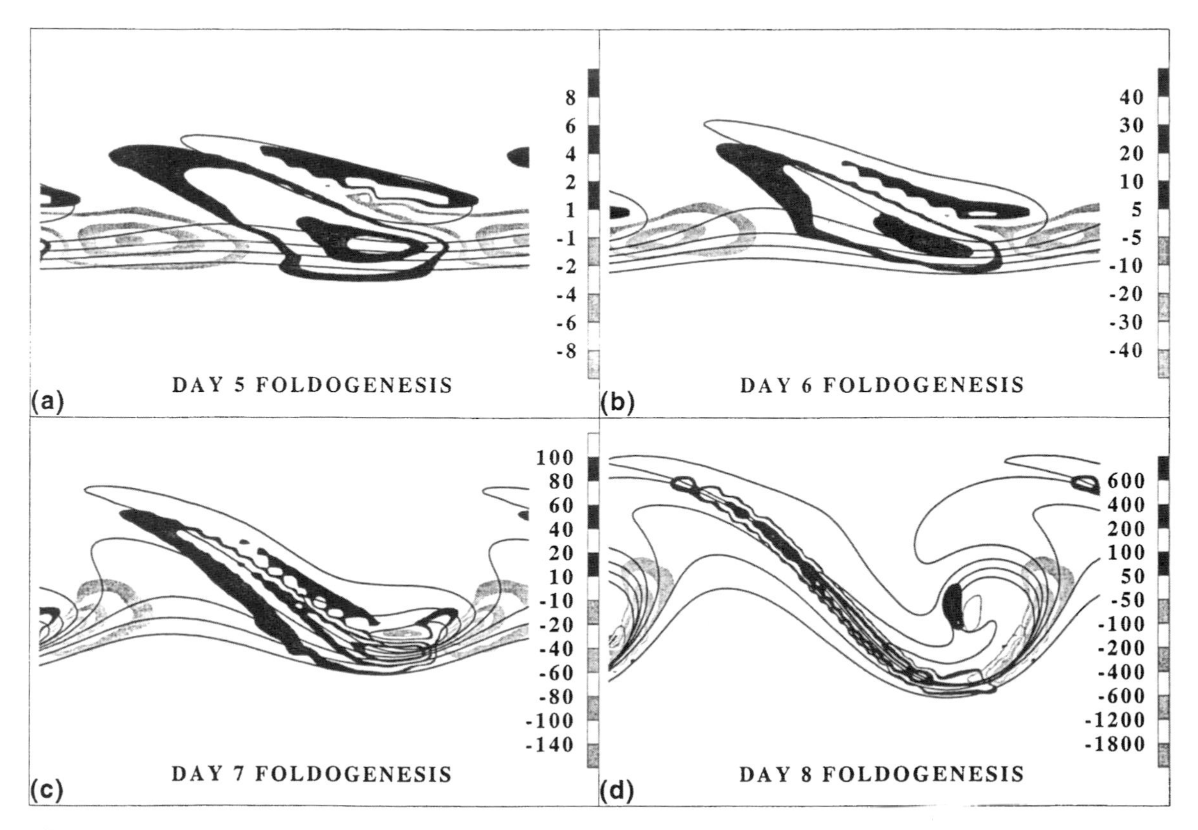

FIG. 2.13. Tropopause height z_T as in Fig. 2.12 (with the addition of the 8875- and 9625-m contours) but including the foldogenesis function (10^{-8} s^{-1}, shaded). Shading intervals are indicated in the panels. Domain is the same as in Fig. 2.12 except $y \in (3000, 6000)$ km. Source: Fig. 9 from Wandishin et al. (2000).

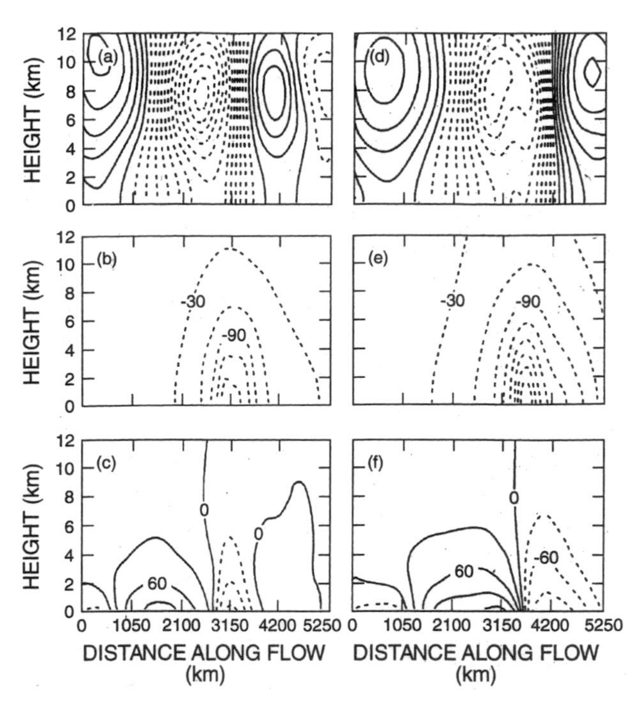

FIG. 2.14. Cross sections (see Fig. 12 of Davis et al. 1996) of geopotential height perturbations associated with (a) upper PV at 1200 UTC 4 Jan, (b) interior PV at 1200 UTC 4 Jan, (c) lower-boundary θ at 1200 UTC 4 Jan, (d) upper PV at 0000 UTC 5 Jan, (e) interior PV at 0000 UTC 5 Jan, and (f) lower-boundary θ at 0000 UTC 5 Jan. The contour interval is 30 m. Source: Fig. 22 from Davis et al. (1996).

intensification computed by Reed et al. (1994) was roughly one-half what they found in their full-physics simulation. Potential vorticity inversion and attribution techniques like those used by Davis et al. (1996) might be employed to address the issue of whether upper-level frontogenesis and tropopause folding play an active or passive role in cyclogenesis.

c. Dynamic tropopause and downstream development

The use of DT maps to visualize global and regional circulations and weather systems has been summarized by Morgan and Nielsen-Gammon (1998) and Nielsen-Gammon (2001). Among the features readily seen on DT maps are jets, baroclinic wave guides, coherent disturbances (Hakim 2000), wave breaking with ridge building at high latitudes and cutoff cyclone development at low latitudes, anticyclonic PV plumes emanating from deep convection in the Tropics, and outflow channels from hurricanes and transitioning tropical cyclones.

Shown in Fig. 2.15 is an example of a baroclinic wave guide and downstream development from a DT perspective in the Northern Hemisphere for the period 5–11 December 1996 as taken from Nielsen-Gammon (2001, his Fig. 4). Evident from Fig. 2.15 is the downstream propagation of high-amplitude troughs and ridges from eastern Asia and the western Pacific on 6 December to the eastern Atlantic and western Europe on 11 December. A common signature of the maps is the strong gradient of potential temperature that marks the location of the jet on the DT. Eruptions of high potential temperature air well poleward occur in conjunction with ridging while excursions of low potential temperature air deep into the subtropics occur with troughing. The poleward and equatorward meanders of the potential temperature gradient associated with the jet and its subsequent evolution are consistent with previous ideas about Rossby wave packets and downstream development (see, e.g., Namias and Clapp 1944; Orlanski and Katzfey 1991; Orlanski and Sheldon 1993, 1995; Bosart et al. 1996; Lackmann et al. 1996; Chang 1993, 1999). The formation of upper-level fronts might be anticipated to occur preferentially in northwesterly flow downstream of amplifying ridges where the potential temperature gradient on the DT becomes very large.

d. Dynamic tropopause and major cyclogenesis

One of the more useful applications of DT maps is to help identify characteristic circulation signatures associated with major cyclogenesis. As an example, displayed in Fig. 2.16 are DT maps for selected times for the Cleveland "Superbomb" of 25–26 January 1978 (e.g., Salmon and Smith 1980; Hakim et al. 1995, 1996) and the "Superstorm" ("Storm of the Century") of 13–14 March 1993 (e.g., Kocin et al. 1995; Huo et al. 1995; Bosart et al. 1996; Dickinson et al. 1997). Both storms illustrate the common major cyclogenesis signature of the formation of a large "PV hook" caused by the cyclonic wrapup of PV (see also Bell and Bosart 1993; Bell and Keyser 1993; Simmons 1999), here manifest as the northward and then westward expansion of higher potential temperature air (with lower PV) on the poleward side of the cyclone while lower potential temperature air (with higher PV) is drawn southward and then eastward. Severe-weather aficionados will recognize the PV hooks shown in Fig. 2.16 as much larger versions of radar reflectivity hooks associated with prominent supercells, illustrating once again how easy it is to get "hooked" on this atmosphere of ours. The theoretical basis for the large-scale PV hook can be understood from Fig. 2.9c,d as part of the development of a cutoff cyclone aloft and attendant upper-level front in the northwesterly flow immediately equatorward of the cutoff cyclone [see also the potential temperature evolution at 4 km for days 6–9 of the idealized baroclinic wave simulation shown in Fig. 15 of Rotunno et al. (1994) and note in particular the intense upper-level front that is apparent by day 8].

e. Dynamic tropopause and nonconservation of PV

Widespread nonconservation of PV is also readily apparent from DT maps. Fig. 2.17 , taken from Bosart (1999, his Fig. 9), shows DT maps from the aforementioned March 1993 Superstorm at 0000 UTC 12–14

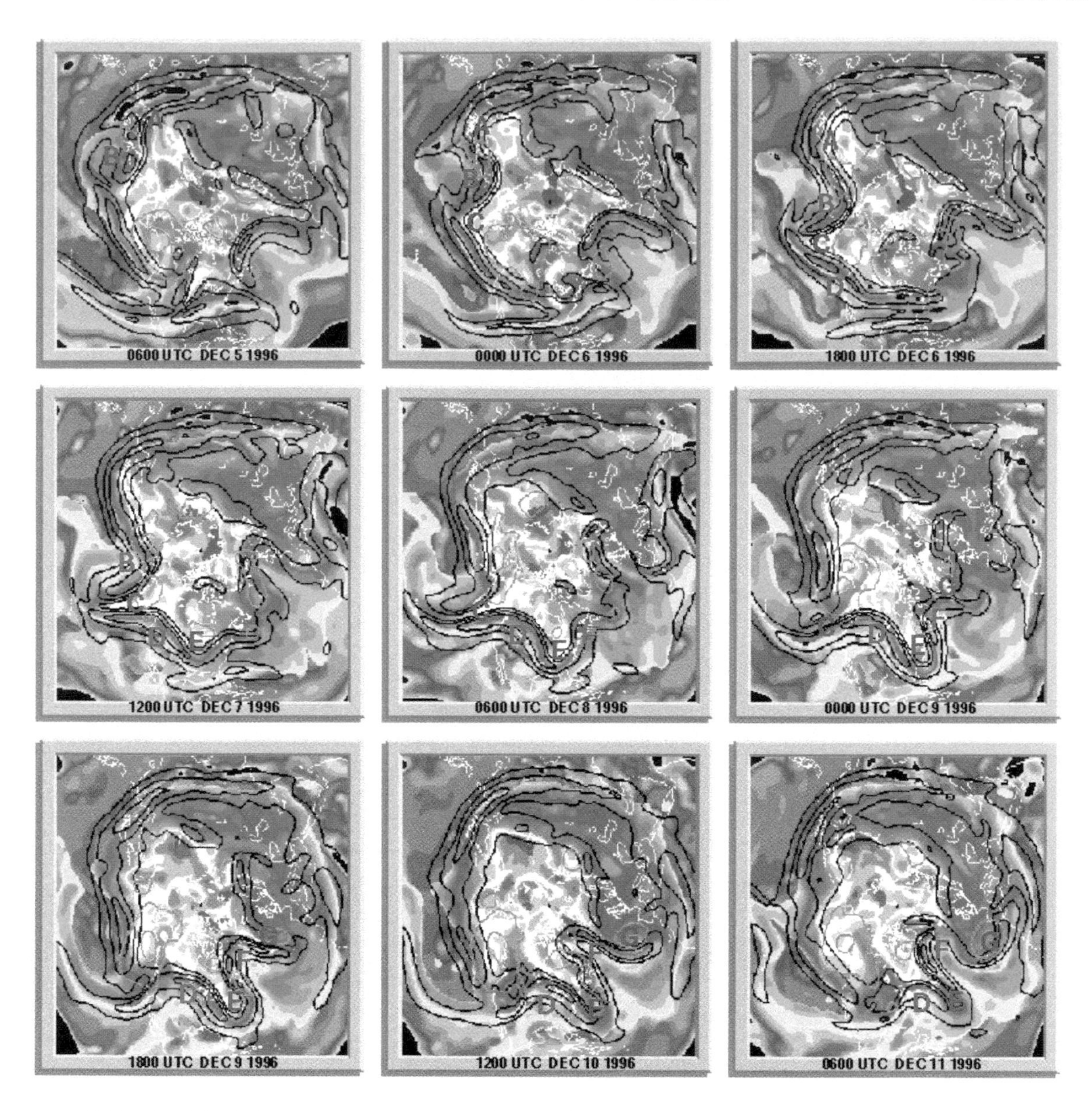

FIG. 2.15. Northern Hemisphere potential temperature on the DT as defined by the 1.5-PVU surface (1 PVU=10^{-6} K m^2 kg^{-1} s^{-1}) is depicted by the colors over the 260–370-K potential temperature range at 5-K intervals for the period 0600 UTC 5 Dec to 0600 UTC 11 Dec 1996 at 18-h intervals. Warm (cool) colors correspond to high (low) values of potential temperature within the 260–370-K band (near the equator, PV on the 370-K surface is colored). Black contours are 250-hPa wind speed at 20 m s^{-1} increments starting at 30 m s^{-1}. Red contours are 1000-hPa heights contoured every 200 m for values of 0 m and below. Labels A-G indicate successive troughs and ridges in a Rossby wave packet. Figure prepared using the "contour-fill overlay method." Source: Fig. 4 from Nielsen-Gammon (2001).

March 1993 and a 24-h forecast DT map verifying at 0000 UTC 13 March 1993 from the National Centers for Environmental Prediction (NCEP) Medium-Range Forecast (MRF) model. The prominent feature to note from the DT analyses, gleaned from the NCEP MRF 2.5° gridded analyses, is the eruption of an enormous ridge and associated jet on the DT over the southeastern United States in the 24-h period ending 0000 UTC 13 March. The building ridge and intensifying jet occur poleward and eastward of a massive eruption of organized deep convection over the northwestern Gulf of Mexico (Bosart et al. 1996; Dickinson et al. 1997). Wind speeds in the southwesterly flow jet double from 30–40 to 70–80 m s^{-1} during this 24-h period. Further ridging over eastern North America continues through 0000 UTC 14 March. The ridging is manifest as an increase in potential temperature and a decrease in pressure on the DT over the southeastern United States.

Inspection of the NCEP-analyzed winds on the DT reveals that these observed changes in potential temperature and pressure on the DT cannot be explained by simple advection. Rather, they must be a reflection of the

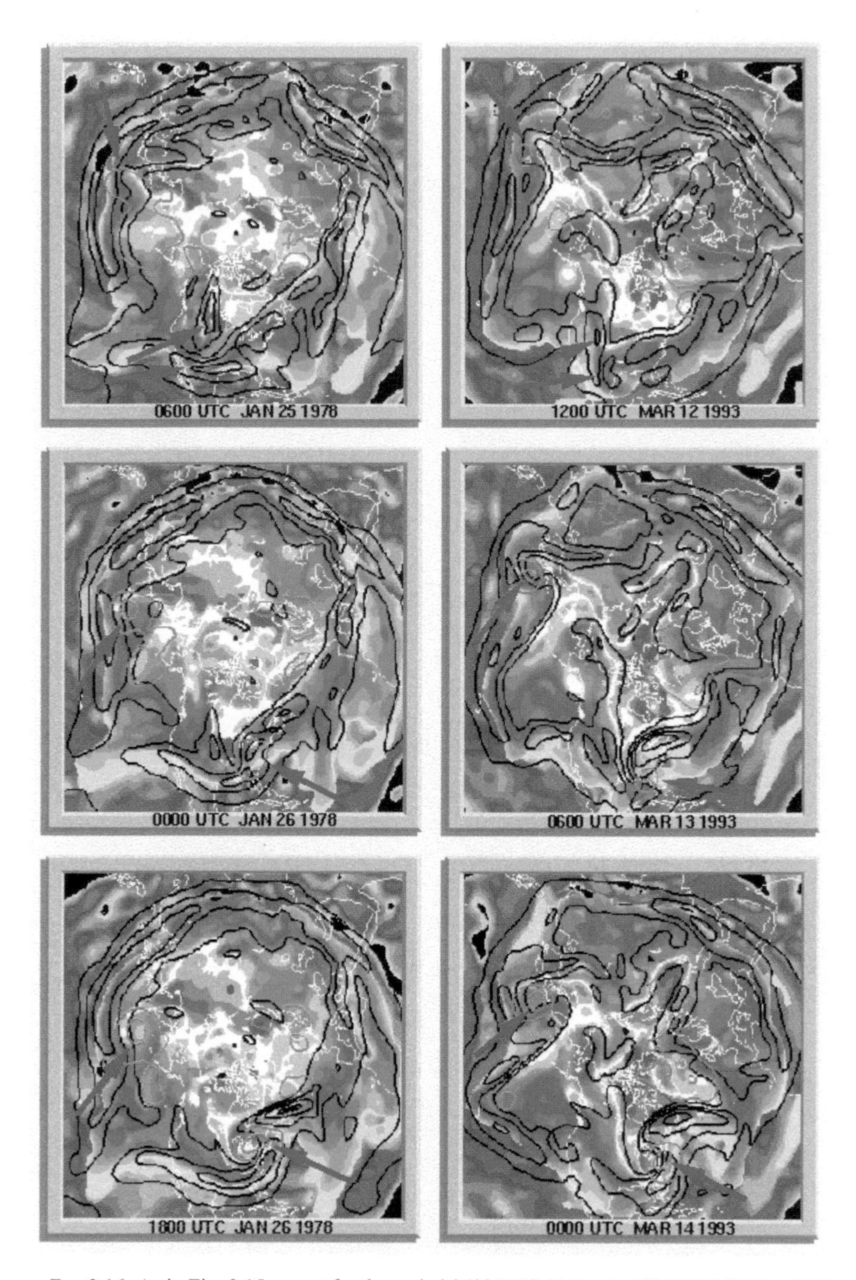

FIG. 2.16. As in Fig. 2.15 except for the period 0600 UTC 25 Jan to 1800 UTC 27 Jan 1978 (left column, the Cleveland Superbomb) and from 1200 UTC 12 Mar to 0000 UTC 14 Mar 1993 (right column, the "Storm of the Century"), at 18-h intervals. Source: Fig. 7 from Nielsen-Gammon (2001).

broad upscale effect of widespread latent heat release from organized deep convection. Proof for this assertion comes from the MRF 24-h DT forecasts shown in Fig. 2.17j–l. Note that the MRF model, which failed to simulate the intensity of the initial cyclogenesis and associated deep convection in the northwestern Gulf of Mexico, also failed to produce the expected downstream theta and pressure ridging on the DT over the southeastern United States as was observed. As shown in Dickinson et al. (1997), the failure of the MRF forecast to replicate the observed nonconservation of PV could likely be attributed to problems with the model convective parameterization scheme. It is also conceivable that the formation of an intense jet downstream of a major trough might also be associated with the development of an upper-level front. Although upper-level fronts tend to occur preferentially in northwesterly flow downstream of a ridge, their formation in southwesterly flow ahead of a trough has recently been documented by Schultz and Doswell (1999). An example of upper-level front formation in southwesterly flow will be presented in section 7.

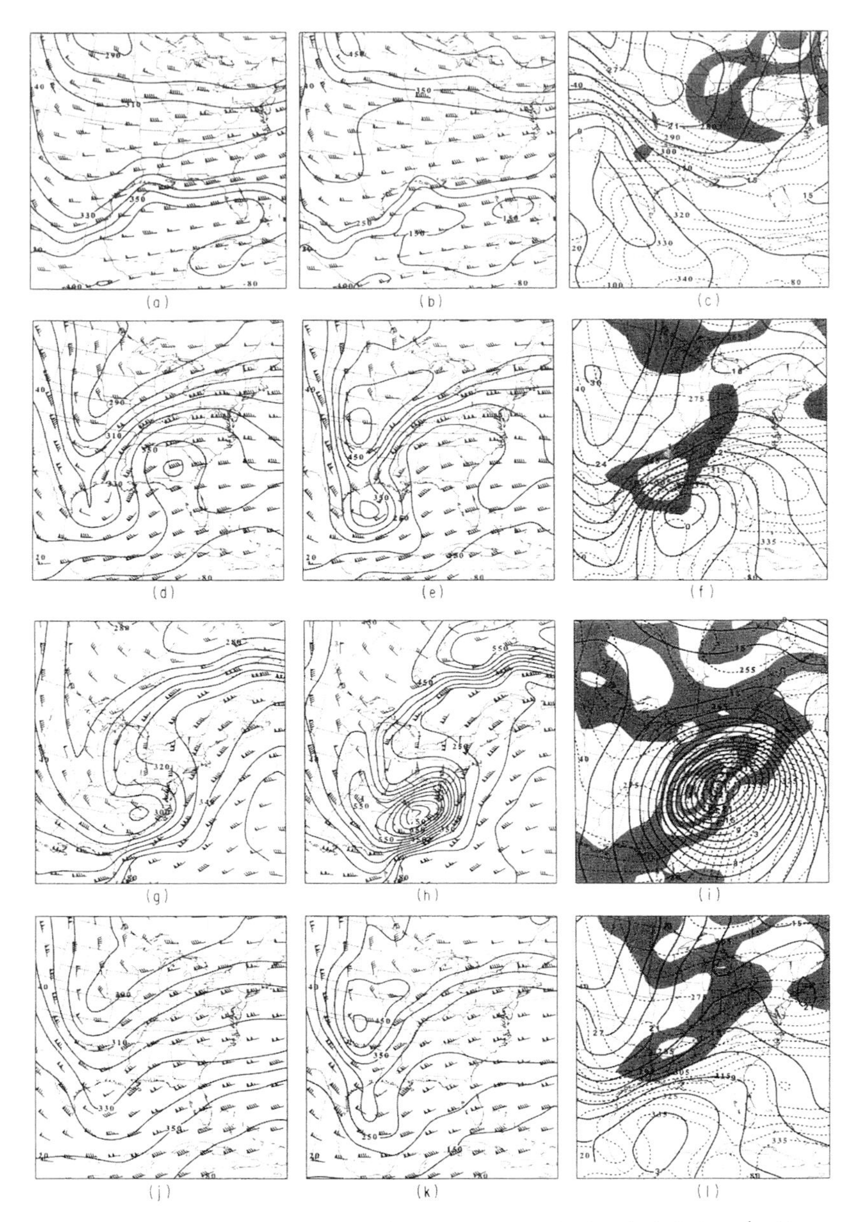

FIG. 2.17. (a): Potential temperature (solid contours every 10 K) and winds (m s^{-1} with one pennant, full barb, and half barb denoting 25, 5, and 2.5 m s^{-1}, respectively) on the dynamic tropopause defined by the 1.5-PVU surface for 0000 UTC 12 Mar. 1993; (b) as in (a) except for pressure (solid contours every 100 hPa) and winds on the dynamic tropopause; (c) as in (a) except for mean sea level pressure (solid contours every 4 hPa) with superimposed 850-700-hPa layer mean PV (shaded according to the scale every 0.5 PVU beginning at 0.5 PVU) and 850-hPa equivalent potential temperature (dashed contours every 5 K); (d)–(f) as in (a)–(c) except for 0000 UTC 13 Mar 1993; (g)–(i) as in (a)–(c) except for 0000 UTC 14 Mar 1993; (j)–(l) as in (a)–(c) except for 24-h MRF forecasts verifying 0000 UTC 13 Mar 1993. Source: Fig. 9 from Bosart (1999).

f. Dynamic tropopause and the "Perfect Storms"

The period 27–31 October 1991 featured the development of twin "Perfect Storms." Most of the press has been reserved for the "Halloween storm" that raked the western Atlantic and coastal regions of New England, New York, New Jersey, and Delaware with high seas and high winds (Cardone et al. 1996; Junger 1997). Equally noteworthy, however, was the remarkable near-simultaneous cyclogenesis event over interior North America that produced record snowfall over parts of the upper Midwest but, as of this writing, has gone

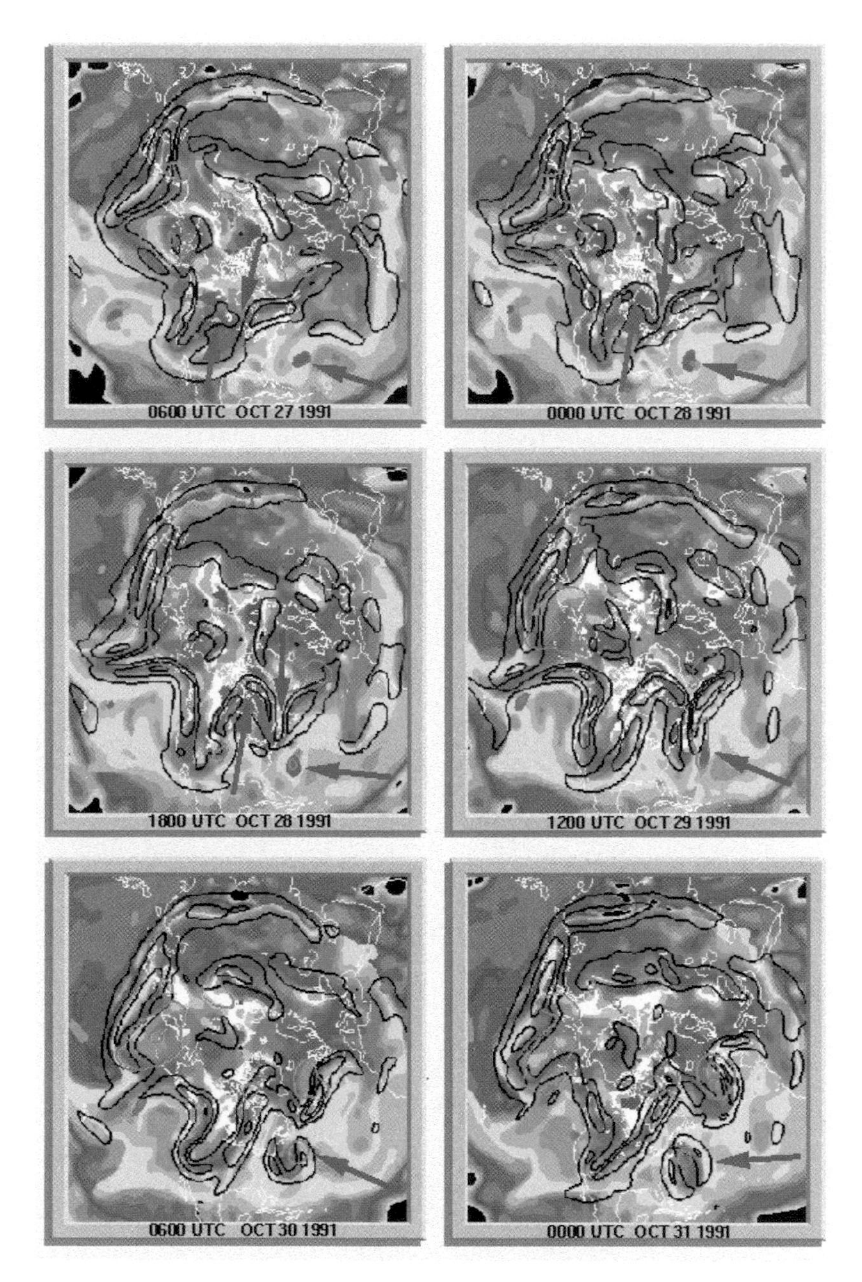

FIG. 2.18. As in Fig. 2.15 except for the period 0600 UTC 27 Oct to 0000 UTC 31 Oct 1991, at 18-h intervals. Red arrow indicates developing trough, pink arrow indicates upstream developing ridge, and blue arrow indicates PV initially associated with Hurricane Grace and later with the "Perfect Storm." Source: Fig. 8 from Nielsen-Gammon (2001).

largely unrecognized. A DT perspective on the twin perfect storms, taken from Fig. 8 in Nielsen-Gammon (2001), is presented in Fig. 2.18. Evident from Fig. 2.18 is that the eastern perfect storm forms subsequent to massive ridging on the DT over eastern Canada that enables a sharp trough to deepen downstream across Labrador into the western Atlantic. As the Labrador trough moves equatorward, it begins to interact with Hurricane Grace, represented by a small-scale quasi-circular region of lowered potential temperature and elevated pressure on the DT, on 29 October. [Nielsen-Gammon (2001) noted that hurricanes are "seen" in these DT analyses "because the stratosphere is assumed to include any air above 650 hPa with an absolute value of PV greater than 1.5 PVU." Mature hurricanes, characterized by deep columns of diabatically generated PV, can often exceed this PVU threshold. However, the coarse resolution of the NCEP/National Center for Atmospheric Research (NCAR) reanalyses used to generate these DT maps results in "hurricanes" that are horizontally too large in terms of PV and horizontally too cool in terms of potential temperature.]

The resulting interaction produces a memorable storm, as Grace is first ingested into the cyclonic circulation envelope and then involved in the transition to an extratropical cyclone that grows in scale and moves westward and equatorward immediately to the east of the North American continent. Very high (low) potential temperature (pressure) air on the DT in the outflow channel poleward of Grace is drawn well poleward and westward, effectively isolating the transitioned cyclone from the westerlies by 31 October. The westward advection of high potential temperature air across eastern Canada also appears to aid ridge and jet development in the southwesterly flow ahead of the western perfect storm over the Midwest. Note also that this Midwest cyclone appears to result from downstream development, as evidenced by massive ridging on the DT over the eastern Pacific on 29 October and equally massive troughing over the western United States 24 h later. The signature of an upper-level front, as evidenced by a strong gradient of potential temperature on the DT, appears initially in the northwesterly flow behind the western U.S. trough and then subsequently in the southwesterly flow ahead of this trough where the gradient of potential temperature appears to be enhanced by the arrival of high potential temperature air moving westward away from the western Atlantic cyclogenesis region. In this regard, the jet development in the southwesterly flow over interior North America bears some resemblance to the aforementioned jet development ahead of the deepening trough associated with the March 1993 superstorm (Fig. 2.17).

g. Extratropical transitions

There has been a renewed interest in the problem of understanding the dynamics of extratropical transitions in general and the rainfall prediction problem in particular associated with transitioning storms that make landfall (see, e.g., Bosart and Lackmann 1995; Browning et al. 1998; Harr and Elsberry 2000; Klein et al. 2000; Thorncroft and Jones 2000; Hart 2003; Hart and Evans 2001; Ritchie and Elsberry 2001; Atallah and Bosart 2003). The use of PV thinking to diagnose transitioning extratropical

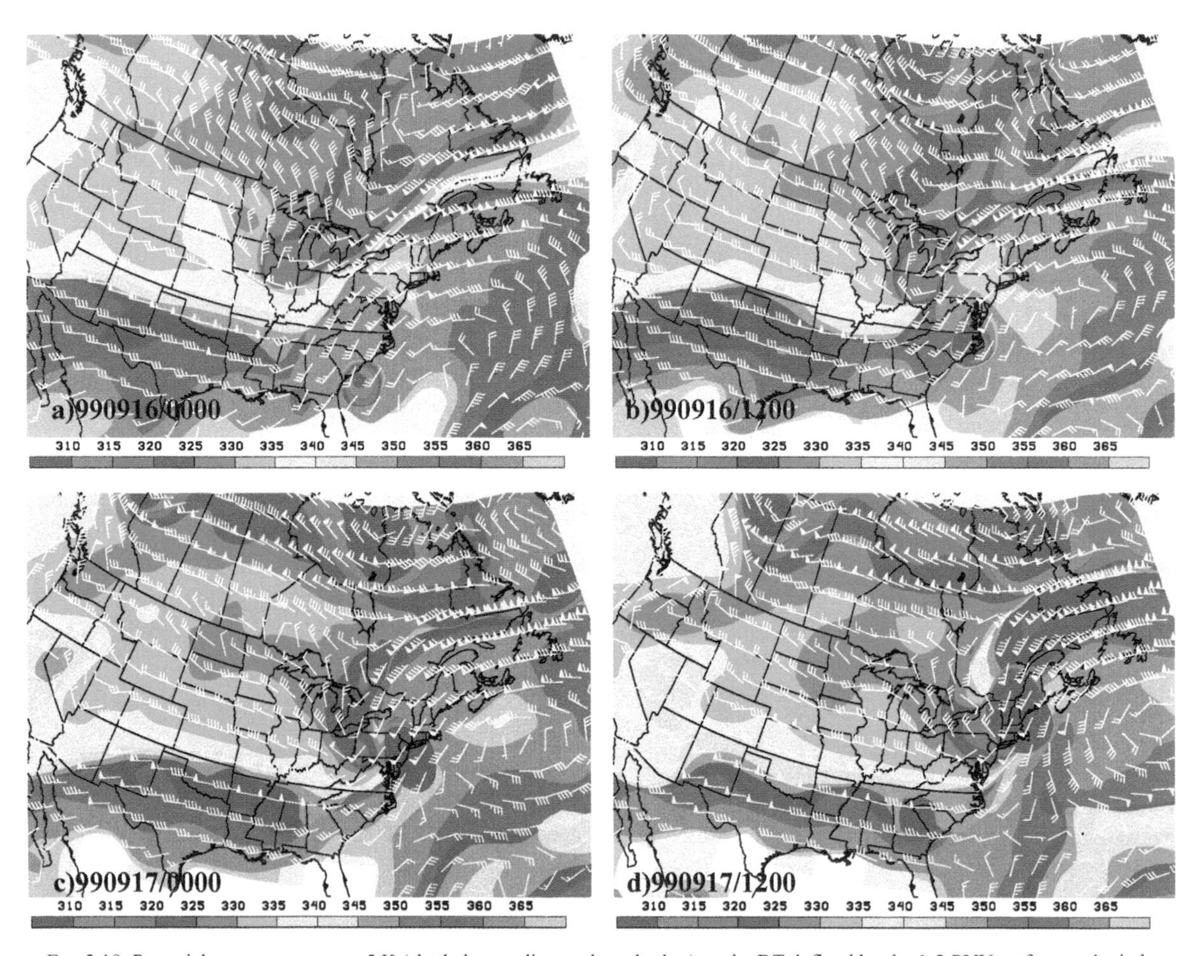

FIG. 2.19. Potential temperature every 5 K (shaded according to the color bar) on the DT defined by the 1.5-PVU surface, and winds on the DT (computed looking down) with one pennant, full barb, and half barb denoting 25, 5, and 2.5 m s^{-1}, respectively, for (a), (b) 0000 and 1200 UTC 16 Sep 1999 and (c), (d) 0000 and 1200 UTC 17 Sep 1999. Plot prepared using the contour-fill overlay method of Morgan and Nielsen-Gammon (1998) and used in Figs. 2.15, 2.16, and 2.18.

cyclones is appealing because (1) the expected nonconservation of PV associated with massive diabatic heating should be reflected in the evolution of the PV signatures, and (2) the development of low- and upper-level baroclinic zones and outflow jets as a poleward-moving tropical cyclone interacts with the midlatitude westerlies should be readily apparent.

A recent example of a significant extratropical transition was Hurricane Floyd over eastern North America on 16–17 September 1999. Figure 2.19, taken from Atallah and Bosart (2003), shows maps of potential temperature and winds on the DT every 12 h for the 36-h period ending 1200 UTC 17 September (the discussion in section 6f about problems with the representation of hurricanes on the DT from coarse-resolution gridded analyses also applies here). The calculations were made using NCEP Aviation Model (AVN) initialized grids available on a 2.5° latitude-longitude resolution. The DT maps were prepared using the contour-fill overlay method (winds were interpolated to the DT looking down) described by Morgan and Nielsen-Gammon (1998). The contour-fill overlay method works as follows. Areas with stratospheric values of PV greater than 1.5 PV units (PVU) are shaded using color 1 starting at some high potential temperature value (365 K in this case). The process is repeated for the next lowest potential temperature surface (360 K) using a slightly different color (color 2). Because the domain of PV values greater than 1.5 PVU will be smaller on the 360-K potential temperature surface as compared to the 365-K surface, a wavy band of varying width containing color 1 will remain after the second iteration. Potential temperature values within this wavy band will be between 360 and 365 K. A third iteration will produce another wavy band containing color 2 for potential temperature values between 355 and 360 K. The iteration process is repeated sequentially at 5-K intervals until a lower potential temperature threshold is reached (305 K in this case). The resulting map shows the distribution of potential temperature on the DT. Likewise, this map can also be viewed as a representation of the distribution of upper-tropospheric PV as noted by Nielsen-Gammon (2001).

Hurricane Floyd made landfall in North Carolina near 0900 UTC 16 September. Prior to landfall at 0000 UTC 16 September, Floyd is marked by an almost circular area of low potential temperature on the DT map off of the North Carolina coast, indicative of a symmetric storm associated with a deep, vertical column of PV (Fig. 2.19a). A prominent outflow jet with winds exceeding 85 m s^{-1} is apparent where the DT is nearly vertical ahead of an advancing trough with low potential temperature on the DT over the Great Lakes.

As Floyd moves up the Atlantic coast, the trough to the northwest advances southeastward toward the storm by 1200 UTC 16 September (Fig. 2.19b). Note also the tropopause fold that develops by this time from West Virginia to north of Maine as the yellow and orange colors "disappear" beneath the red colors. The tropopause fold is likely the result of 1) the monconservation of potential temperature (and PV) in response to diabatic heating and associated upper-tropospheric outflow originating from heavy precipitation in the vicinity and poleward of Floyd, and 2) confluent frontogenesis. Evidence for the former is found in the westward expansion of high potential temperature air on the DT despite winds blowing parallel to potential temperature contours at both 0000 and 1200 UTC 16 September (Fig. 2.19a,b), and in previous studies of Hurricane David (1979) by Bosart and Lackmann (1995) and the 12–14 March 1993 Superstorm by Dickinson et al. (1997; see also Fig. 2.16). Evidence for the latter is found in the strong confluent flow associated with a migratory short-wave trough depicted in the tropopause fold region northeast of Maine (Fig. 2.19b). An anonymous reviewer also suggested that ageostrophic vertical shear associated with the strong divergent outflow from the storm could act to produce a tropopause fold as depicted in Fig. 2.11c.

By 0000 UTC 17 September, the upper-level trough begins to overspread Floyd as the extratropical transition process proceeds while the tropopause fold region extends downstream past Newfoundland as the aforementioned short-wave trough reaches the North Atlantic (Fig. 2.19c). The tropopause remains folded in the anticyclonic outflow region immediately westward and poleward of Floyd. Although horizontal confluence is not evident close in to Floyd at this time, it is apparent poleward of New York where strong southwesterly flow at higher potential temperatures in the storm outflow channel is forced over air with lower potential temperatures on the east side of the trough located over the Great Lakes (seen as reds adjacent to greens). By 1200 UTC 17 September the DT begins to unfold just to the northwest of Floyd (note reapparance of the yellow and orange colors) while the DT remains folded well downstream along the outflow jet channel where the approach of a second trough and its associated confluent flow likely contributes to continued folding as well (Fig. 2.19d). Note also the signature of the continuing extratropical transition of Floyd as a PV hook appears over the storm as evidenced by the emergence of an area of blue colors on the northeast side of the deep PV column and associated low potential temperature air on the DT that marks the storm (Fig. 2.19d). The signature of the Floyd tropopause fold begs the question as to what extent tropopause folds that form in the southwesterly flow along the poleward side of tropical storm outflow jet channels are manifestations of PV nonconservation due to widespread storm-induced diabatic heating and whether tropopause folds that form under these circumstances can contribute appreciably to tropospheric–stratospheric exchange processes, given that the associated deep convection itself could also be associated with significant stratospheric–tropospheric exchange.

To what extent the downstream jet is diabatically generated and/or is a manifestation of larger-scale deformation

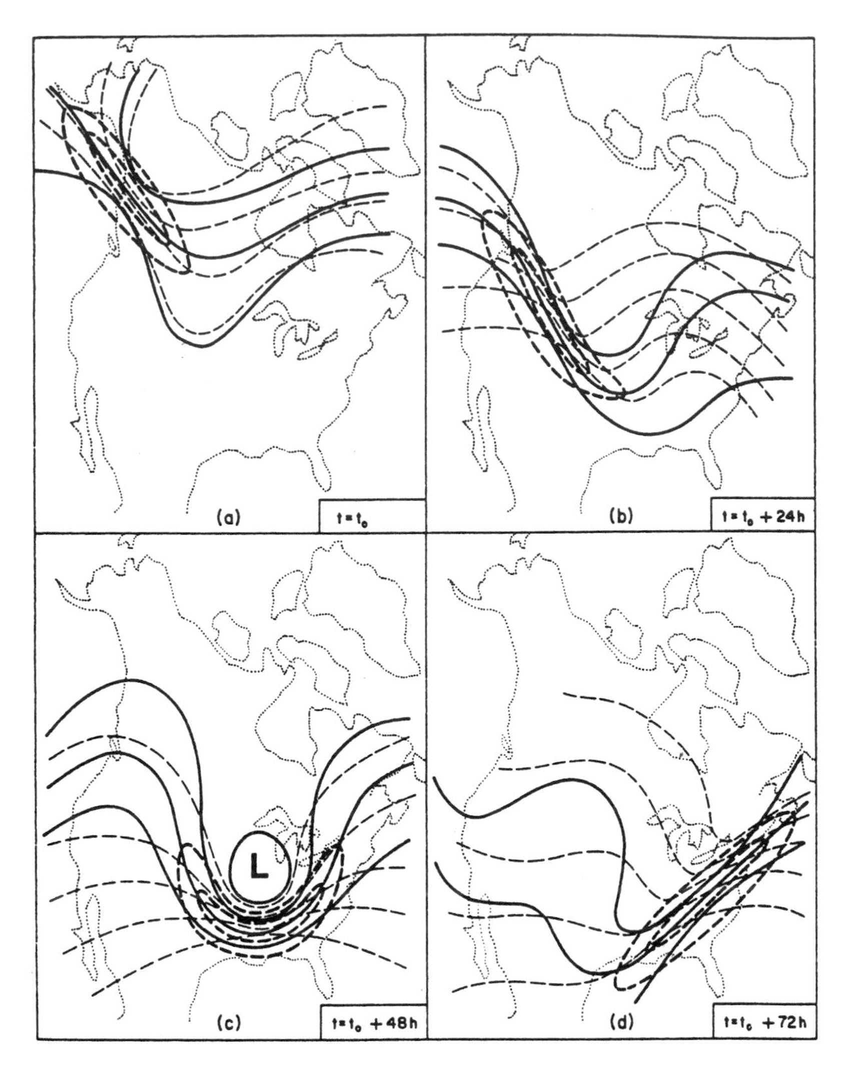

FIG. 2.20. Idealized schematic depiction on a constant pressure surface of the propagation of an upper-tropospheric jet-front system through a midlatitude baroclinic wave over a 72-h period: (a) formation of jet front in the confluence between mid- and high-latitude currents, (b) jet front situated in the northwesterly flow inflection of amplifying wave, (c) jet front at the base of the trough of fully developed wave, and (d) jet front situated in the southwesterly flow inflection of damping wave. Geopotential height contours, thick solid lines; isotachs, thick dashed lines; isentropes or isotherms, thin dashed lines. From Shapiro (1983). Source: Fig. 19 from Keyser and Shapiro (1986).

processes associated with the hurricane-trough-jet interaction remains to be determined [a similar question was raised in the case of Hurricane Opal in October 1995; see, e.g., Bosart et al. (2000)]. Clearly, in some circumstances upper-level frontogenesis may perhaps be possible when a deep, warm tropical air mass comes into juxtaposition with a much cooler polar air mass along the forward flank of a major midlatitude trough that is intercepting a poleward-moving hurricane. Again, there is good reason to conduct further investigations of the extent to which upper-level frontogenesis might occur in southwesterly flow, and to compare and contrast the relevant dynamics with the more common occurrences of northwesterly flow upper-level frontogenesis. Based on the previous discussion, the use of PV thinking and a DT perspective are likely to be helpful in thisendeavor.

7. An example of upper-level frontogenesis in southwesterly flow

a. Conceptual models of frontal life cycles

Although case studies of upper-level fronts over North America since the pioneering investigations by Reed and

Sanders (1953) and Reed (1955) have uncovered many important aspects of the structure of these fronts, attempts to document complete frontal life cycles have been more difficult. Prominent upper-level fronts are a product of the cold season when thermal gradients and vertical wind shears are large. Strong winds aloft ensured that upper-level fronts associated with migratory short-wave troughs could not be sampled completely by the North American radiosonde network for more than three or four consecutive 12-h upper-air sampling periods, precluding the construction of complete frontal life cycles from the available observations. A notable exception was a study by Sanders et al. (1991) of an upper-level front from October 1963. This front formed in northwesterly flow over western North America and underwent most of its life cycle over North America because winds aloft were weaker than would be the case deeper into the cold season.

Occasionally, the sounding sampling frequency was increased to 3 h in support of special research projects. On one such occasion in February 1964, soundings were taken over the southeastern United States every 3 h in anticipation of a major cyclogenesis event. Although the expected cyclogenesis failed to materialize, an intense upper-level front was inadvertently captured by the special sounding network. Bosart (1970) took advantage of this serendipitous opportunity to study the structure of the upper-level front in considerable detail. He showed that frontogenesis occurred in a short-wave trough that was migrating through a major long-wave trough. The front intensified initially in northwesterly flow upstream of the long-wave trough, mostly through tilting. Subsequently, the front migrated around the base of the long-wave trough where it reached its greatest intensity due to horizontal confluence (tilting contributed to frontolysis at this stage). The changeover from tilting-driven frontogenesis to confluence-driven frontogenesis occurred as the upper-level front rounded the base of the long-wave trough into the southwesterly flow. Bosart (1970) also showed that individual air parcels that underwent intense frontogenesis as they passed through the frontal zone carried their frontal properties from the northwesterly flow well downstream in the southwesterly flow. A life cycle study was not possible, however, because the special sounding domain was limited to the southeastern United States.

Shapiro (1982) synthesized the results from disparate case studies, including Bosart (1970), into schematic snapshots (Fig. 2.20) depicting the movement of an upper-level jet-front system through a larger-scale trough–ridge system over North America. The jet-front system forms initially in confluent northwesterly flow downstream of a ridge axis with intensification occurring in northwesterly flow farther downstream in the presence of cold-air advection that allows tilting to contribute positively to frontogenesis on the mesoscale and positively to the amplification of the baroclinic wave on the synoptic scale. Subsequently, the jet-front system rounds

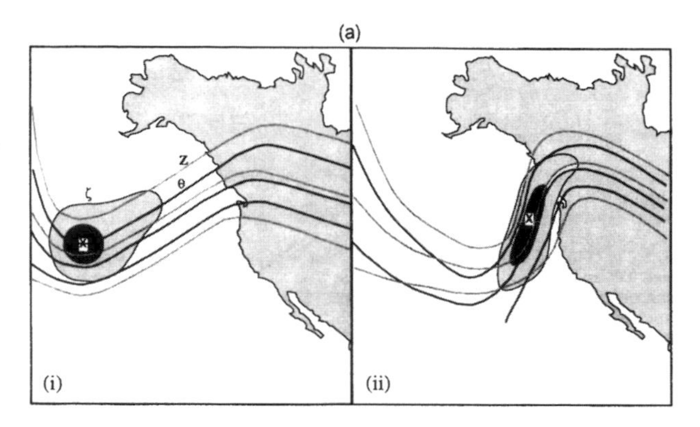

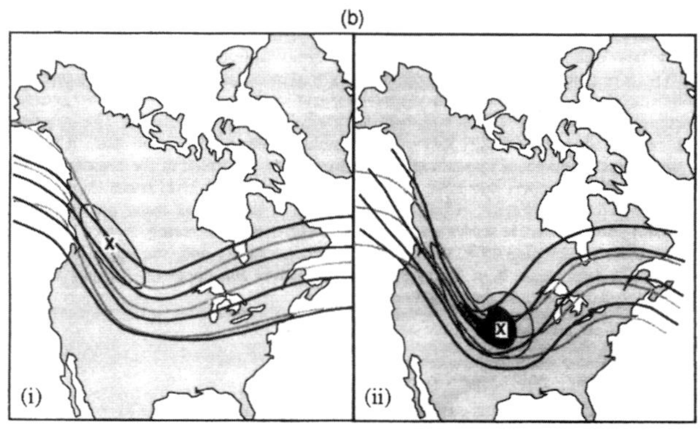

Fig. 2.21. Revised conceptual model: idealized schematic depiction, on an upper-tropospheric isobaric surface, of the early evolution of an upper-level jet-front system through a midlatitude baroclinic wave over a 12-24-h period: (a) southwesterly flow case and (b) northwesterly flow case. Geopotential height contours (solid gray lines) are labeled Z in (a) (i), isentopes (solid black lines) are labeled θ in (a) (i), and relative vorticity (shaded) is labeled ζ in (a) (i). Source: Fig. 16 from Schultz and Doswell (1999).

the base of the trough near peak intensity as amplification of the synoptic-scale baroclinic wave ceases as the thermal and geopotential height troughs come into phase. Finally, the jet-front system migrates downstream into the southwesterly flow where confluence in the presence of warm-air advection contributes to frontogenesis. A possible difficulty with the Shapiro (1982) schematic representation in Fig. 2.20, however, is that individual upper-level fronts are seldom observed to migrate around the base of the baroclinic wave trough from the northwesterly to the southwesterly flow [the Bosart (1970) case is atypical].

Schultz and Doswell (1999), while acknowledging the general success of the Shapiro (1982) conceptual model shown in Fig. 2.20 in representing properly the movement of a jet-front system through a synoptic-scale baroclinic wave, argued that his conceptual model was an oversimplification of the structure of the associated upper-level frontogenesis. Based upon further studies of upper-level frontogenesis over the eastern Pacific and North America and a climatology of upper-level fronts associated with eastern Pacific cyclones that reached western North America, they prepared a modified schematic (Fig. 2.21) that illustrates a jet-front system

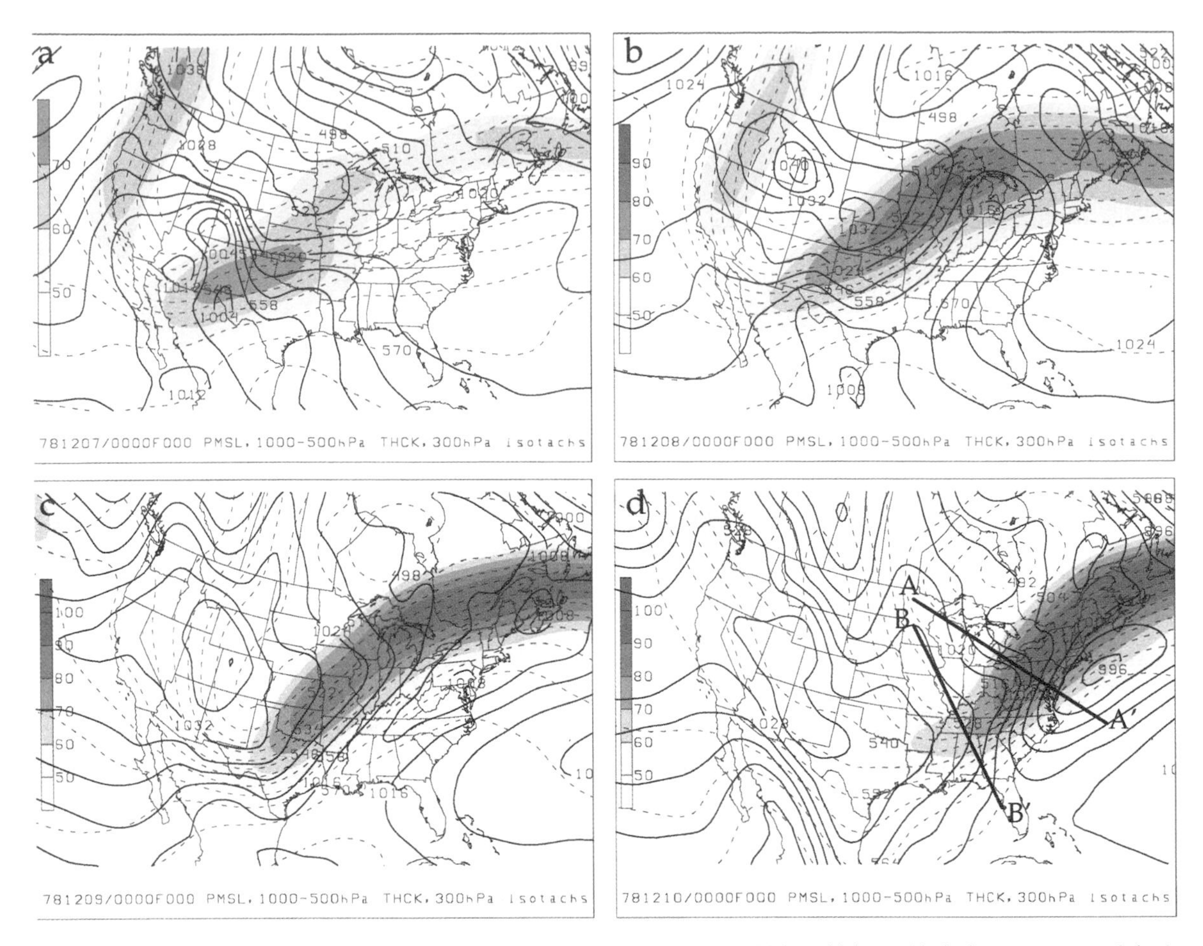

FIG. 2.22. (a)–(d). Mean sea level pressure (solid contours every 4 hPa), 1000–500-hPa thickness (dashed contours every 6 dam), and 300-hPa isotachs (beginning at 50 m s^{-1} and shaded according to the color bar) for 0000 UTC 7–10 Dec 1978. Cross-section lines $A-A^1$ and $B-B^1$ in Figs. 2.27 and 2.28, respectively, are indicated in (d).

interacting with a synoptic-scale baroclinic wave in both southwesterly and northwesterly flow.

The southwesterly flow interaction case is shown in Figs. 2.21a,b. Initially, the isotherms and height contours are quasi-parallel from the trough eastward toward the downstream ridge. As a potent vorticity maximum migrates through the trough axis and interacts with the larger-scale baroclinic wave, curvature vorticity is converted to shear vorticity, the vorticity maximum elongates poleward in the southwesterly flow, and cold-air advection develops in the southwesterly flow behind the vorticity maximum. In the view of Schultz and Doswell (1999), the interaction of the vorticity maximum with the larger-scale baroclinic wave is responsible for the development of the cold-air advection and the associated strengthening of the upper-level frontal zone.

The contrasting northwesterly flow upper-level frontal evolution shown in Figs. 2.21c,d bears resemblance to the first two stages of the Shapiro (1982) conceptual model shown in Fig. 2.20a,b. However, the Schultz and Doswell (1999) schematic differs in two important aspects. First, there is substantial thermal advection, warm (cold) upstream (downstream) in the northwesterly flow, and second, the leading cold-air advection region becomes concentrated in the base of the deepening trough as the vorticity increases and compacts (Lackmann et al. 1996, 1997), indicative of a growing baroclinic wave (see also Fig. 2.9d).

b. Upper-level front of 7–10 December 1978

The period 6–11 December 1978 featured a very strong west-southwesterly flow over North America along with a gradual eastward movement of a major large-scale trough initially located over western North America. The mean level pressure, 1000-500-hPa thickness, and 300-hPa isotachs for 0000 UTC 7–10 December 1978 are shown in Fig. 2.22. [these and the following figures are based upon the NCEP-NCAR gridded reanalyses described in Kalnay et al. (1996) and Kistler et al. (2001)]. Over the 72-h period ending 0000 UTC 10 December a cyclone develops over the lower Mississippi River valley (Fig. 2.22a) and moves northeastward while deepening modestly to 1007 hPa to a position over Nova

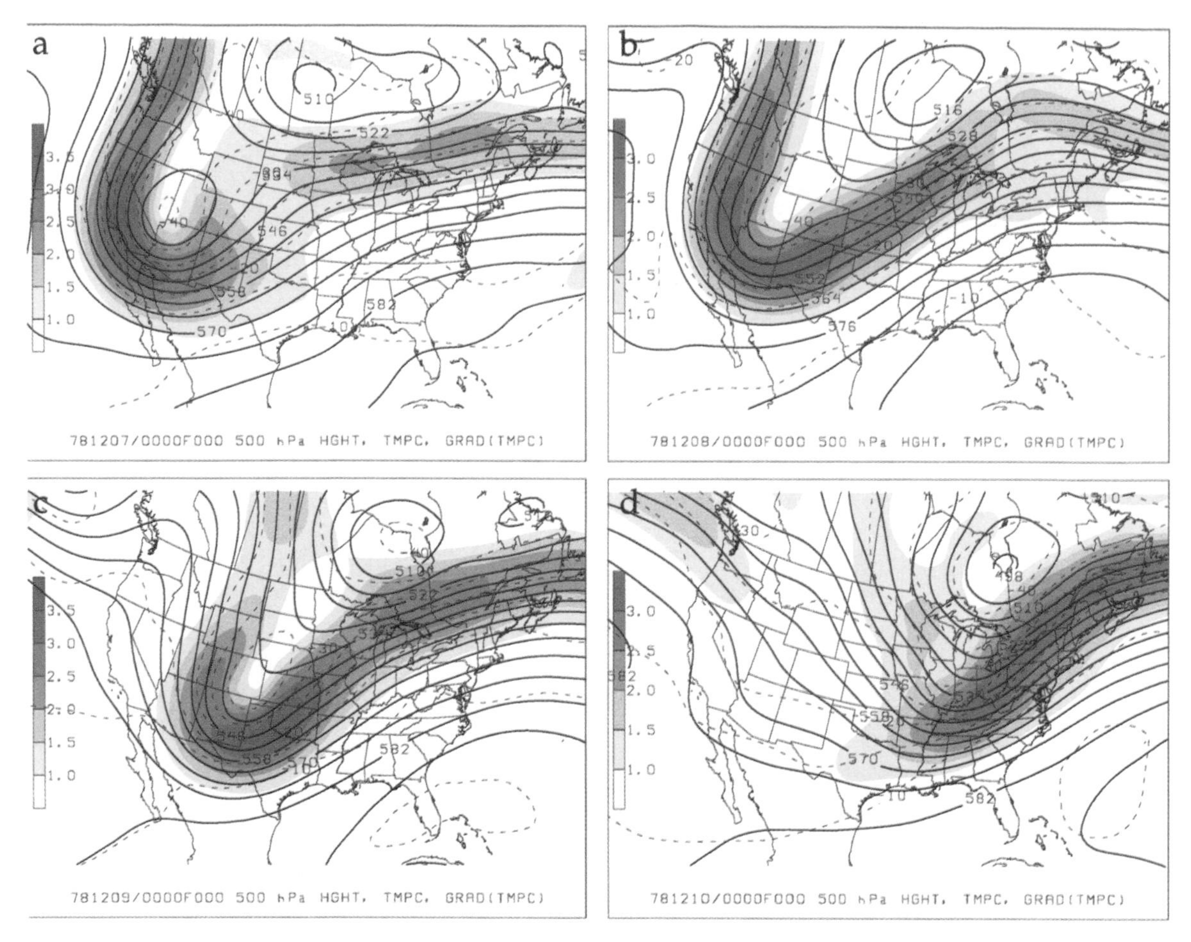

FIG. 2.23. (a)–(d) The 500-hPa heights (solid contours every 6 dam), 500-hPa temperatures (dashed contours every 5°C), and horizontal temperature gradient (beginning at 1×10^{-5}°C m^{-1} and shaded according to the color bar) for 0000 UTC 7–10 Dec 1978.

Scotia by 0000 UTC 9 December (Fig. 2.22c). A second cyclone forms along the trailing cold front over the eastern Ohio River valley near 0000 UTC 9 December and deepens moderately to 994 hPa just southeast of New England by 0000 UTC 10 December (Fig. 2.22d). Over the next 36 h (not shown) this second cyclone deepens strongly to 960 hPa as it races northeastward to a position well east of Labrador. A very strong jet was observed in the southwesterly flow with speeds in excess of 100 m s^{-1} analyzed at 0000 UTC 9–10 December (Fig. 2.22c,d). The upper-level front of interest is embedded in this southwesterly flow.

Shown in Fig. 2.23 are the 500-hPa heights, temperatures, and temperature gradient for 0000 UTC 7–10 December. At 0000 UTC 7 December the strongest temperature gradient resides in the northerly flow along the west coast of the United States from where it curls around the base of the cutoff cyclone while a second, weaker, baroclinic zone stretches from the upper Midwest eastward to north of New England in a confluent flow region (Fig. 2.23a). By 0000 UTC 8 December the zone of strongest temperature gradient has both migrated around the base of the deep trough in the southwestern United States and redeveloped northeastward toward the western Great Lakes (Fig. 2.23b). The northeastward strengthening of the 500-hPa temperature gradient occurs in conjunction with confluent flow over the northern Plains as a weak trough lifts northeastward from the southwestern U.S. trough (Fig. 2.23b). The northeastward movement of this weak trough supports the surface inverted trough formation and 300-hPa jet development seen in Fig. 2.22b. By 0000 UTC 9 December the downstream end of the region of maximum 500-hPa temperature gradient has passed through the ridge and reached the east coast of Canada (Fig. 2.23c). Confluent flow to the east of the Rockies strengthens the 500-hPa temperature gradient over northern Texas and Oklahoma at 0000 UTC 9 December as a short-wave trough located to the southwest of a cutoff cyclone situated near James Bay elongates equatorward and then phases with the main trough lifting northeastward in the 24 h subsequent to 0000 UTC 9 December (Fig. 2.23c,d).

The corresponding 500-hPa maps for 0000 UTC 7–10 December with total frontogenesis [computed following Miller (1948)] superimposed are displayed in Fig. 2.24. At 0000 UTC 7 December frontogenesis is mostly confined to the northerly flow along the West Coast and in the confluent jet entrance region over the Dakotas

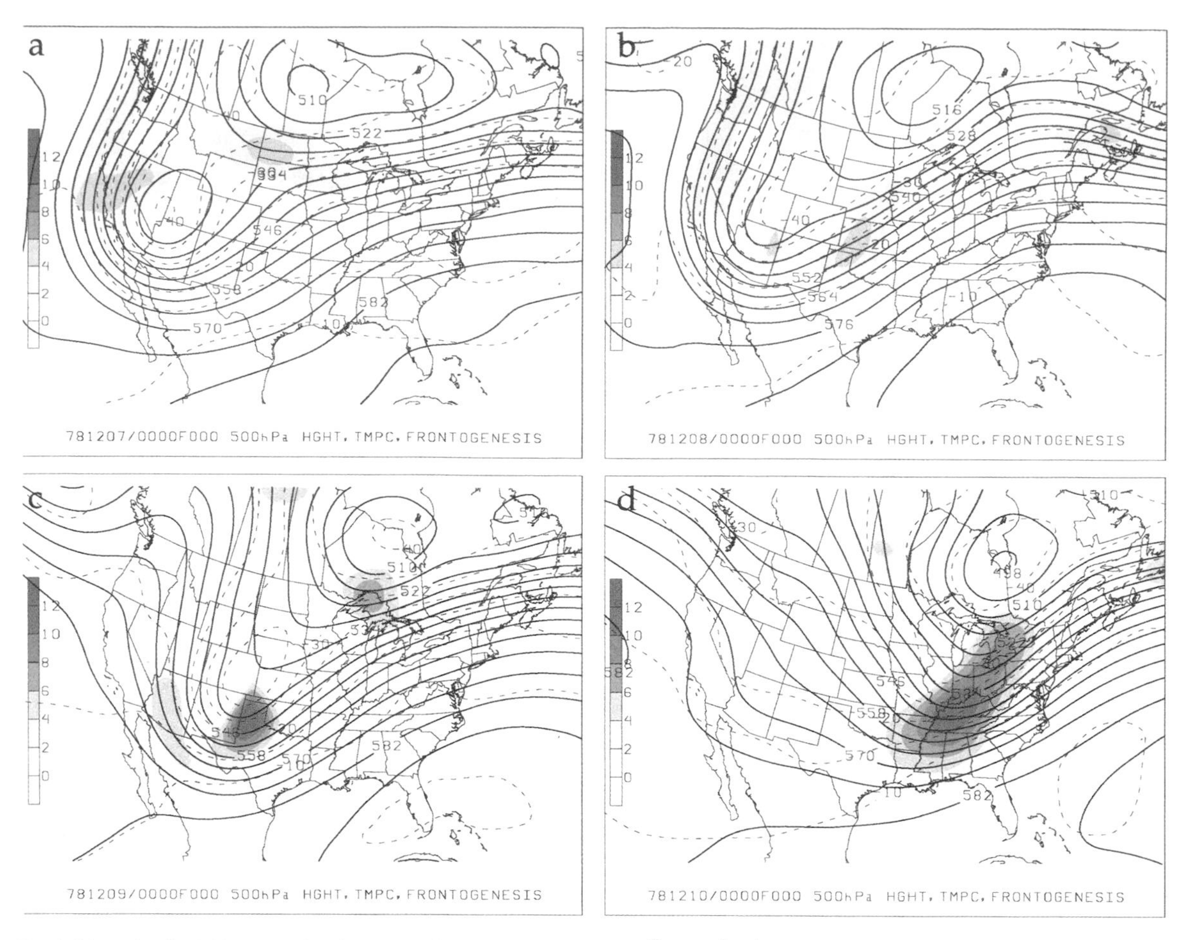

FIG. 2.24. As in Fig. 2.23 except that frontogenesis (beginning at 1×10^{-10}°C m^{-1} s^{-1} and shaded according to the color bar) replaces the horizontal temperature gradient.

(Fig. 2.24a). By 0000 UTC 8 December, frontogenesis develops in the Oklahoma–Kansas region, driven mostly by horizontal confluence (not shown), as the aforementioned weak short-wave trough moves northeastward (Figs. 2.23b, 2.24b). A new frontogenesis maximum develops over the Texas panhandle where the flow remains confluent as the Great Lakes trough elongates southwestward by 0000 UTC 9 December (Figs. 2.23c, 2.24c). This frontogenesis maximum strengthens further and expands northeastward to the eastern Great Lakes by 0000 UTC 10 December (Fig. 2.24d). The expansion and strengthening of the 500-hPa frontogenesis maximum in the 24 h ending 0000 UTC 10 December is driven almost exclusively by horizontal confluence (not shown). The twisting term contribution to frontogenesis is weakly negative in response to large-scale ascent maximizing along the warm side of the main baroclinic zone (not shown). Comparison of Figs. 22–24 shows that the frontogenesis maximum at 0000 UTC 10 December is situated near the southwestern end of the maximum 500-hPa temperature gradient and the entrance region of the 100 m s^{-1} 300-hPa jet. Taken together, these figures suggest that air parcels are undergoing frontogenesis in a confluent jet entrance region and then carrying their frontal properties well downstream, analogous to what Bosart (1970) found in his observational study of an intense upper-level front.

To help give an additional perspective on the frontogenesis calculation, maps of potential temperature on the DT for 0000 UTC 7–10 December, patterned after Morgan and Nielsen-Gammon (1998), are presented in Fig. 2.25. Although the structure of the DT is poorly resolved in the 2.5° resolution NCEP–NCAR gridded reanalysis datasets, it is apparent from Fig. 2.25 that the DT is quite steep over much of the United States. Minor tropopause folds are apparent over the Texas–Oklahoma panhandle area and east of Florida at 0000 UTC 7 December (Fig. 2.25a). By 0000 UTC 8 December the tropopause fold over the Texas and Oklahoma panhandle area has moved northeastward to southeastern Nebraska (Fig. 2.25b). This tropopause fold region can be associated with the weak short-wave trough lifting out of the deep trough in the southwestern United States (Fig. 2.23b) and a small region of frontogenesis at 500 hPa just upstream (Fig. 2.24b). A "spot" of tropopause folding west of the Baja Peninsula at 0000 UTC

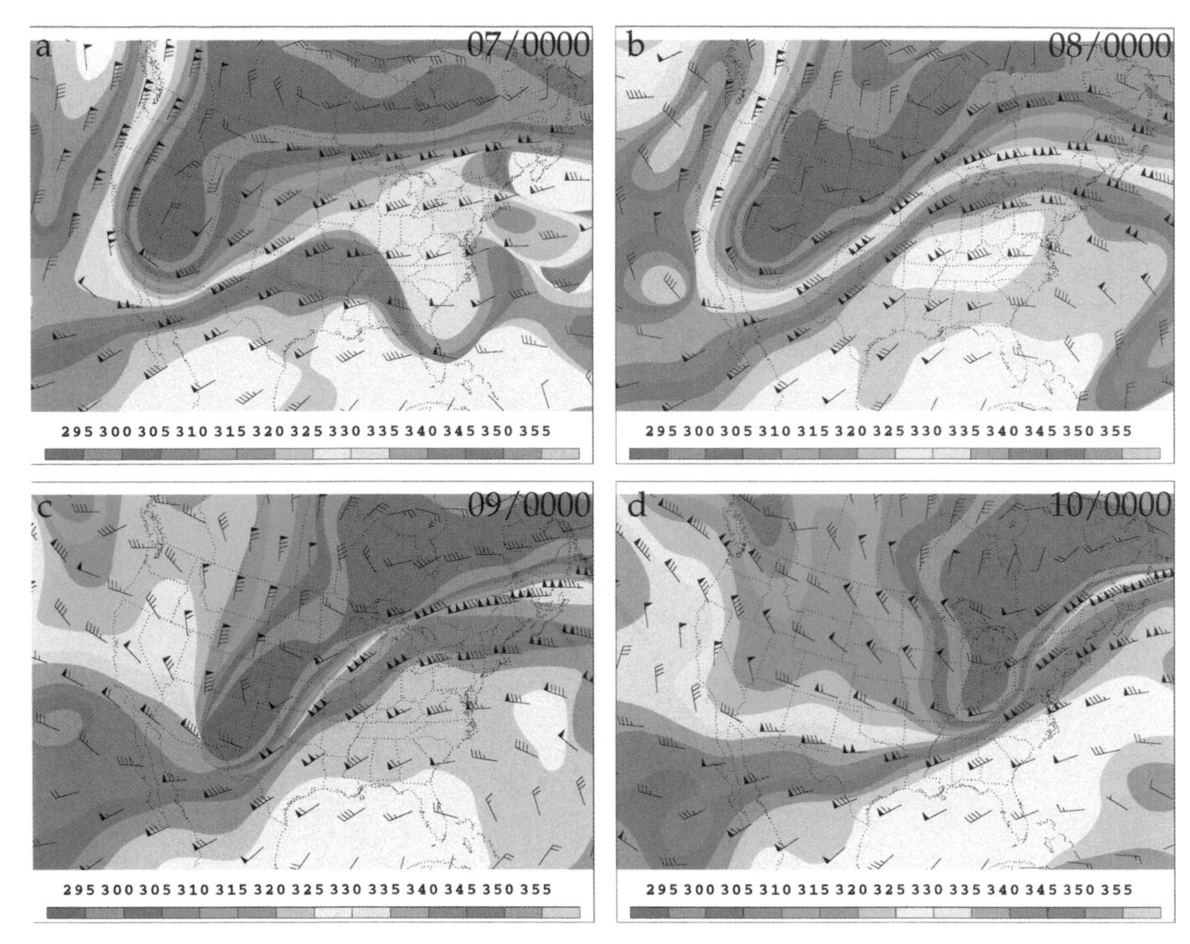

Fig. 2.25. (a)–(d) As in Fig. 2.19 except at 0000 UTC 7–10 Dec 1978.

8 December develops into a significant tropopause fold to the east of the major trough axis over western Texas by 0000 UTC 9 December in conjunction with a strengthening of the 500-hPa temperature gradient (Fig. 2.23c) and an increase in the 500-hPa frontogenesis (Fig. 2.24c). Meanwhile, the tropopause fold that had been situated over southeastern Nebraska now extends from the western Great Lakes to north of Maine as the previous weak short-wave trough rides over the ridge in confluent flow over the eastern United States (cf. Figs. 2.23 2.24 2.25c). By 0000 UTC 10 December significant tropopause folding is concentrated over the lower Ohio valley and the central Appalachians ahead of the major trough lifting out of the southwestern United States (Fig. 2.25d). The now well-defined region of tropopause folding at 0000 UTC 10 December is concentrated at the leading edge of an area of strengthening 500-hPa frontogenesis that is dominated by horizontal confluence (not shown) (Fig. 2.24d). Likewise, tropopause folding lies in the entrance region of the 300-hPa jet (Figs. 2.22d, 2.25d).

A computation of total foldogenesis from (3) and the contributions to foldogenesis from horizontal confluence from (4) and from differential vertical motion from (5) for 1200 UTC 9 December and 0000 UTC 10 December are shown in Fig. 2.26. On the coarse-resolution scale of the NCEP–NCAR reanalysis, foldogenesis (computed by looking upward to the DT from below) is dominated by horizontal confluence (Figs. 2.26c–f), is maximized on the upstream side of the tropopause fold region (Figs. 2.25d and Fig. 2.26d), and is situated above the region of maximum 500-hPa frontogenesis (Fig. 2.24d). The signature suggests that air parcels that undergo frontogenesis and foldogenesis carry their frontal and foldal properties well downstream, in agreement with the findings of Bosart (1970).

A northwest–southeast-oriented cross section of PV and potential temperature along the leading edge of the tropopause fold and frontogenesis region at 0000 UTC 10 December is shown in Fig. 2.27a (cross-section line appears in Fig. 2.22b) while soundings for Pittsburgh, Pennsylvania (PIT) and Sterling, Virginia (IAD), are plotted in Fig. 2.27b. Despite the relatively coarse resolution of the NCEP-NCAR reanalysis dataset, a prominent tropopause fold extending downward to near 600 hPa is apparent. The winds in the plane of the cross

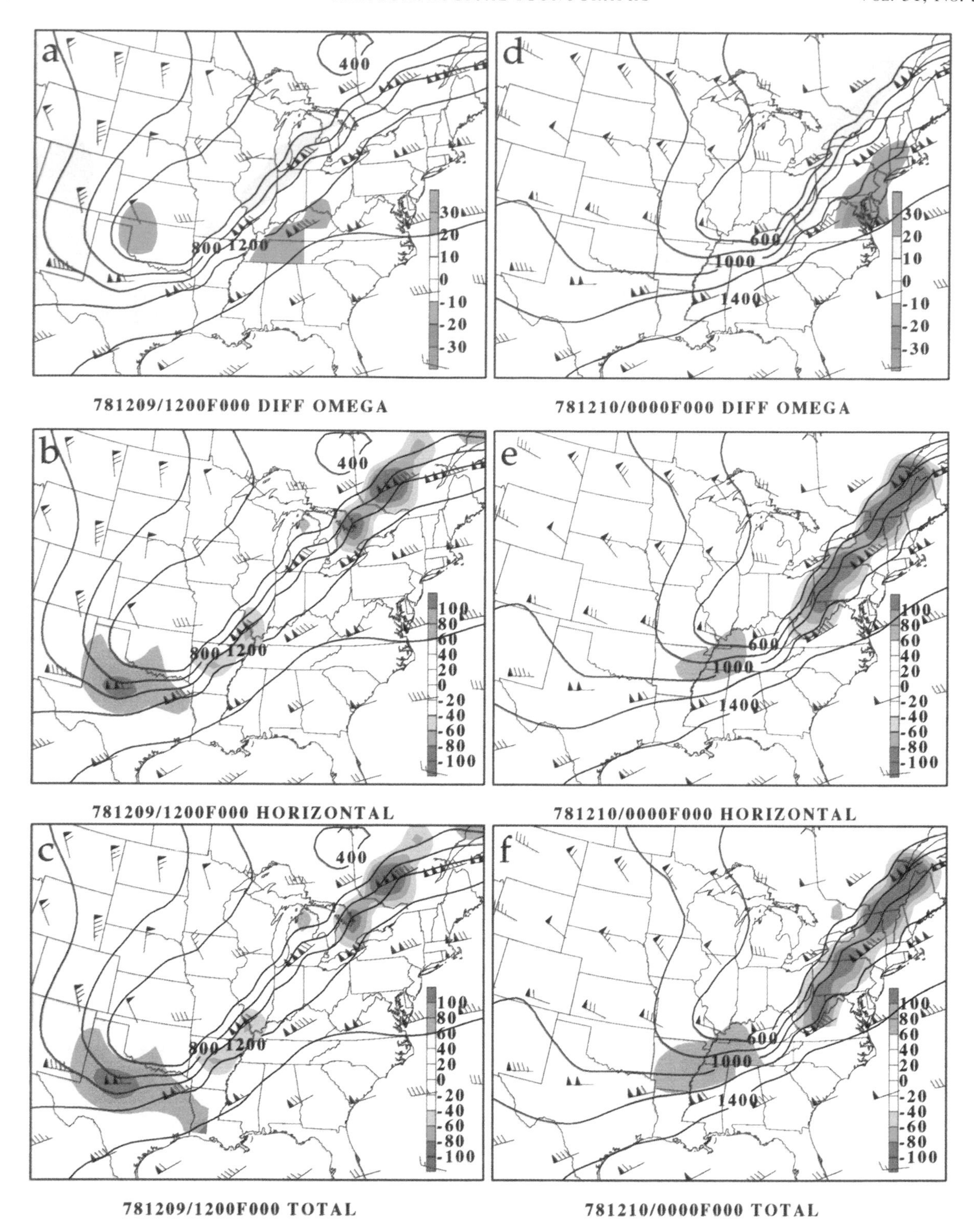

FIG. 2.26. Height of the dynamic tropopause as viewed from below (solid contours in dam) and foldogenesis in units of 10^{-8} s^{-1} shaded according to the color bar with warm (cool) colors corresponding to foldogenesis (foldolysis) (left): 1200 UTC 9 Dec 1978 and (right) 0000 UTC 10 Dec 1978. (a), (d) Contribution to foldogenesis from differential vertical motion [Eq. (5) in text] at 1200 UTC 9 (0000 UTC 10) Dec 1978 on the left (right). (b), (e) As in the top row except for horizontal confluence [Eq. (4) in text]. (c), (f) As in the top row except for total foldogenesis [Eq. (3) in text].

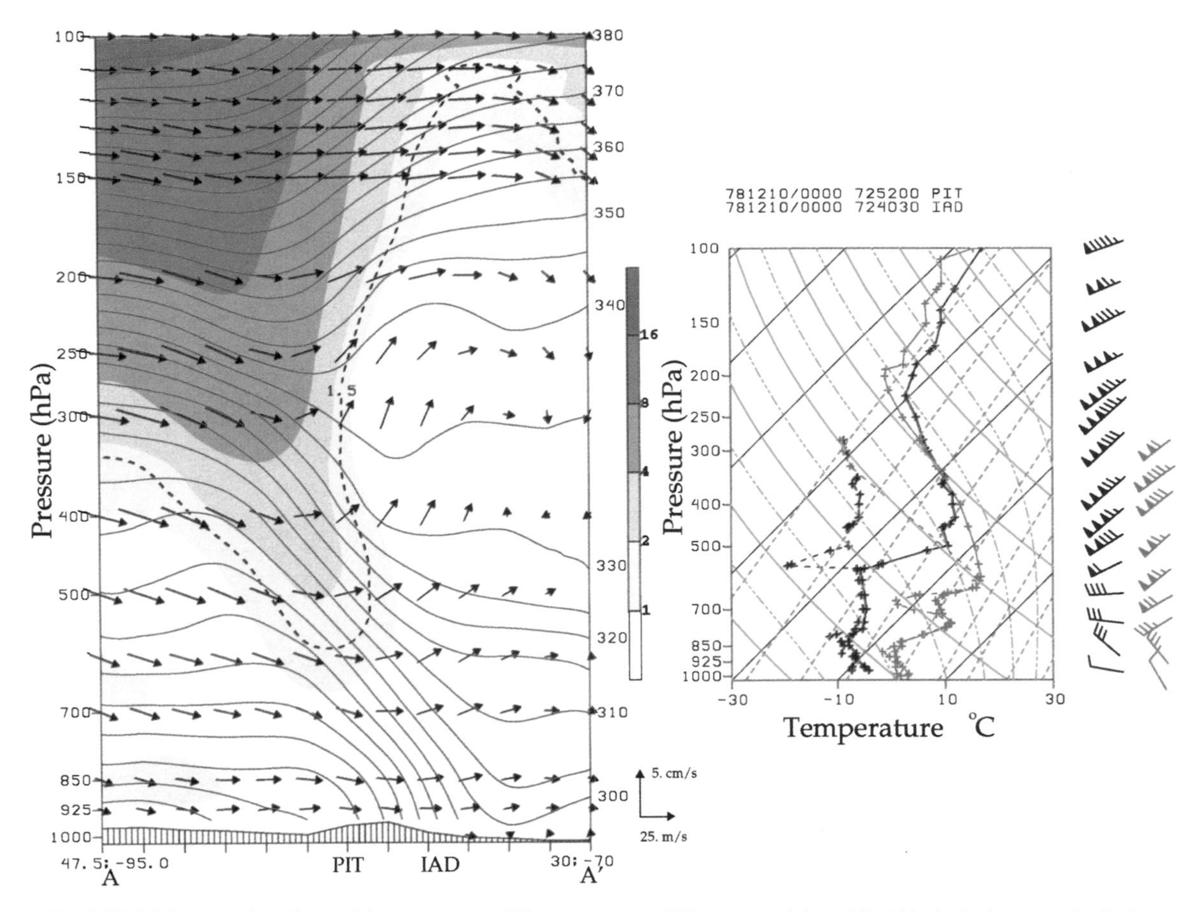

FIG. 2.27. (a) Cross section of potential temperature (solid contours every 5 K) and potential vorticity (thin dashed contours beginning at 0.25 PVU, 0.50 PVU, and then shaded according to the color bar beginning at 1 PVU), for 0000 UTC 10 Dec 1978. The bold dashed line indicates the 1.5-PVU contour. Winds (m s^{-1}) in the plane of the cross section are indicated by arrows and are scaled according to the reference vector. Smoothed terrain is shaded along the abscissa. Locations of the Pittsburgh, PA, and Sterling, VA, soundings shown in (b) are indicated by PIT and IAD, respectively, along the abscissa. Line A–A′ in Fig. 2.22d marks the NW-SE-oriented cross section. (b) PIT (black) IAD (red) temperature (solid) and dewpoint temperature (dashed) soundings with winds (in m s^{-1} as in Fig. 2.19) to the right.

section appear to be confluent across the tropopause fold at 500 and 400 hPa, consistent with the domination of frontogenesis (Fig. 2.24d) by confluence (not shown). At PIT the tropopause fold is associated with an increase of potential temperature of 45 K and an increase in wind speed of almost 50 m s^{-1}, respectively, over the 575-500-hPa layer. Comparison of Fig. 2.27b with Fig. 2.22d reveals that PIT, where steady light snow was falling (not shown) lies well to the west of a departing cyclone. A remnant of the surface frontal inversion is apparent near the 800-hPa level. At IAD the tropopause fold is weaker and lower (675–625 hPa) while the surface frontal zone slightly higher and stronger (850–750 hPa). Clearly, the upper- and lower-level frontal zones are not part of a continuous and deep frontal zone separating tropical and polar air, a common characteristic originally noted by Reed (1955).

A second cross section of frontogenesis and vertical motion, taken along a Minneapolis, Minnesota (MSP), to Tampa, Florida (TPA), line is shown in Fig. 2.28. The orientation of this cross section was chosen to sample the frontogenesis maximum shown in Fig. 2.24d and to "nick" the western edge of the folded DT as shown in Fig. 2.25d. Note how frontogenesis is maximized over roughly the 500–350 hPa layer where the flow is confluent along the cold boundary just upstream of where the DT is folded to the east (Fig. 2.28a). Strong sinking motion is indicated along the cold boundary and the base of the PV extrusion into the troposphere (Fig. 2.28b). Comparison with Fig. 2.23b shows that this sinking motion is occurring in a region of cold-air advection in the confluent flow area along the cold side of the primary baroclinic zone. Comparison of Figs. 2.23, 2.24, 2.25d, and 2.28a,b with Fig. 2.26f reveals that the eastern half of the zone where the foldogenesis function is positive corresponds well with the region of frontogenesis, sinking air, and cold-air advection. Collectively, these figures suggest that the tropopause fold originates as air parcels situated on the DT where the DT is closer to the ground are forced to sink in conjunction with

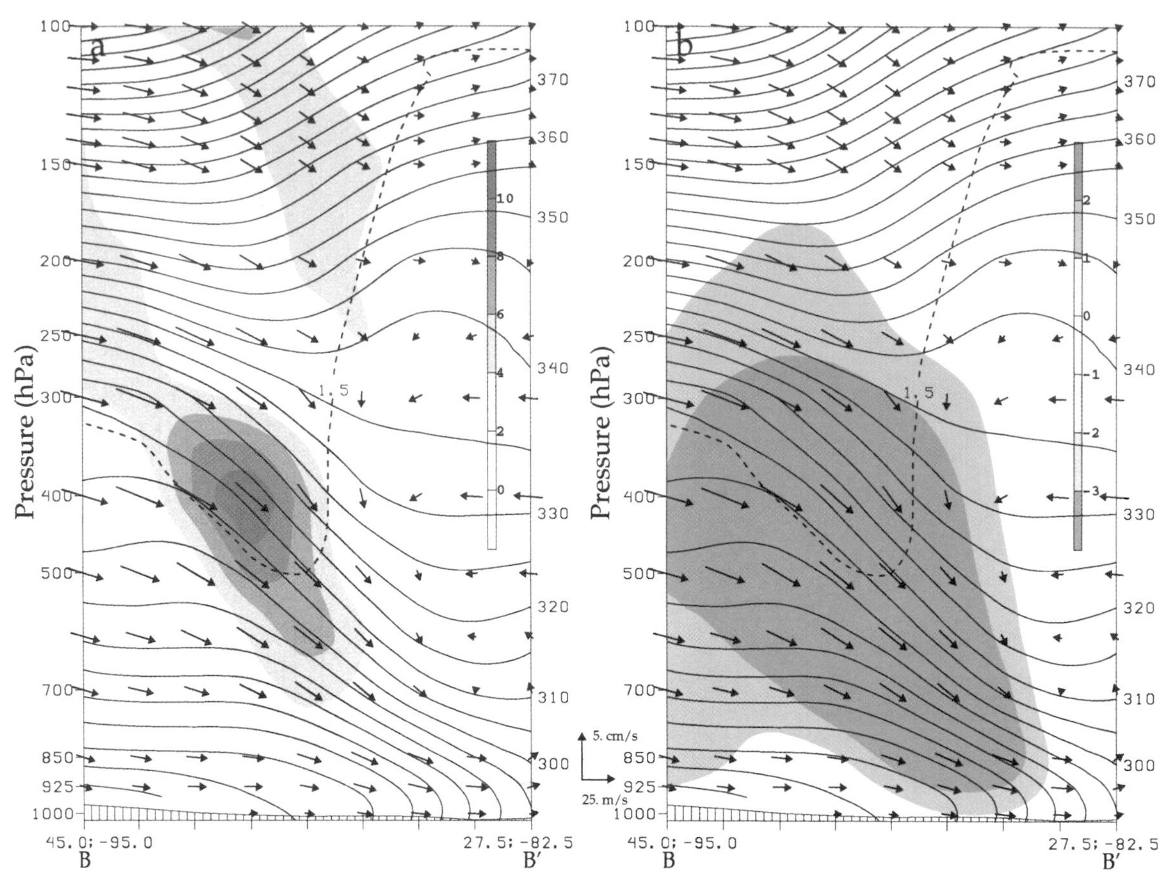

FIG. 2.28. (a) Cross section of potential temperature (solid contours every 5 K) and frontogenesis [shaded according to the color bar beginning at 2.0×10^{-10} C m^{-1} s^{-1} taken along a Minneapolis, Minnesota (MSP), to Tampa, Florida (TPA), line as indicated by line B–B′ in Fig. 2.22d. Bold dashed line denotes the 1.5-PVU surface as defined in Fig. 2.27a. Winds as in Fig. 2.27a. (b) As in (a) except for potential temperature and vertical motion (shaded according to the color bar beginning at $\pm -1.0 \times 10^{-3}$ hPa s^{-1}). Reference scale vector for winds as in Fig. 2.27.

cold-air advection and confluence-driven frontogenesis and foldogenesis.

Although the upper-level front case of 7–10 December 1978 occurred in southwesterly flow, it differed from the Shapiro (1982) schematic shown in Fig. 2.20 in that warm-air advection was absent in the frontogenesis region. In the Schultz and Doswell (1999) schematic of southwesterly flow upper-level frontogenesis, cold-air advection develops behind the front in response to upstream baroclinic wave intensification and vorticity compaction in the trailing trough. In contrast, the upper-level frontogenesis in the 7–10 December 1978 case is driven by horizontal confluence that maximizes along the poleward side of the dominant baroclinic zone in response first to a weak short-wave trough lifting northeastward toward a jet entrance region in the northern branch of the westerlies and second to the phasing of a trough in the northern branch of the westerlies with the dominant trough in the southern branch. The findings from this brief case study, along with the tropopause folding associated with the divergent outflow from Hurricane Floyd (1999) and confluence ahead of migratory short-wave troughs in the polar westerlies shown in Fig. 2.19, illustrate that tropopause folding can occur in conjunction with flow regimes that depart significantly from the classic cold-air advection in northwestly flow situation as studied by Reed (1955). As will be discussed in the next section, these newly documented tropopause folding situations open up new opportunities for future research.

8. Future research directions

In the 1950s and 1960s advances in our knowledge and understanding of upper-level fronts and frontogenesis were derived largely from observational studies typified by Reed (1955). Numerical investigations began in the 1970s (e.g., Bleck 1973, 1974) and continue to the present. Theoretical investigations of baroclinic wave cyclone life cycles and associated upper-level fronts (e.g., Rotunno et al. 1994; Wandishin et al. 2000) have dominated research in the last decade. The ascendancy of numerical and theoretical investigations of upper-level frontogenesis over observational studies in the last 20 years is likely a reflection of the inability of the current observational database to provide critical meteorological measurements on the mesoscale. These changes in research emphasis over the last 50 years likely reflect historical "push–pull" relationships between observational, theoretical, and numerical studies of upper-level fronts and frontogenesis.

Further observational advances in our knowledge and understanding of upper-level fronts and frontogenesis must await improvements to our observational database to include enhanced mesoscale observations from a mix of conventional and remotely sensed observations. Kuo et al. (1998) provided an example of how a global positioning system (GPS) receiver carried aboard a satellite and modified to sound the atmosphere could be used to construct atmospheric soundings with sufficient vertical resolution to reveal stable layers associated with upper-level fronts (their Fig. 7) through a GPS limb-sounding technique. More generally, Anthes et al. (2000) have discussed how applications of this GPS/MET (meteorology) sounding technique could be used to map the global dynamic tropopause (DT) with considerable accuracy. The ability to map the global DT accurately and to delineate stable layers associated with upper-level fronts will facilitate the global calculation of PV and permit more comprehensive observational investigations of the structure of upper-level fronts, especially on the mesoscale, and baroclinic wave cyclone life cycles. Recent papers by Donnadille et al. (2001a,b) on a study of upper-level frontogenesis and associated tropopause folding during the Fronts and Storm Track Experiment (FASTEX) are representative examples of new research that takes advantage of enhanced measurements from special field programs.

Danielsen (1964, 1966) suggested that dry air with a previous history of descent that characterized the dry slot of an extratropical cyclone could result in extensive air mass destabilization as it began to ascend near the tip of the dry slot and continued to ascend above the surface warm frontal zone. In Danielsen's view the resulting destabilization in association with ascent could trigger deep convection in the dry slot of a continental spring cyclone and/or in the vicinity of the surface warm front. An interpretation of Danielsen's argument in terms of PV thinking is that organized deep convection can be triggered by ascent ahead of a mobile disturbance on the DT provided that the "coupling index," defined as the difference between the potential temperature on the DT and the equivalent potential temperature at 850 hPa (Bosart and Lackman 1995), is negative. Bosart et al. (1996) and Dickinson et al. (1997), for example, showed that the massive deep convection that erupted in the northwest Gulf of Mexico in association with the initial cyclogenesis that defined the 12–14 March 1993 Superstorm occurred as a coherent disturbance on the DT with potential temperature near 330 K approached and overspread low-level air with an equivalent potential temperature of 340–345 K over the northwest Gulf of Mexico.

More recently, Griffiths et al. (2000), in a study of convective destabilization by a tropopause fold diagnosed using PV inversion, concluded that, "The importance of tropopause folds, or indeed of tropopause depressions in general in the context of convection, is that the small horizontal scale has the potential to focus and amplify any convective destabilization locally." These and the previous findings suggest that the debate as to whether tropopause folding plays an active or passive role in cyclogenesis should perhaps be redirected to ask to what extent the triggering of organized deep convection in the tip of the dry slot, near the surface warm front, or ahead of a coherent disturbance on the DT plays an active or passive role in the development of vertical circulations and jet-front systems. Idealized, adiabatic simulations suggest that tropopause folding is a passive response to the evolution of a baroclinic wave embedded in a synoptic scale disturbance. In the real atmosphere the bulk effects of an outbreak of organized deep convection may result in a significant upscale impact through the intensification of a downstream jet and associated baroclinic zone in response to massive ridging downstream of the deep convection such as happened in the 12–14 March 1993 Superstorm. In this context the extent to which an upper-level front and associated tropopause fold plays an active or passive role in the development of mesoscale vertical circulations within the cyclogenetic region could be a subject of further research.

9. Conclusions

The pioneering observational studies of upper-level fronts and tropopause folding by Reed and Sanders (1953) and Reed (1955) showed that upper-level fronts with a typical horizontal width and depth of 100 and 1 km, respectively, formed in northwesterly flow downstream of a ridge in conjunction with cold-air advection. Frontal horizontal temperature gradients intensified by an order of magnitude to $10°–20°C\ (100\ km)^{-1}$ on timescales of less than 1 day in conjunction with cold-air advection-driven subsidence that maximized along the warm boundary of the baroclinic zone. Furthermore, the Reed studies demolished the idea that the polar front was a continuous and deep structure that encircled the Northern Hemisphere and separated tropical air from polar air. Through the use of PV as a tracer, the Reed studies also showed that stratospheric air could penetrate deep into the troposphere in association with a tropopause fold. The results from the Reed studies led to the discovery of the dry slot in extratropical cyclones, stimulated a new interest in identifying stratospheric–tropospheric exchange processes, and established the value of using PV as a tracer in studies of frontogenesis and cyclogenesis.

The bulk of this paper was devoted to a discussion of how Reed's early ideas gleaned from studies of upper-level fronts and tropopause folding seeded future research that is still bearing fruit today. Emphasis was placed on the fruits of studies, such as those of Shapiro (1982), that showed the connection of upper-level jet-front systems with low-level baroclinic zones and low-level jets through a qualitative interpretation of the Sawyer-Eliassen equation. Emphasis was also placed on the results of idealized numerical and theoretical studies of upper-level frontogenesis within a developing wave cyclone (e.g., Shapiro and Keyser 1990; Keyser 1999; Rotunno et al., 1994; Wandishin et al. 2000; Hakim and Keyser 2001) that quantified the "Shapiro effect" (cold-air advection-induced subsidence leading to tilting-generated frontogenesis) and demonstrated that upper-level frontogenesis and tropopause folding was an attribute of baroclinic wave cyclone formation that could also be understood from a PV perspective (foldogenesis), and culminated in the identification of canonical idealized frontal circulation patterns for the Sawyer-Eliassen equation.

Additional emphasis was placed on how applications of "PV thinking" (Hoskins et al. 1985) and dynamic tropopause maps (Nielsen-Gammon 2001), a natural extension of the early use of PV as a tracer by Reed and others, can be used to diagnose extratropical cyclones and tropical cyclones that transition to extratropical cyclones. Finally, results were presented of a brief case study of upper-level frontogenesis that occurred from 7 to 10 December 1978. The case was noteworthy because cases of upper-level frontogenesis in southwesterly flow are comparatively rare (Schultz and Doswell 1999). The large-scale pattern featured a deep trough over the southwestern United States with an elongated and intense baroclinic zone extending northeastward to the Great Lakes and Canada. Peak winds exceeded 100 m s^{-1} in the downstream 300-hPa jet. As a leading short-wave trough in the northern branch of the westerlies settled southward and eastward toward the Great Lakes, a confluent flow regime and associated cold-air advection became established over the north-central United States eastward to the western Great Lakes. Confluent frontogenesis and resulting foldogenesis in this region resulted in an exceptionally intense upper-level front by 0000 UTC 10 December 1978. Despite its great intensity, atmospheric soundings showed that the upper-level front was a distinct feature from a lower-level cold front, in agreement with Reed's original result that disproved the assumption that the polar front was a continuous and deep boundary that circumnavigated the Northern Hemisphere and separated tropical air from polar air. The case study also illustrated that future idealized studies of upper-level frontogenesis, tropopause folding, and cyclone life cycles must consider the effects of multiple trough interactions.

Acknowledgments. Richard J. Reed is thanked for inspiring a generation of students and researchers through his many pioneering studies and for helping to make synoptic meteorology an exciting field for scientific studies. The two anonymous reviewers are thanked for their suggestions and comments that helped to improve the manuscript. Eyad Atallah and Mike Dickinson of the University at Albany are thanked for providing stimulating scientific discussions and critical technical assistance. Mike Dickinson also assisted the author with the December 1978 case study. Celeste Iovinella is thanked for providing additional technical assistance that enabled this manuscript to be completed. This research was supported by National Science Foundation Grants ATM-9912075, ATM-9413012, ATM-9912075, and ATM-000673.

REFERENCES

Atallah, E., and L. F. Bosart, 2003: Extratropical transition and precipitation distribution of Hurricane Floyd (1999). *Mon. Wea. Rev.*, **131**, 1063–1081.

Anthes, R. A., C. Rocken, and Y.-H. Kuo, 2000: Applications of COSMIC to meteorology and climate. *Terr. Atmos. Oceanic Sci.*, **11**, 115–156.

Bell, G. D., and L. F. Bosart, 1993: A case study diagnosis of the formation of an upper-level cutoff cyclonic circulation over the eastern United States. *Mon. Wea. Rev.*, **121**, 1635–1655.

——, and D. Keyser, 1993: Shear and curvature vorticity and potential-vorticity interchanges: Interpretation and application to a cutoff cyclone event. *Mon. Wea. Rev.*, **121**, 76–102.

Berggren, R., 1952: The distribution of temperature and wind connected with active tropical air in the higher troposphere and same remarks concerning clear air turbulence at high altitude. *Tellus*, **4**, 45–53.

Bjerknes, J., and E. Palmén, 1937: Investigations of selected European cyclones by means of serial ascents. *Geofys. Publ., Norske Videnskaps-Akad. Oslo*, **12**, (2), 1–62.

Bleck, R., 1973: Numerical forecasting experiments based on the conservation of potential vorticity on isentropic surfaces. *J. Appl. Meteor.*, **12**, 737–752.

——, 1974: Short-range prediction in isentropic coordinates with filtered and unfiltered numerical models. *Mon. Wea. Rev.*, **102**, 813–829.

——, and C. Mattocks, 1984: A preliminary analysis of the role of potential vorticity in alpine lee cyclogenesis, *Beitr. Phys. Atmos.*, **57**, 357–368.

Bluestein, H. B., 1986: Fronts and jet streaks: A theoretical perspective. *Mesoscale Meteorology and Forecasting*, P. S. Ray, Ed., Amer. Meteor. Soc., 173–215.

Bosart, L. F., 1970: Mid-tropospheric frontogenesis. *Quart. J. Roy. Meteor. Soc.*, **96**, 442–471.

——, 1981: The Presidents' Day snowstorm of February 1979: A subsynoptic scale event. *Mon. Wea. Rev.*, **109**, 1542–1566.

——, 1999: Observed cyclone life cycles. *The Life Cycles of Extratropical Cyclones*, C. W. Newton and S. Grønås, Eds., Amer. Meteor. Soc., 187–213.

——, and O. Garcia, 1974: Gradient Richardson number profiles and changes within an intense mid-tropospheric baroclinic zone. *Quart. J. Roy. Meteor. Soc.*, **100**, 593–607.

——, and G. M. Lackmann, 1995: Postlandfall tropical cyclone reintensification in a weakly baroclinic environment: A case study of Hurricane David (September 1979). *Mon. Wea. Rev.*, **123**, 3268–3291.

——, G. J. Hakim, K. R. Tyle, M. A. Bedrick, W. E. Bracken, M. J. Dickinson and D. M. Schultz, 1996: Large-scale antecedent conditions associated with the 12–14 March 1993 cyclone ("Superstorm '93") over eastern North America. *Mon. Wea. Rev.*, **124**, 1865–1891.

——, C. S. Velden, W. E. Bracken, J. Molinari and P. G. Black, 2000: Environmental influences on the rapid intensification of Hurricane Opal (1995) over the Gulf of Mexico. *Mon. Wea. Rev.*, **128**, 322–352.

Boyle, J. S. and L. F. Bosart, 1986: Cyclone–anticyclone couplets over North America. Part II: Analysis of a major cyclone event over the eastern United States. *Mon. Wea. Rev.*, **114**, 2432–2465.

Briggs, J., and W. T. Roach, 1963: Aircraft observations near jet streams. *Quart. J. Roy. Meteor. Soc.*, **89**, 225–247.

Browning, K. A., 1971: Radar measurements of air motion near fronts. *Weather*, **26**, 320–340.

——, 1990: Organization of clouds and precipitation in extratropical cyclones. *Extratropical Cyclones The Erik Palmén Memorial Volume*, C. W. Newton and E. O. Holopainen, Eds., Amer. Meteor. Soc., 129–153.

——, 1999: Mesoscale aspects of extratropical cyclones: An observational perspective. *The Life Cycles of Extratropical Cyclones*, C. W. Newton and S. Grønås, Eds., Amer. Meteor. Soc., 265–283.

——, and C. D. Watkins, 1970: Observations of clear air turbulence by high power radar. *Nature*, **227**, 260–263.

——, G. Vaughan and P. Panagi, 1998: Analysis of an ex-tropical cyclone after its reintensification as a warm-core extratropical cyclone. *Quart. J. Roy. Meteor. Soc.*, **124**, 2329–2356.

Cardone, V. J., R. E. Jensen, D. T. Resio, V. R. Swail and A. T. Cox, 1996: Evaluation of contemporary ocean wave models in rare extreme events: The "Halloween storm" of October 1991 and the "storm of the century" of March 1993. *J. Atmos. Oceanic Technol.*, **13**, 198–230.

Carr, F. H., and J. P. Millard, 1985: A composite study of comma clouds and their association with severe weather over the Great Plains. *Mon. Wea. Rev.*, **113**, 370–387.

Chang, E. K. M., 1993: Downstream development of baroclinic waves as inferred from regression analysis. *J. Atmos. Sci.*, **50**, 2038–2053.

——, 1999: Characteristics of wave packets in the upper troposphere. Part II: Seasonal and hemispheric variations. *J. Atmos. Sci.*, **56**, 1729–1747.

Danielsen, E. F., 1959: The laminar structure of the atmosphere and its relation to the concept of a tropopause. *Arch. Meteor. Geophys. Bioklimatol.*, **A11**, 294–332.

——, 1961: Trajectories: Isobaric, isentropic and actual. *J. Atmos. Sci.*, **18**, 479–486.

——, 1964: Project Springfield report. Tech. Rep. DASA 1517, Defense Atomic Support Agency, Washington DC, 97 pp. [NTIS AD-607980.]

——, 1966: Research in four-dimensional diagnosis of cyclonic storm cloud systems. The Scientific Rep. 1, Pennsylvania State University, 1–53.

——, 1967: Transport and diffusion of stratospheric radioactivity based on synoptic hemispheric analyses of potential vorticity. Final Rep. Part 3, The Pennsylvania State University, 1–91.

——, 1968: Stratospheric-tropospheric exchange based on radioactivity, ozone and potential vorticity. *J. Atmos. Sci.*, **25**, 502–518.

——, 1990: In defense of Ertel's potential vorticity and its general applicability as a meteorological tracer. *J. Atmos. Sci.*, **47**, 2013–2020.

——, and R. Bleck, 1967: Research in four-dimensional diagnosis of cyclonic storm cloud systems. Final Scientific Rep., The Pennsylvania State University, 1–96.

Davies, H. C., 1999: Theories of frontogenesis. *The Life Cycles of Extratropical Cyclones*, C. W. Newton and S. Grønås, Eds., Amer. Meteor. Soc., 215–238.

Davis, C. A., 1992: Piecewise potential vorticity inversion. *J. Atmos. Sci.*, **49**, 1397–1411.

——, M. T. Stoelinda and Y.-H. Kuo, 1993: The integrated effect of condensation in numerical simulations of extratropical cyclogenesis. *Mon. Wea. Rev.*, **121**, 2309–2330.

——, and K. A. Emanuel, 1991: Potential vorticity diagnostics of cyclogenesis. *Mon. Wea. Rev.*, **119**, 1929–1953.

——, E. D. Grell, and M. Shapiro, 1996: The balanced dynamical nature of a rapidly intensifying oceanic cyclone. *Mon. Wea. Rev.*, **124**, 3–26.

Dickinson, M. J., M. A. Bedrick, L. F. Bosart, W. E. Bracken, G. J. Hakim, D. M. Schultz, and K. R. Tyle, 1997: The March 1993 Superstorm cyclogenesis: Incipient phase synoptic- and convective-scale flow interaction and model performance. *Mon. Wea. Rev.*, **125**, 3041–3072.

Donnadille, J., J.-P. Cammas, P. Mascart, D. Lambert, and R. Gall, 2001a: FASTEXIOP18: A very deep tropopause fold. I: Synoptic description and modelling. *Quart. J. Roy. Meteor. Soc.*, **127**, 2247–2268.

——, ——, ——, ——, and ——, 2001b: FASTEXIOP18: A very deep tropopause fold. Part II: Quasi-geostrophic omega diagnoses. *Quart. J. Roy. Meteor. Soc.*, **127**, 2269–2286.

Eliassen, A., 1962: On the vertical circulation in frontal zones. *Geofys. Publ.*, **24** (4), 147–160.

——, 1990: Transverse circulations in frontal zones. *Extratropical Cyclones: The Erik Palmén Memorial Volume*, C. W. Newton and E. O. Holopainen, Eds., Amer. Meteor. Soc., 155–164.

Farrell, B. F., 1999: Advances in cyclogenesis theory: Toward a generalized theory of baroclinic development. *The Life Cycles of Extratropical Cyclones*, C. W. Newton and S. Grønås, Eds., Amer. Meteor. Soc., 111–122.

Gettelman, A., and A. H. Sobel, 2000: Direct diagnoses of stratosphere-troposphere exchange. *J. Atmos. Sci.*, **57**, 3–16.

Godson, W. L., 1951: Synoptic properties of frontal surfaces. *Quart. J. Roy. Meteor. Soc.*, **77**, 633–653.

Griffiths, M., A. J. Thorpe, and K. A. Browning, 2000: Convective destabilization by a tropopause fold diagnosed using potential-vorticity inversion. *Quart. J. Roy. Meteor. Soc.*, **126**, 125–144.

Gyakum, J. R., 1983a: On the evolution of the *QE II* storm. I: Synoptic aspects. *Mon. Wea. Rev.*, **111**, 1137–1155.

——, 1983b: On the evolution of the *QE II* storm. II: Dynamic and thermodynamic structure. *Mon. Wea. Rev.*, **111**, 1156–1173.

——, 1991: Meteorological precursors to the explosive intensification of the *QE II* storm. *Mon. Wea. Rev.*, **119**, 1105–1131.

Hakim, G. J., 2000: Climatology of coherent structures on the extratropical tropopause. *Mon. Wea. Rev.*, **128**, 385–406.

——, and D. Keyser, 2001: Canonical frontal circulation patterns in terms of Green's functions for the Sawyer-Eliassen equation. *Quart. J. Roy. Meteor. Soc.*, **127**, 1795–1814.

——, L. F. Bosart, and D. Keyser, 1995: The Ohio valley wave-merger cyclogenesis event of 25–26 January 1978. Part I: Observations. *Mon. Wea. Rev.*, **123**, 2663–2692.

——, D. Keyser, and L. F. Bosart, 1996: The Ohio valley wave-merger cyclogenesis event of 25–26 January 1978. Part II: Diagnosis using quasigeostrophic potential vorticity inversion. *Mon. Wea. Rev.*, **124**, 2176–2205.

Harr, P. A., and R. L. Elsberry, 2000: Extratropical transition of tropical cyclones over the western North Pacific. Part I: Evolution of structural characteristics during the transition process. *Mon. Wea. Rev.*, **128**, 2613–2633.

Hart, R. E., 2003: A cyclone phase space derived from thermal wind and thermal asymmetry. *Mon. Wea. Rev.*, **131**, 585–616.

——, and J. L. Evans, 2001: A climatology of the extratropical transition of Atlantic tropical cyclones. *J. Climate*, **14**, 546–564.

Haynes, P. H., and M. E. McIntyre, 1990: On the conservation and impermeability theorems for potential vorticity. *J. Atmos. Sci.*, **47**, 2021–2031.

Holopainen, E. O., 1990: Role of cyclone-scale eddies in the general circulation of the atmosphere: A review of recent observational studies. *Extratropical Cyclones: The Erik Palmén Memorial Volume*, C. W. Newton and E. O. Holopainen, Eds., Amer. Meteor. Soc., 48–60.

——, 1999: Cyclone climatology and its relationship to planetary waves: A review of recent observational studies. *The Life Cycles of Extratropical Cyclones*, C. W. Newton and S. Grønås, Eds., Amer. Meteor. Soc., 89–99.

Holton, J. R., P. H. Haynes, M. E. McIntyre, A. R. Douglass, R. B. Rood, and L. Pfister, 1995: Stratosphere-troposphere exchange. *Rev. Geophys.*, **33**, 403–439.

Hoskins, B. J., 1990: Theory of extratropical cyclones. *Extratropical Cyclones: The Erik Palmén Memorial Volume*, C. W. Newton and E. O. Holopainen, Eds., Amer. Meteor. Soc., 63–80.

——, and F. P. Bretherton, 1972: Atmospheric frontogenesis models: Mathematical formulation and solution. *J. Atmos. Sci.*, **29**, 11–37.

——, and P. Berrisford, 1988: A potential vorticity perspective of the storm of 15–16 October 1987. *Weather*, **23**, 122–129.

——, M. E. McIntyre and A. W. Robertson, 1985: On the use and significance of isentropic potential vorticity maps. *Quart. J. Roy. Meteor. Soc.*, **111**, 877–946.

Huo, Z., D.-L. Zhang, J. Gyakum and A. Staniforth, 1995: A diagnostic analysis of the superstorm of March 1993. *Mon. Wea. Rev.*, **123**, 1740–1761.

Juckes, M. N., 2000: The descent of tropospheric air into the stratosphere. *Quart. J. Roy. Meteor. Soc.*, **126**, 317–337.

Junger, S., 1997: *The Perfect Storm: A True Story of Men Against the Sea*. Norton, 227 pp.

Kalnay, E., and Coauthors, 1996: The NCEP/NCAR 40-Year Reanalysis Project. *Bull. Amer. Meteor. Soc.*, **77**, 437–471.

Keyser, D., 1986: Atmospheric fronts: An observational perspective. *Mesoscale Meteorology and Forecasting*, P. S. Ray, Ed., Amer. Meteor. Soc., 216–258.

——, 1999: On the representation and diagnosis of frontal circulations in two and three dimensions. *The Life Cycles of Extratropical Cyclones*. C. W. Newton and S. Grønås, Eds., Amer. Meteor. Soc., 239–264.

——, and M. J. Pecnick, 1985: Diagnosis of ageostrophic circulations in a two-dimensional primitive equation model of frontogenesis. *J. Atmos. Sci.*, **42**, 1283–1305.

——, and M. A. Shapiro, 1986: A review of the structure and dynamics of upper-level frontal zones. *Mon. Wea. Rev.*, **114**, 452–499.

——, and R. Rotunno, 1990: On the formation of potential-vorticity anomalies in upper-level jet-front systems. *Mon. Wea. Rev.*, **118**, 1914–1921.

——, B. D. Schmidt, and D. G. Duffy, 1992a: Quasigeostrophic diagnosis of three-dimensional ageostrophic circulations in an idealized baroclinic disturbance. *Mon. Wea. Rev.*, **120**, 698–730.

——, B.D. Schmidt, and D. G. Duffy, 1992b: Quasigeostrophic vertical motions diagnosed from along- and cross-isentrope components of the Q vector. *Mon. Wea. Rev.*, **120**, 731–741.

Kistler, R. and Coauthors, 2001: The NCEP-NCAR 50-year reanalysis: Monthly means CD-ROM and documentation. *Bull. Amer. Meteor. Soc.*, **82**, 247–267.

Klein, P. M., P. A. Harr and R. L. Elsberry, 2000: Extratropical transition of western North Pacific tropical cyclones: An overview and conceptual model of the transformation stage. *Wea. Forecasting*, **15**, 373–396.

Kleinschmidt, E., 1950: On the structure and origin of cyclones (Part 1). *Meteor. Rundsch.*, **3**, 1–6.

Kocin, P. J., P. N. Schumacher, R. F. Morales Jr., and L. W. Uccellini, 1995: Overview of the 12–14 March 1993 superstorm. *Bull. Amer. Meteor. Soc.*, **76**, 165–182.

Kuo, Y.-H. and Coauthors, 1998: A GPS/MET sounding through an intense upper-level front. *Bull. Amer. Meteor. Soc.*, **79**, 617–626.

Lackmann, G. M., L. F. Bosart, and D. Keyser, 1996: Planetary- and synoptic-scale characteristics of explosive wintertime cyclogenesis over the western North Atlantic Ocean. *Mon. Wea. Rev.*, **124**, 2672–2702.

——, D. Keyser, and L. F. Bosart, 1997: A characteristic life cycle of upper-tropospheric cyclogenetic precursors during the Experiment on Rapidly Intensifying Cyclones over the Atlantic (ERICA). *Mon. Wea. Rev.*, **125**, 2729–2758.

Lamarque, J.-F. and P. G. Hess, 1994: Cross-tropopause mass exchange and potential vorticity budget in a simulated tropopause folding. *J. Atmos. Sci.*, **51**, 2246–2269.

Mattocks, C., and R. Bleck, 1986: Jet streak dynamics and geostrophic adjustment processes during the initial stage of lee cyclogenesis. *Mon. Wea. Rev.*, **114**, 2033–2056.

McIntyre, M. E., 1999: Numerical weather prediction: A vision of the future, updated still further. *The Life Cycles of Extratropical Cyclones*, C. W. Newton and S. Grønås, Eds., Amer. Meteor. Soc., 337–355.

Miller, J. E., 1948: On the concept of frontogenesis. *J. Meteor.*, **5**, 169–171.

Morgan, M. C., and J. W. Nielsen-Gammon, 1998: Using tropopause maps to diagnose midlatitude weather systems. *Mon. Wea. Rev.*, **126**, 2555–2579.

Namias, J. and P. F. Clapp, 1944: Studies of the motion and development of long waves in the westerlies. *J. Meteor.*, **1**, 57–77.

Newton, C. W., 1954: Frontogenesis and frontolysis as a three-dimensional process. *J. Meteor.*, **11**, 449–461.

——, 1958: Variations in frontal structure of upper level troughs. *Geophysica*, **6**, 357–375.

——, and A. Trevisan, 1984: Clinogenesis and frontogenesis in jet-stream waves. Part II: Channel model numerical experiments. *J. Atmos. Sci.*, **41**, 2735–2755.

Nielsen-Gammon, J. W., 2001: A visualization of the global dynamic tropopause. *Bull. Amer. Meteor. Soc.*, **82**, 1151–1167.

——, and R. J. Lefevre, 1996: Piecewise tendency diagnosis of dynamical processes governing the development of an upper-tropospheric mobile trough. *J. Atmos. Sci.*, **53**, 3120–3142.

Orlanski, I., and J. Katzfey, 1991: The life cycle of a cyclone wave in the Southern Hemisphere: Part I: Eddy energy budget. *J. Atmos. Sci.*, **48**, 1972–1988.

——, and J. Sheldon, 1993: A case of downstream baroclinic development over western North America. *Mon. Wea. Rev.*, **121**, 2929–2950.

——, and J. Sheldon, 1995: Stages in the energetics of baroclinic systems. *Tellus*, **47A**, 605–628.

Palmén, E., 1949: On the origin and structure of high-level cyclones south of the maximum westerlies. *Tellus*, **1**, 22–31.

——, and K. M. Nagler, 1949: The formation and structure of a large-scale disturbance in the westerlies. *J. Meteor.*, **6**, 227–242.

——, and C. W. Newton, 1969: *Atmospheric Circulation Systems: Their Structure and Physical Interpretation*. Academic Press, 603 pp.

Petterssen, S., 1956: *Weather Analysis and Forecasting*, 2d ed., Vol. I, McGraw-Hill, 428 pp.

Reed, R. J., 1955: A study of a characteristic type of upper-level frontogenesis. *J. Meteor.*, **12**, 226–237.

——, 1990: Advances in knowledge and understanding of extratropical cyclones during the past quarter century: An overview. *Extratropical Cyclones: The Erik Palmén Memorial Volume*, C. W. Newton and E. O. Holopainen, Eds., Amer. Meteor. Soc., 27–45.

——, and E. F. Danielsen, 1959: Fronts in the vicinity of the tropopause. *Archiv. Meteor. Geophys. Bioklimatol.*, **A11**, 1–17.

——, and F. Sanders, 1953: An investigation of the development of a mid-tropospheric frontal zone and its associated vorticity field. *J. Meteor.*, **10**, 338–349.
——, Y.-H. Kuo and S. Low-Nam, 1994: An adiabatic simulation of the ERICA IOP 4 storm: An example of quasi-ideal frontal cyclone development. *Mon. Wea. Rev.*, **122**, 2688–2708.
Ritchie, E. A., and R. L. Elsberry, 2001: Simulations of the transformation stage of the extratropical transition of tropical cyclones. *Mon. Wea. Rev.*, **129**, 1462–1480.
Rotunno, R., W. C. Skamarock, and C. Snyder, 1994: An analysis of frontogenesis in numerical simulations of baroclinic waves. *J. Atmos. Sci.*, **51**, 3373–3398.
Salmon, E. W., and P. J. Smith, 1980: A synoptic analysis of the 25–26 January 1978 blizzard cyclone in the central United States. *Bull. Amer. Meteor. Soc.*, **61**, 453–460.
Sanders, F., 1955: An investigation of the structure and dynamics of an intense surface frontal zone. *J. Meteor.*, **12**, 542–552.
——, L. F. Bosart and C.-C. Lai, 1991: Initiation and evolution of an intense upper-level front. *Mon. Wea. Rev.*, **119**, 1337–1367.
Sawyer, J. S., 1956: The vertical circulation at meteorological fronts and its relation to frontogenesis. *Proc. Roy. Soc. London*, **A234**, 346–362.
Schultz, D. M. and C. A. Doswell III, 1999: Conceptual models of upper-level frontogenesis in south-westerly and north-westerly flow. *Quart. J. Roy. Meteor. Soc.*, **125**, 2535–2562.
Shapiro, M. A., 1970: On the applicability of the geostrophic approximation to upper-level frontal-scale motions. *J. Atmos. Sci.*, **27**, 408–420.
——, 1976: The role of turbulent heat flux in the generation of potential vorticity in the vicinity of upper-level jet stream systems. *Mon. Wea. Rev.*, **104**, 892–906.
——, 1978: Further evidence of the mesoscale and turbulent structure of upper level jet stream-frontal zone systems. *Mon. Wea. Rev.*, **106**, 1100–1111.
——, 1981: Frontogenesis and geostrophically forced secondary circulations in the vicinity of jet stream-frontal zone systems. *J. Atmos. Sci.*, **38**, 954–973.
——, 1982: Mesoscale weather systems of the central United States. CIRES/NOAA Tech. Rep., University of Colorado, 78 pp.
——, 1983: Mesoscale weather systems of the central United States. *The National STORM Program: Scientific and Technological Bases and Major Objectives*, R. A. Anthes, Ed., University Corporation for Atmospheric Research, 3.1–3.77.
——, and D. Keyser, 1990: Fronts, jet streams and the tropopause. *Extratropical Cyclones: The Erik Palmén Memorial Volume*, C. W. Newton and E. O. Holopainen, Eds., Amer. Meteor. Soc., 167–191.
——, and Coauthors, 1999: A planetary-scale to mesoscale perspective of the life cycles of extratropical cyclones: The bridge between theory and observations. *The Life Cycles of Extratropical Cyclones*, C. W. Newton and S. Grønås, Eds., Amer. Meteor. Soc., 139–185.
Simmons, A., 1999: Numerical simulations of cyclone life cycles. *The Life Cycles of Extratropical Cyclones*, C. W. Newton and S. Grønås, Eds., Amer. Meteor. Soc., 123–137.
Staley, D. O., 1960: Evaluation of potential-vorticity changes near the tropopause and the related vertical motions, vertical advection of vorticity, and transfer of radioactive debris from stratosphere to troposphere. *J. Meteor.*, **17**, 591–210.
Sutcliffe, R. C., 1952: Principles of synoptic weather forecasting. *Quart. J. Roy. Meteor. Soc.*, **78**, 291–320.
Thorncroft, C., and S. C. Jones, 2000: The extratropical transitions of Hurricanes Felix and Iris in 1995. *Mon. Wea. Rev.*, **128**, 947–972.
Thorpe, A. J., 1999: Dynamics of mesoscale structure associated with extratropical cyclones. *The Life Cycles of Extratropical Cyclones*, C. W. Newton and S. Grønås, Eds., Amer. Meteor. Soc., 285–296.
Uccellini, L. W., 1986: The possible influence of upstream upper-level baroclinic processes on the development of the *QE II* storm. *Mon. Wea. Rev.*, **114**, 1019–1027.
——, 1990: Processes contributing to the rapid development of extratropical cyclones. *Extratropical Cyclones: The Erik Palmén Memorial Volume*, C. W. Newton and E. O. Holopainen, Eds., Amer. Meteor. Soc., 167–191.
——, P. J. Kocin, R. A. Petersen, C. H. Wash, and K. F. Brill, 1984: The Presidents' Day cyclone of 18–19 February 1979: Synoptic overview and analysis of the subtropical jet streak influencing the pre-cyclogenetic period. *Mon. Wea. Rev.*, **112**, 31–55.
——, D. Keyser, K. F. Brill, and D. H. Wash, 1985: The Presidents' Day cyclone of February 1979: Influence of upstream trough amplification and associated tropopause folding on rapid cyclogenesis. *Mon. Wea. Rev.*, **113**, 962–988.
Wandishin, M. S., J. W. Nielsen-Gammon, and D. Keyser, 2000: A potential vorticity diagnostic approach to upper-level frontogenesis within a developing baroclinic wave. *J. Atmos. Sci.*, **57**, 3918–3938.
Wei, M.-Y., 1987: A new formulation of the exchange of mass and trace constituents between the stratosphere and troposphere. *J. Atmos. Sci.*, **44**, 3079–3086.
Whittaker, J. S., L. W. Uccellini, and K. F. Brill, 1988: A model-based diagnostic study of the rapid development phase of the Presidents' Day cyclone. *Mon. Wea. Rev.*, **116**, 2337–2365.
Wirth, V., 2000: Thermal versus dynamical tropopause in upper-tropospheric balanced flow anomalies. *Quart. J. Roy. Meteor. Soc.*, **126**, 299–317.
——, and J. Egger, 1999: Diagnosing extratropical synoptic-scale stratosphere-troposphere exchange: A case study. *Quart. J. Roy. Meteor. Soc.*, **125**, 635–655.

Chapter 3

Back to Frontogenesis

BRIAN HOSKINS

Department of Meteorology, University of Reading, Reading, United Kingdom

"...the increase in horizontal temperature gradient is due to a steepening of the slope of isentropic surfaces as the result of an indirect solenoidal circulation. Here it is found that the upper-level frontal zone is a dynamically produed phenomenon that intensifies as the storm intensifies, and that the circulation within it is reverse of that required for energy release in the classical sense."—(Reed 1955)

1. Introduction

It is a great pleasure for me to contribute to this symposium and volume in honor of Dick Reed. My Ph.D. research on frontogenesis owed much to the papers by Dick and his collaborators. Since then I have always found scientific meetings with him to be exhilarating, and my wife and I have found social meetings with Joan and him to be very enjoyable. I suspect that my main contribution to Dick's life is that I guided him in the art of driving around roundabouts in England!

The overriding impact of Dick's research reflects his love of what atmospheric weather systems look like and his pleasure in telling others about his insights, especially when they go against prevailing dogma! Another crucial side of his scientific character has always been his willingness to take on board new theories or techniques when he sees that they may be relevant.

In recent years there have been a number of excellent review articles on fronts and related topics, particularly in the volumes that have arisen from the Erik Palmen meeting (Newton and Holopainen 1990) and the Bergen meeting (Shapiro and Gronås 1999). The former included very interesting contributions from Dick himself (Reed 1990) and from Arnt Eliassen (Eliassen 1990). In the latter in particular there was a superb review on frontogenesis by Huw Davies (Davies 1999), as well as other excellent and related articles (e.g., Keyser 1999; Browning 1999; Thorpe 1999; Simmons 1999; Shapiro et al. 1999). Because of this and because I have not worked actively on frontogenesis for some 20 years, this article is not intended to provide a comprehensive review of the topic. Instead it is written from a rather personal perspective, and much of it reflects the time when I was active in the area.

In section 2, I will recall some of the research that appeared to me to be influential in the late 1960s. This is followed in section 3 by a summary of frontogenesis theory. In section 4 I will give a brief discussion of where the study of frontogenesis is today and of some questions that still appear to be in need of being answered.

2. Some research influential in the late 1960s

In 1967 when I began my research on frontogenesis the body of observational studies that I found to be most influential was that associated with Dick Reed and collaborators: Reed and Sanders (1953), Reed (1955), Sanders (1955), and Reed and Danielsen (1959), hereafter RS, R55, S55, and RD, respectively. Two basic questions were what was the nature of fronts and which came first, the cyclone or the front. The opening paragraph of the final section of R55 was typically unequivocal on both questions:

> "The position is taken that the classical polar-front model of cyclone formation does not provide an adequate description of the structure and behavior of frontal zones in the real atmosphere and of the relationship between frontogenesis and cyclogenesis. It is proposed that the frontal problem be re-examined with recognition, first, that many so-called fronts are not fronts according to any definition involving discontinuity surfaces but are merely zones of relatively strong temperature contrast and, secondly, that, intermittently in time and space, true frontal zones do appear in the atmosphere in connection with cyclonic development. The main task of frontal theory is then to distinguish characteristic types of frontal zones and to study their origin and development."

As referenced in R55 the notion that fronts were not preexisting discontinuities was consistent with those of Sutcliffe (1952), another meteorologist who always made his views very clear! The primacy of the cyclogenesis on a preexisting relatively broad temperature contrast was consistent with the setup and results of the baroclinic instability studies of Charney (1947) and Eady (1949). Later it was lent strong support by the numerical integrations performed by Edelmann (1963) and Williams (1967) using the primitive equations. The only reservation associated with the formation of intense, realistic-looking fronts in the cyclone simulation by Edelmann (1963) was that his initial conditions already had a strong, idealized cyclone structure. The 2D Eady wave integration by Williams (1967) had the advantages of a very simple setup and no question that any initial transients or other numerical artifacts were important. It gave a strong indication that there was an internal dynamical mechanism that led to a tendency for frontal collapse.

In R55, Dick credits Fred Sanders with the realization that strong fronts were observed to form only near the surface and in the upper troposphere. The surface front shown in S55, and reproduced here in Fig. 3.1 had large velocity and temperature gradients close to the surface but these weakened rapidly with height. In contrast to the classic horizontal confluence model, vertical velocity was seen as order one important for the strong frontal gradients. The picture of an upper-tropospheric front

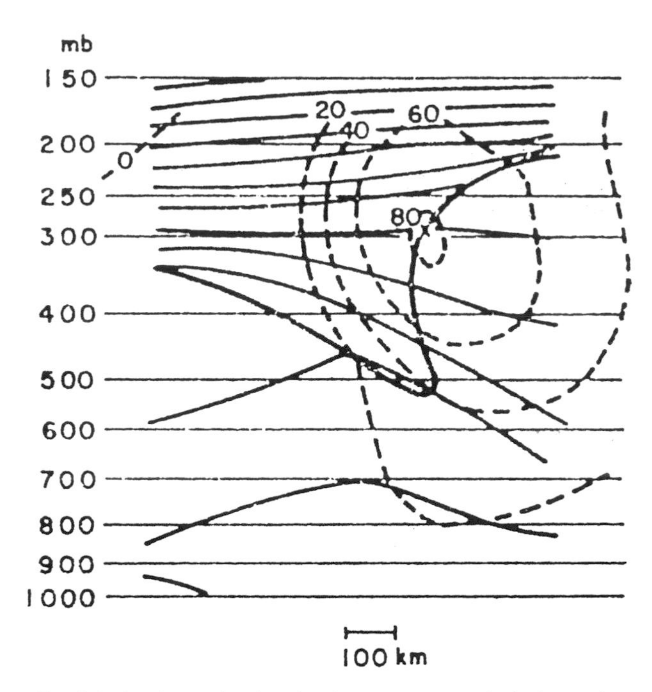

FIG. 3.2. An observational study of an upper-tropospheric front, taken from R55. This picture is the middle one of the three showing the formation of a very strong upper front. A slightly thicker continuous line denotes the tropopause, continuous lines are volumes of potential temperature (contour interval is 10K), and broken lines are isolines of long-front velocity (contour interval is 20 m s^{-1}).

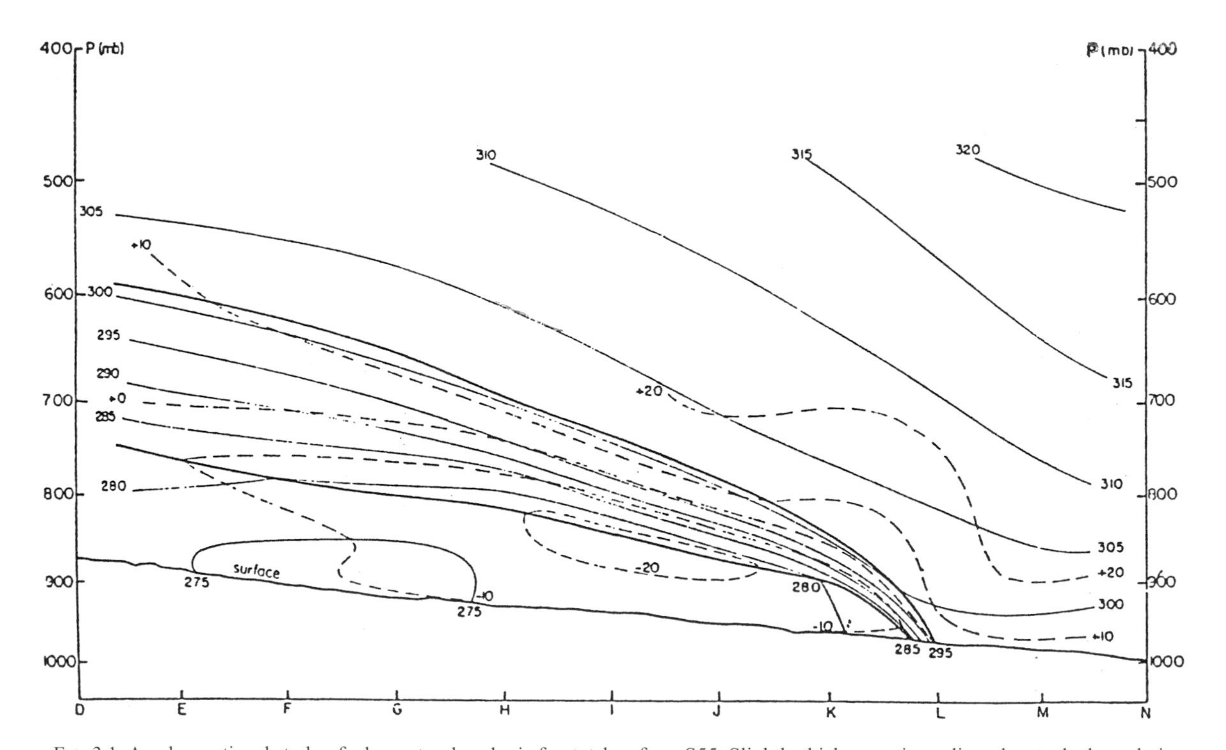

FIG. 3.1. An observational study of a lower trophospheric front, taken from S55. Slightly thicker continous lines denote the boundaries of the region of very strong gradients, continous lines are isolines of potential temperature (contour interval, 5 K), and broken lines are isolines of long-front velocity (contour interval, 10 m s^{-1}).

painted in R55 (see Fig. 3.2 for the middle stage), and of many such fronts in RD, was one in which stratospheric air descended deep into the troposphere in thin tongues. They were not transitions between subtropical and polar air. A most challenging aspect given to the theoretician was the importance ascribed to descent in forming gradients and the indirect nature of this vertical circulation, the strongest descent being on the warm side.

For me, starting my frontogenesis research in a mathematics department, it was vital to have a clear description of fronts and frontogenesis in the real atmosphere in which I could have confidence. These papers provided this for me.

Another vital aspect introduced in RS and exploited in R55 and RD was the use of Rossby–Ertel potential vorticity (PV) as a tracer of the motion. The convincing description in R55 and RD of upper-tropospheric fronts being associated with descent of stratospheric air in folds was made using PV. This was a striking use of such an advanced theoretical concept in an observational study. The material conservation of PV and the related identification of stratospheric air were questioned by Staley (1960) but the results and technique stood.

As discussed in Hoskins (1999), Sutcliffe (1938, 1939) considered the differential acceleration required at two levels suggested by the requirement of maintaining geostrophic balance. Sutcliffe's arguments led to the necessity of the existence of a direct cross-frontal ageostrophic circulation in the confluence and shear cases shown in Fig. 3.3 and discussed below, where the wind and temperature fields suggest that the front is intensifying. The nature of the geostrophic assumptions in this early work was not explicit. However RS showed that the twisting terms were important in the vorticity equation and R55 gave evidence that quasigeostrophic approximations to PV were not valid.

Sawyer (1956), building on the work of Sutcliffe, and then more formally Eliassen (1962) showed that it was possible to approximate the wind along an almost straight front by its geostrophic value. This "frontal geostrophy" approximation enabled Sawyer (1956) and Eliassen (1962) to derive an elliptic equation for the vertical circulation in a plane across the front. Another aspect of these studies that was crucial for my later work was the introduction by Eliassen (1962) of a coordinate transformation in the horizontal that turned the cross-frontal circulation into normal form.

Coming from a mathematical background it was easier for me that the observational studies of Dick Reed and collaborators, as well as the theoretical studies, stressed the fluid and dry thermodynamical aspects of fronts and frontogenesis. However Eliassen (1959) thought that latent heat release could indeed be a dominant mechanism.

The other ingredient that was present in 1967 was the time-dependent quasigeostrophic (QG) confluence model of Stone (1966). This was interesting in both its successes and failures. It showed frontogenesis in an analytic model based on the classic confluence mechanism. However there was no sign of frontal collapse in a finite time and the structures lacked some realism, consistent with the fact that observations had shown that the approximations underlying the theory were not valid for a strong front.

3. Frontogenesis

I will now summarize the basic theory of frontogenesis as seen from my perspective. The account has its basis in the papers discussed above and in my own Ph.D. research published in Hoskins (1971) and Hoskins and

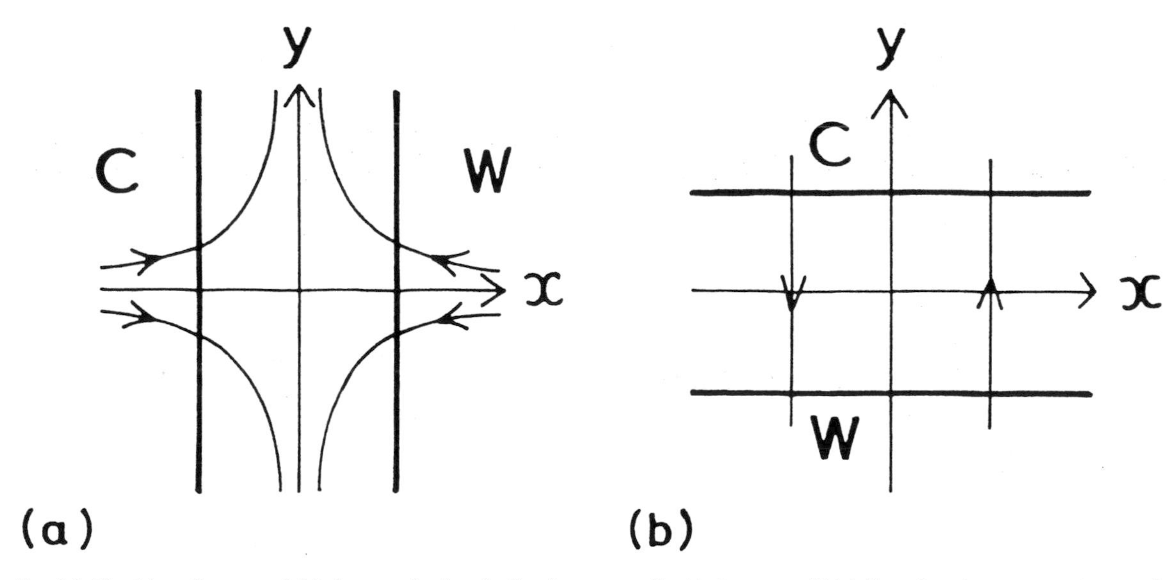

FIG. 3.3. The (a) confluence and (b) shear mechanism for forming an x gradient in buoyancy. Thick lines show buoyancy contours and warm and cold air are indicated by W and C, respectively. Thin lines with arrows denote streamlies. [Taken from Hoskins (1982).]

Bretherton (1972). Although the time-dependent theory is not most neatly expressed this way, I find that the most natural approach to frontogenesis is through the cross-frontal circulation equation. This will be done first in the brief account here as well as in Hoskins (1982), to which reference should be made for more mathematical details.

a. Cross-frontal circulation

We use as thermodynamic variable the buoyancy $b = g\theta/\theta_0$, where θ is the potential temperature and θ_0 a standard value. If b is advected horizontally by the geostrophic wind, then simple differentiation gives

$$D_g b_x = Q_1, \quad (1)$$

$$\text{where } Q_1 = -u_{gx} b_x - v_{gx} b_y, \quad (2)$$

Here, D_g is the geostrophic advection operator, the suffix g refers to geostrophic values, and the suffices x and y denote derivatives. The increasing buoyancy gradients in the x direction in the confluence and shear mechanisms, which are shown in ideal form in Fig. 3.3, are described by the first and second terms in Q_1, respectively. The vector form may be written

$$D_g \nabla \boldsymbol{b} = \boldsymbol{Q}, \quad (3)$$

$$\text{where } \boldsymbol{Q} = -(\boldsymbol{k} \times \partial \boldsymbol{v}_g / \partial s) \nabla b. \quad (4)$$

Here, s is a coordinate along b contours with warm air on the right. The geostrophic frontogenesis function is

$$D_g \mid \nabla b \mid^2 = 2\boldsymbol{Q} \cdot \nabla b. \quad (5)$$

It is positive if **Q** points toward warmer air. As can be seen from (4), this requires that the component of the wind in the horizontal along the buoyancy contour increases with s.

At the level of QG theory, thermal wind balance may be written $b_x = f v_{gz}$. It is easily shown that, following the motion, as well as giving a tendency $+Q_1$ to b_x, the geostrophic motion gives a tendency $-Q_1$ to $f v_{gz}$. Geostrophic motion therefore tends to destroy its own balance and the role of the ageostrophic motion is to maintain it. It is easily found that to do this the ageostrophic cross-front velocity (u_a, w) must satisfy

$$N^2 w_x - f^2 u_{az} = 2Q_1, \quad (6)$$

where the stratification measure, N^2, is at most a function of height. If we use a cross-front streamfunction such that

$$(u_a, w) = (\psi_z, -\psi_x),$$

then (6) gives

$$N^2 \psi_{xx} + f^2 \psi_{zz} = -2Q_1. \quad (7)$$

If buoyancy gradients are tending to increase with time in the x direction (Q_1 positive), then balance is maintained by a direct ageostrophic circulation like that shown in Fig. 3.4a. This acts to reduce the increasing buoyancy gradients through ascent that cools adiabatically on the warm side and descent that warms adiabatically on the cold side. It also acts to increase the vertical shear in v through ageostrophic flow toward the warm air at low levels and away from it at upper levels.

Such a circulation was implicit in the quasigeostrophic frontogenesis studies of Stone (1966); Williams

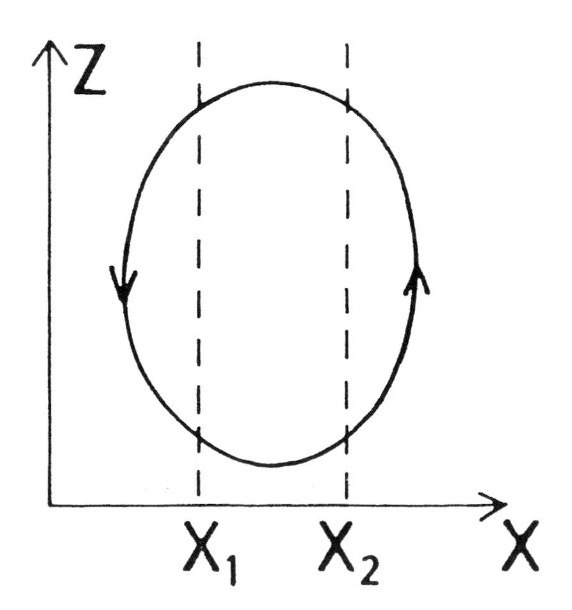

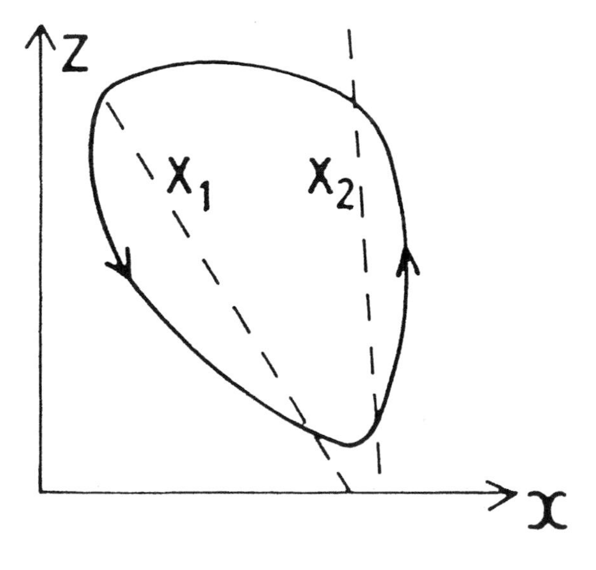

FIG. 3.4. Cross-frontal ageostrophic circulations for a region in which Q_1 is positive. (a) The quasigeostrophic circulation in physical space or the semigeostrophic circulation in X space. (b) The semigeostrophic circulation in physical space. The dashed lines are lines of constant X, which are close together near the surface in physical space in regions of large vorticity. [Taken from Hoskins (1982).]

and Plotkin (1968); and Williams (1968). Given the vorticity equation

$$D_g \xi = f w_z, \tag{8}$$

where ξ is the vertical component of relative vorticity, it is clear that the best that can be hoped for, and is indeed found in the QG studies, is linear growth in buoyancy gradients and in both positive and negative vorticity, and no real frontal slope of maximum gradients.

In semigeostrophic theory we make only the frontal geostrophy approximation that the long-front wind is geostrophic. The same analytical approach can be shown to yield the more complete form of the Sawyer–Eliassen cross-front circulation equation:

$$L(\psi) \equiv N^2 \psi_{xx} - 2S^2 \psi_{xz} + F^2 \psi_{zz} = -2Q_1, \tag{9}$$

where $N^2 = b_z$, $S^2 = b_x = f v_z$, and $F^2 = f(f + v_{gx})$ are now all in general functions of position and time. The ellipticity of this equation depends on the quantity

$$F^2 N^2 - S^4 = fP \equiv f^2 \mathcal{N}^2, \tag{10}$$

where P is proportional to the PV. Thus the material conservation of PV means that whatever gradients are formed the elliptic nature of (9) is maintained. However it also means that as larger and larger gradients are formed the cancellation between the terms in P becomes larger. Mathematically, this means that the ellipticity of the equation becomes weaker and it becomes more nearly parabolic. As it does so, the solution becomes distorted.

This distortion is most easily seen using the coordinate transformation of Eliassen (1962), replacing x by $X = x + v_g/f$ as independent variable in the x direction. Since

$$\mathcal{N}^2 = \partial(X, b)/\partial(x, z), \tag{11}$$

as gradients increase in a frontal region, the angle between "absolute momentum" lines, X = constant, and buoyancy lines becomes smaller. However the cross-front circulation equation (9) using X coordinates has the form

$$(\mathcal{N}^2 \psi_X)_X + f^2 \psi_{ZZ} = -2Q_1/J, \tag{12}$$

where Z = z and

$$J = \partial X/\partial x = 1 + f^{-1} v_{gx} = \zeta/f \tag{13}$$

is the Jacobian of the transformation. Here, $\zeta = f + v_{gx}$ is the vertical component of absolute vorticity, the $-u_y$ term being omitted, which is consistent with the frontal geostrophy approximation. The analogy with the quasigeostrophic form in (7) is clear. However the fact that it is in the transformed coordinate is crucial. As shown in Fig. 3.4, near the surface where vorticity is large, X lines crowd together and the implied ageostrophic convergence and vortex stretching both become larger, even for the same Q_1. (Over most of the domain J is close to 1 and the reduction in the forcing term is small.)

At the surface, where $w = 0$,

$$w_z = -\psi_{xz} = -J\psi_{XZ}.$$

Therefore at the surface front we may write

$$w_z = \gamma \xi, \tag{14}$$

where $\gamma = -\psi_{XZ}/f$ is a dimensionless number.

We now have all the ingredients for frontal collapse. The relevant surface vorticity equation may be written

$$D\xi/Dt = (f + \xi) w_z \tag{15}$$

and includes the nonlinear vortex stretching term. Substituting (14) in (15) gives an equation that may be solved subject to assuming that on the short timescale of frontal collapse the air remains in the front and γ may be taken as constant. Taking t = 0 as the time when the relative vorticity is equal to f, the solution is

$$\xi = f / [2exp(-\gamma f t) - 1]. \tag{16}$$

Infinite vorticity and buoyancy gradients are predicted at $t = (\gamma f)^{-1} ln2$. Taking γ = 0.2 predicts a frontal collapse in only about 10 h. The large gradients tend to be oriented along the sloping X surfaces, giving the frontal slope. However, these gradients rapidly weaken away from the boundary, consistent with the observations.

Following Ooyama (1966) and Hoskins (1974, 1978), other insights are possible through further analysis of (9). When the equations of motion are linearized about a balanced flow in the y direction with uniform F, S, and N, the perturbation streamfunction satisfies the following equation:

$$\frac{\partial^2}{\partial t^2}(\psi_{xx} + \psi_{zz}) = -L(\psi). \tag{17}$$

If we define the maximum and minimum oscillation frequencies to be σ_{max} and σ_{min}, respectively, then it can be shown that the product of the squares of the two frequencies is proportional to fP. Apart from changes in latitude, the product of the frequencies is thus materially conserved in adiabatic motion. In this way the PV is a conserved measure of the stiffness of the fluid. In a situation in which the hydrostatic approximation is valid, the first term in the brackets on the left-hand side of (17) is negligible.

In this case the maximum frequency and strongest stability are in the vertical and correspond to gravitational oscillations with $\sigma_{max} = N$. The minimum frequency is along buoyancy/isentropic surfaces and corresponds to inertial oscillations with $\sigma_{min} = (f\xi_\theta)^{1/2}$, where ξ_θ is the vertical component of absolute vorticity with the derivatives evaluated on buoyancy/isentropic surfaces.

As a surface front becomes stronger, the enhanced static stability means that resistance to quasi-vertical motion orthogonal to isentropic surfaces becomes stronger. However, at the same time the resistance to motion along these surfaces must become weaker. Thus the motion must

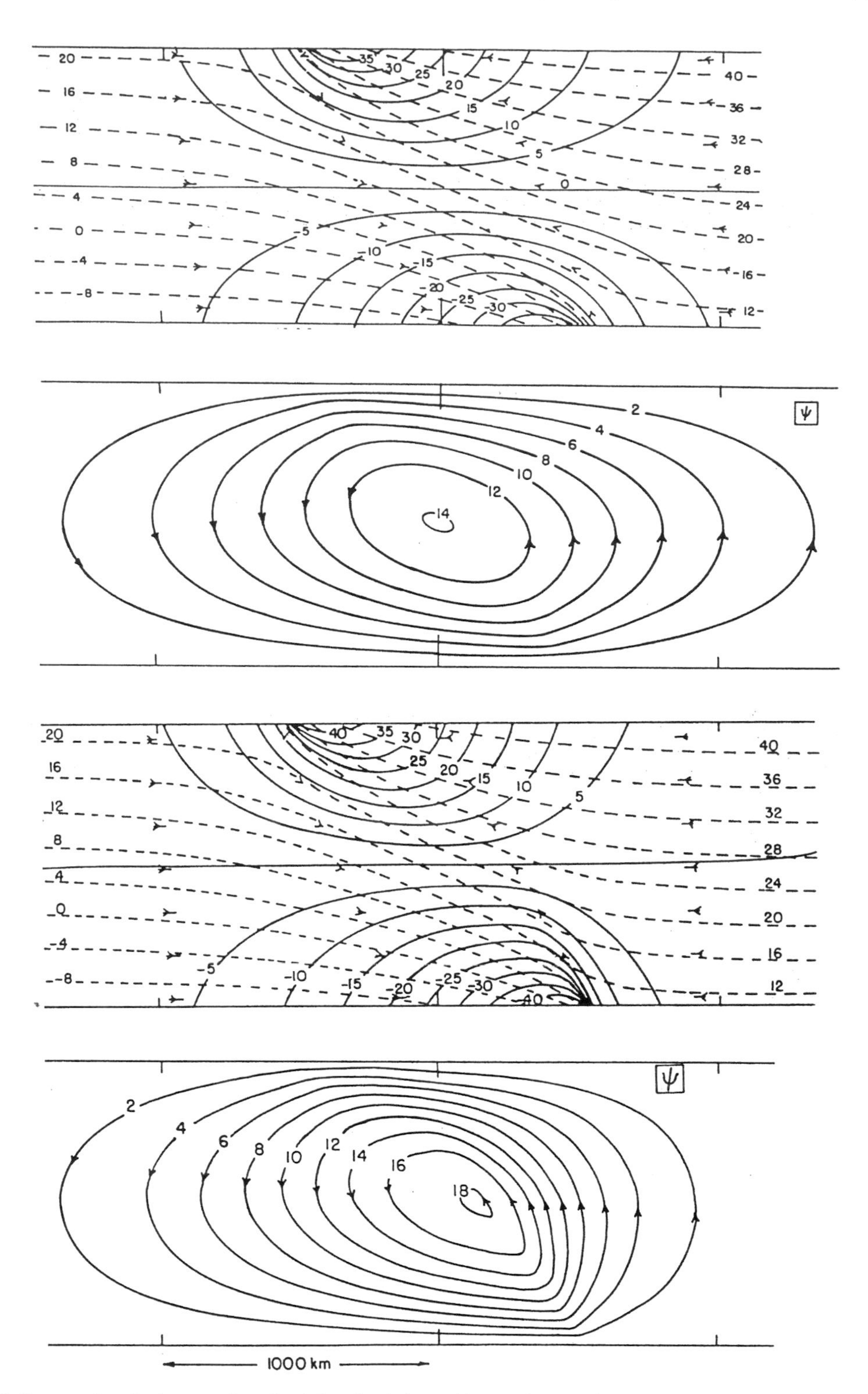

FIG. 3.5. Cross sections in the x–z plane for 1-day simulations with a semigeostrophic confluence frontogenesis model with an initially uniform PV atmosphere between rigid surfaces 10 km apart. (a) Long-front velocity (m s^{-1}, solid), potential temperature (°C, dashed), and total velocity in the plane (arrows) for the dry calculation. (b) Cross-frontal streamfunction (10 kg $m^{-1}s^{-1}$) for the dry calculation. (c) and (d) As in (a) and (b) but for a moist calculation in which, in the region of ascent, a moist PV is conserved and a very small effective PV is used in the circulation equation. [Taken from Thorpe and Emmanuel (1985).]

become more oriented along the isentropic direction of least "resistance." Given that, as frontal gradients become stronger absolute momentum surfaces become almost parallel to isentropic surfaces, the motion must also become more oriented along them. This explains the distortion of the ageostrophic circulation sketched in Fig. 3.4b and leads to frontal collapse in a finite time.

If the tropopause behaved like a rigid surface, then the same process would occur there. However the high PV of the stratosphere is reflected in large resistance only to motions almost orthogonal to isentropic surfaces. Its resistance to motion along isentropic surfaces is similar to that in the troposphere so that the response to increasing temperature gradients in the troposphere is for stratospheric air to descend in a sheet down isentropic surfaces: the tropopause fold. This process can only occur when there are potential temperature gradients along the tropopause. The arguments given here and the 2D model solutions, such as those given in Hoskins 1972, are generally consistent with (R55) and RD, except that the fold is occurring in a thermally direct vertical circulation. This disagreement suggests that the indirect circulation inferred by Dick and colleagues must owe its existence to a fully 3D development. Keyser and Pecnick (1985) have mimicked a 3D situation by mixing the confluence and shear mechanisms in a 2D model and shown how the descent can be moved to the warm side. In any developing weather system the overall circulation must be direct. However, locally it can be indirect, with the along buoyancy contour flow decreasing downstream. As discussed in Heckley and Hoskins (1982), this happens routinely downstream of the upper ridge. It can occur more spectacularly, for example as in the case of Shapiro (1981), when the buoyancy contours cross the jet core, implying a rotation of the jet with height.

If the PV is negative in the Northern Hemisphere, then $\sigma_{\min^2}$ is negative and the fluid is unstable. This so-called symmetric instability is in the hydrostatic case merely inertial instability on isentropic surfaces. However, away from the equator it can only occur if some diabatic process makes the PV negative. When the ascending atmosphere becomes saturated, the effective PV may be near zero or even negative. Emanuel (1983) found this to be the case in observations and considered the slantwise convective available potential energy along absolute momentum surfaces. Earlier, Sawyer (1956) solved the cross-frontal circulation equation for various idealized situations, including some with reduced static stability in the warm air. He found that this reduction in stability was necessary in order to get values of vertical velocity consistent with significant rainfall. Under Eliassen's guidance, in 1964 Todsen (see Eliassen 1990) included a near-zero PV representation of moist processes in a solution of the cross-frontal circulation equation for a real situation. Bennetts and Hoskins (1979) explored the possibility that frontal rainbands could be the result of negative effective PV in the ascending air. They also considered that this negative PV would suddenly be realized when saturation was achieved and termed this conditional symmetric instability (CSI). Recent high-resolution Doppler radar

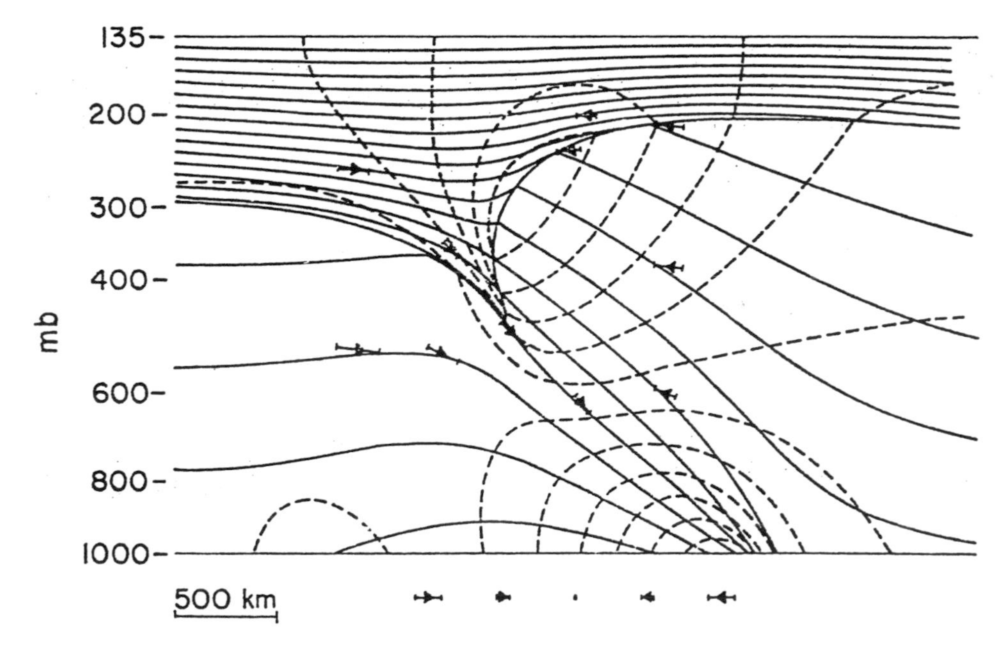

FIG. 3.6. Cross section in the x–z plane for a simulation with a semigeostrophic confluence frontogenesis model with two uniform PV regions, the higher value representing the stratosphere and the lower value the troposphere. Continuous lines are isolines of potential temperature drawn every 7.8 K, broken lines are isolines of long-front velocity drawn every 10.5 m s^{-1}, and the arrows indicate the motion in the palne. [Taken from Hoskins (1972).]

observations of cold fronts by Browning et al. (2001) reveal multiple slantwise circulations with small vertical scale, which they attribute to CSI.

b. Time-dependent models

The crucial linkage in my work was made when I tried to estimate the forcing term for the cross-frontal circulation equation in X space, (12), in my attempt to understand the numerical primitive equation integration solution for the 2D Eady wave produced by Terry Williams (1967). Enabled by his excellent figures, I found that the forcing seemed to be very nearly sinusoidal in X. This led me to consider the time-dependent solution using this coordinate. I will now summarize the essence of such a time-dependent frontogenesis analysis.

With the frontal geostrophy approximation the motion in the y direction is approximated by its geostrophic value. The y-momentum equation then implies that $DX/Dt = u_g$, so that the ageostrophic motion in the x direction becomes implicit. The geostrophic and hydrostatic relations hold in X coordinates with there existing a variable Φ such that

$$fv = \Phi_X, \quad b = \Phi_Z. \tag{18}$$

The Jacobian of the transformation and the vertical component of absolute vorticity, (13), can be shown to be given by

$$J = \zeta/f = (1 - f^{-2}\Phi_{XX})^{-1}. \tag{19}$$

From (11), the PV and N^2 may be written

$$f^{-1}P = \mathcal{N}^2 = \frac{\partial(X, b)}{\partial(X, Z)}\frac{\partial(X, Z)}{\partial(x, z)} = Jb_Z. \tag{20}$$

Substituting for b from (18) and for J from (19), (20) may be rewritten in the following form:

$$f^{-2}\Phi_{XX} + \mathcal{N}^{-2}\Phi_{ZZ} = 1. \tag{21}$$

There is the same analogy between the semigeostropic and quasigeostrophic forms as there is for the cross-front circulation equation. Equation (21) can be solved for Φ and hence for all the balanced variables if P is uniform in regions in the interior and with $b = \Phi_z$ geostrophically advected on horizontal boundaries. From (19), discontinuities in the coordinate transformation and thus in v and b in physical space are predicted when $v_X = f$, that is, when the quasigeostrophic solution would give a relative vorticity of f. Again the absolutely central role of PV in determining the motion and the importance of its material conservation is apparent.

The physical reality check I always applied when determining solutions from this theory for both the confluence and shear models was, "did they agree with the observations summarized by Dick Reed and collaborators?" The theory worked in suggesting surface frontal collapse and strong upper-air fronts. The 2D surface fronts in a single PV troposphere, such as the one in the confluence model given in Figs. 3.5a,b compared reasonably enough with the observational

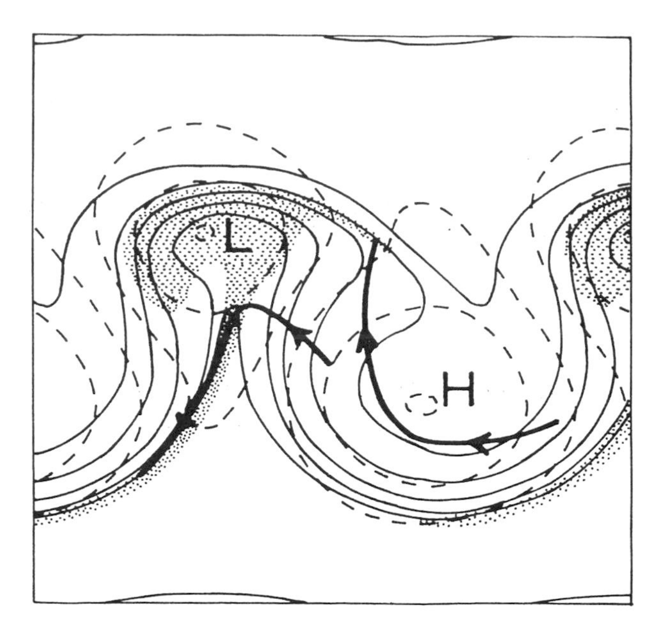

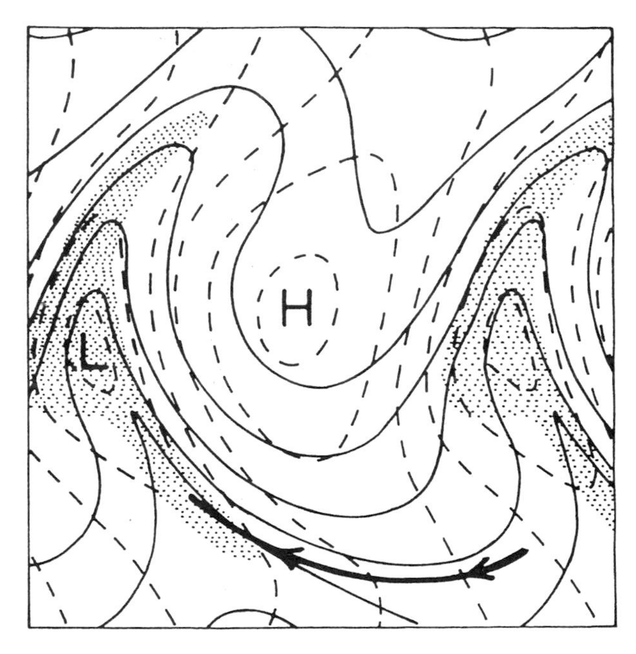

FIG. 3.7. Surface maps for two 3D semigeostrophic uniform PV baroclinic wave simulations in a doubly periodic domain. (a) Day 6.3 for a basic flow with zero surface wind and at the lid a sinusoidal westerly going from zero at the "edges" to a maximum of 29.4 m s^{-1} at the "center" of the "channel." (b) Day 5.5 for a basic flow at the surface varying from 4.4 m s^{-1} westerlies at the edges to 4.4 m s^{-1} easterlies at the center of the channel and at the lid uniform 25 m s^{-1} westerlies. Temperature contours every 4 K in (a) and 8 K in (b) are continuous lines, pressure contours are dashed lines, and the regions of relative vorticity greater than f/2 are shaded. The bold lines in each case indicate two trajectories relative to the system from day 3.

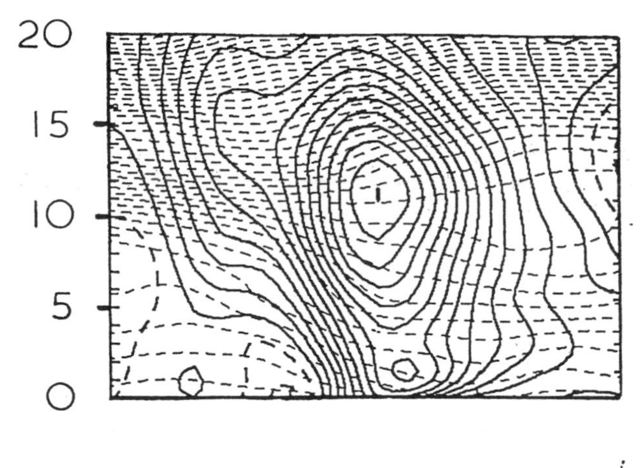

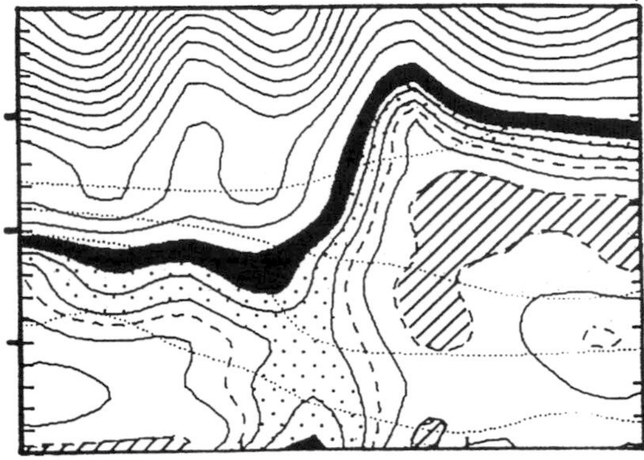

FIG. 3.8. Vertical sections across the storm near the United Kingdom at 0000 UTC 16 Oct 1987. (a) Velocity into the section with isotachs every 5 m s^{-1} (negative contours heavy dashed) and potential temperature contours every 5K. (b) PV is shown by hatching for regions with values less than ¼, dashed contour at 3/4 and continuous contours at ½, 1,2, 3 and then every 2. The region with values between 1 and 2 is stippled and that between 2 an 3 blackened. Also shown are the 300, 315, 330, and 350 K isentropes. [Taken from Hoskins and

structures like that in Fig. 3.1. However, particularly in the confluence model, they give little indication of the spatial or temporal relationship to a weather system. The upper-tropospheric fronts, such as that in Fig. 3.6 obtained adding a second-higher PV region on the top to represent the stratosphere, inspired by the observational studies, compared well with the middle time observational structure shown in R55 and reproduced in Fig. 3.2. However they did not show the dramatic fold and stratospheric tongue discussed in R55 and RD. Moore (1993) later showed that such tongues could be produced if the model was modified so that it could be integrated for an unrealistically long period. Even then, as indicated before, the sense of the implied vertical circulation was opposite to that inferred from observations.

The main hope for increasing realism was to make the structures fully 3D. We found (Hoskins 1975, 1976; Hoskins and West 1979; Hoskins and Draghici 1977; Heckley and Hoskins 1982) that the geostrophic momentum approximation analysis of Eliassen (1948) and Fjortoft (1962) could be used as the basis for a 3D version of the semigeostrophic theory described above. McWilliams and Gent (1980) in their very thorough analysis of the accuracy of approximations to the primitive equations raised some questions over the geostrophic momentum and semigeostrophic equations. I always realized that they were no better than quasigeostrophic theory in terms of a Rossby number asymptotic approximation. To me the crucial aspects were that, unlike the quasigeostrophic equations, they had a full PV that is advected in three dimensions and that they can describe the formation of strong fronts and jet streams with arbitrary orientations. These qualities were recognized by McWilliams and Gent (1980). I was reassured when errors in particular baroclinic wave life cycle simulations exhibited by Snyder et al. (1991) were interpreted by Davies et al. (1991) as merely reflecting sensitive behavior near a bifurcation point in the parameter space of the initial state.

The semigeostropic equations enabled us to look at the formation of fronts in baroclinic waves growing on westerly flows. Cold fronts produced in these models, such as that reproduced in Fig. 3.7a, seemed to pass the Dick Reed test. I remember him approving of the descent that dominated the southern portion of the front. The thermal circulations are direct on average. Air following a trajectory like that shown in the cold front region in Fig. 3.7a experiences strong frontogenesis with the vectors **Q** pointing toward the warm air near the cyclone center (Hoskins and Pedder 1980). Farther along the front the "Q vectors" are almost parallel to temperature contours. Dick was also happy to see the general lack of a strong warm front ahead of the warm sector, the only strong feature being on the north and northeastern boundary of the warm air.

Before the time shown here, a discontinuity formed at the rigid lid in the model downstream from the ridge and on the cold side of the thermal contrast. This again indicates that tropopause-level frontogenesis would be stronger if the tropopause was rigid. However the strong gradients descending into the troposphere would not occur. Because the model was simple enough that x and z could be reversed, I realized that this upper-air discontinuity would become a very strong warm front with this transformation. In this way it was found that strong warm fronts could be generated, but only in baroclinic waves on flows with stronger surface westerlies on their southern flanks. With such a horizontal shear, although the large-scale forcing of vertical motion is similar, the details of the frontogenetic regions change. The Q vectors now point toward the warm air ahead of the warm sector and along the thermal contours behind it. As in the example shown in Fig. 3.7b, strong warm frontogenesis now occurs to the east of the warm sector.

Adrian Simmons and I used these ideas as the basis for designing the zonal flows we later used in our primitive equation studies of baroclinic wave growth (see, e.g., Hoskins 1990; Simmons 1999). In particular the two life cycles LC1 and LC2 were based on the usual and the strong warm front case, respectively, and yielded frontogenesis and frontal structures very similar to those given by semigeostrophic theory.

Adding the 3D growing baroclinic wave context clearly gives extra realism to the 2D time-dependent frontogenesis models. The other aspect that is clearly lacking is a representation of the enhanced frontogenesis due to large-scale latent heat release. As discussed above, one representation is in terms of reduced or even zero PV. One of the cases that we included in Hoskins and Bretherton (1972) was the zero PV confluence model, which can be solved very easily. The frontogenesis is very similar to that in the finite-PV case but occurs quicker because there is no resistance to motion along isentropes. Thorpe and Emanuel (1985) included in the 2D confluence model the near-zero PV representation of the impact of latent heat release in ascending air. As shown in Fig. 3.5c,d, this gives the anticipated enhanced frontal gradients (e.g., about a factor of 6 in the maximum vorticity) and circulation (the maximum frontal ascent is 3.7 cm s^{-1} as opposed to 1.1 cm s^{-1} in the dry case).

An alternative perspective on latent heat release is through its impact on the material change in PV and the consequent influence on the time-dependent frontogenesis. Latent heat release leads to increasing PV below and decreasing PV above. The trajectories of the air determine the resulting impact on the PV distribution at a later time. In a realistic synoptic development it appears that the reduced PV in the upper troposphere is advected rapidly downstream, acting to decrease the already low PV on the anticyclonic side of the jet stream or increase the strength of the downstream ridge. However the lower-tropospheric air below the latent heating maximum in general tends to stay in the frontal region, leading to a significant PV maximum there. This then contributes to much enhanced vorticity and stability in the frontal region. In PV inversion terms it adds to the cyclonic circulation induced by the warm air ahead of the cold front. Many observational studies have shown such a positive PV feature. As a personal example (Hoskins and Berrisford 1988) in Fig. 3.8 are shown vertical cross sections of (a) wind and potential temperature and (b) PV for the October 1987 storm that uprooted a significant proportion of the trees in southern England. The lowered tropopause and the large near-surface PV, together with the surface warmth, give a "PV tower." Associated with this is a very strong cyclonic circulation throughout the troposphere, and particularly near the surface.

4. Some concluding comments

The discussion here has been personal and somewhat limited: a much wider one is given in Davies (1999) and in the other review papers mentioned previously. As summarized in Shapiro et al. (1999), semigeostrophic and primitive equation integrations show a range of realistic developments. Indeed routine high-resolution weather forecasts exhibit a wide range of structures that in general verify quite well. Whether this fulfills the characterization task set in the last sentence of my Dick Reed quote is perhaps less clear. From a more observational perspective, Keith Browning and others (see, e.g., Browning 1999) have continued to enlighten in this regard.

One question that still does not appear to have an answer is what limits the collapse process in frontal regions. Two-dimensional semigeostrophic theory has as its small number

$$(U/V)^2 \times (\xi/f),$$

where U and V are characteristic velocities across and along the front in a frame moving with the front. The relative magnitudes of U and V will depend on the rate of increase in buoyancy gradients implied by the geostrophic motion. However it would be expected that the theory for an almost-straight front will be valid until vorticities of at least 5f are produced. Since it can be shown that the Richardson number is approximately f/ξ, fronts with increasingly large vorticity will have decreasing Richardson numbers. Consequently in Hoskins and Bretherton (1972) we said that mixing due to Kelvin-Helmholtz instability may limit the frontogenetic process. Indeed Browning et al. (1970) had earlier presented observational evidence that frontal zones are Richardson number limited. Recently, Chapman and Browning (2001), using Doppler radar observations, have estimated that Kelvin-Helmholtz billows in frontal zones act to dissipate the zones on a timescale of a few hours.

Cullen and Purser (1984) for zero PV and Koshyk and Cho (1992) for finite PV have been happy to carry semigeostrophic theory even beyond the surface discontinuity and allow the fluid in contact with the boundary and the frontal discontinuity to both propagate into the interior. When frontal curvature is significant, the 3D semigeostrophic theory that is then required does not have an accuracy that suggests validity beyond $\xi = f$, but such curvatures are not often present. The breakdown of semigeostrophic theory may lead to the end of frontal collapse, but this is not clear at present. Various authors, such as Snyder et al. (1993) have used very high resolution primitive equation integrations and seen the generation of gravity waves but not a clear end to the collapse process other than with the inclusion of diffusion. It is, however, possible that strictly 2D fronts are very special and that in reality the tendency to produce 3D structures may often provide a limit to the frontogenesis.

Very recent simulations of one of the Fronts and Atlantic Storm Tracks Experiment (FASTEX) storms by H. Lean (2002, personal communication) using the new U.K. Met Office forecast model run at a range of horizontal and vertical resolutions provide suggestive information on frontal collapse and its limitation. Lean has run a 45-layer model with 60-, 24-, 12-, 4-, and 2-km horizontal resolutions. The 800-hPa vertical velocity in the frontal region rises with decreasing resolution when viewed at the grid scale of each model. However, when averaged to the 60-km grid it shows very good signs of convergence at the higher resolutions. Spectral energy density plots show increased energy near the truncation scale except with the 2-km grid. This suggestion that the 2-km grid is resolving the frontal structure is supported by plots of 800-hPa vertical motion. At 2 km the scale of the frontal structure appears to be above the grid scale, whereas with the larger grids, this is not the case. This may be a sign that other mechanisms come into play on such kilometer scales. However it may, alternatively, be a sign that the vertical resolution is spuriously limiting the horizontal collapse.

The fact that the strongest gradients in lower-tropospheric fronts are in the boundary layer strongly suggests that its dynamics and physics are of more importance than is usually recognized. A recent surprise for me was a PV diagnosis of the results of including a boundary layer in a dry baroclinic wave life cycle (D. Adamson 2002, personal communication). The expected Ekman pumping and mechanically reduced PV in the region of the surface cyclone were indeed found. However, there were also important increases in PV ahead of the warm sector and this air moved over and around the low center in the bent-back occlusion region. The high PV acted to strengthen this front somewhat but the associated enhanced static stability acted to decrease the development of the whole system.

Observational experiments such as FASTEX have shown that frontal regions have a wealth of different modes of precipitation organization rather than the simple large-scale latent heat release mimicked in frontogenesis models. Only some of these are easily categorized as associated with processes like CSI.

We are entering a very interesting period when modern observational systems will be giving us even more detail of frontogenesis and frontal regions, and it will be possible to simulate cyclone development using numerical models with a resolution of a kilometer or so in the horizontal. This will open up new opportunities for examining the detailed interplay of the semigeostrophic frontal collapse mechanism, boundary layer and internal turbulence processes, and moist processes on many scales. It will need people of the caliber and breadth of Dick Reed to fully realize the tremendous opportunity for increasing both our knowledge of frontogenesis and our ability to predict the detailed weather associated with it.

Acknowledgments. I am pleased to acknowledge the extremely helpful comments of Kerry Emanuel and another reviewer, and also those of Keith Browning and Alan Thorpe.

REFERENCES

Bennetts, D. A., and B. J. Hoskins, 1979: Conditional symmetric instability—a possible explanation for frontal rainbands. *Quart. J. Roy. Meteor. Soc.,* **105,** 945–962.

Browning, K. A., 1999: Mesoscale aspects of extratropical cyclones: An observational perspective. *The Life Cycles of Extratropical Cyclones,* M. A. Shapiro and S. Gronås, Eds., Amer. Meteor. Soc., 265–283.

——, T. W. Harrold, and J. R. Starr, 1970: Richardson number limited shear zones in the free atmosphere. *Quart. J. Roy. Meteor. Soc.,* **96,** 40–49.

——, D. Chapman, and R. S. Dixon, 2001: Stacked slantwise convective circulations. *Quart. J. Roy. Meteor. Soc.,* **127,** 2513–2536.

Chapman, D., and K. A. Browning, 2001: Measurement of dissipation in frontal zones. *Quart. J. Roy, Meteor. Soc.,* **127,** 1939–1959.

Charney, J. G., 1947: The dynamics of long waves in a baroclinic westerly current. *J. Meteor.,* **4,** 135–162.

Cullen, M. J. P., and R. J. Purser, 1984: An extended Lagrangian theory of semigeostrophic frontogenesis. *J. Atmos. Sci.,* **41,** 1477–1497.

Davies, H. C., 1999: Theories of frontogenesis. *The Life Cycles of Extratropical Cyclones,* M. A. Shapiro and S. Gronås, Eds., Amer. Meteor. Soc., 215–238.

——, C. Schär and H. Wernli, 1991: The palette of fronts and cyclones within a baroclinic wave development. *J. Atmos. Sci.,* **48,** 1666–1689.

Eady, E. T., 1949: Long waves and cyclone waves. *Tellus,* **1** (3), 33–52.

Edelmann, W., 1963: On the behaviour of disturbances in a baroclinic channel. Summary Rep. 2, Research in Objective Weather Forecasting, Part F, Contract AF61 (052)-373, Deutscher Wetterdienst, Offenbach/Main, Germany, 35 pp.

Eliassen, A., 1948: The quasi-static equations of motion. *Geofys. Publ.,* **17** (3), 1–44.

——, 1959: On the formation of fronts in the atmosphere. *The Atmosphere And Sea in Motion (Rossby Memorial Volume),* B. Bolin, Ed., Rockefeller Institute Press, 277–287.

——, 1962: On the vertical circulation in frontal zones. *Geofys. Publ.,* **24,** 147–160.

——, 1990: Transverse circulations in frontal zones. *Extratropical Cyclones: The Erik Palmén Memorial Volume,* C. W. Newton and E. O. Holopainen, Eds., Amer. Meteor. Soc., 155–165.

Emanuel, K. A., 1983: On assessing local conditional symmetric instability from atmospheric soundings. *Mon. Wea. Rev.,* **111,** 2016–2033.

Fjortoft, R., 1962: On the integration of a system of geostrophically balanced prognostic equations. *Proc. Int. Symp. on Numerical Weather Prediction,* Tokyo, Japan, Meteorological Society of Japan, 153–159.

Heckley, W. A., and B. J. Hoskins, 1982: Baroclinic waves and frontogenesis in a non-uniform potential vorticity semi-geostrophic model. *J. Atmos. Sci.,* **39,** 1999–2016.

Hoskins, B. J., 1971: Atmospheric frontogenesis models: Some solutions. *Quart. J. Roy. Meteor. Soc.,* **97,** 139–153.

——, 1972: Non-Boussinesq effects and further development in a model of upper tropospheric frontogenesis. *Quart. J. Roy. Meteor. Soc.,* **98,** 532–541.

——, 1974: The role of potential vorticity in symmetric stability and instability, *Quart. J. Roy. Meteor. Soc.,* **100,** 480–482.

——, 1975: The geostrophic momentum approximation and the semigeostrophic equations. *J. Atmos. Sci.,* **32,** 233–242.

——, 1976: Baroclinic waves and frontogensis. Part 1: Introduction and Eady waves. *Quart. J. Roy. Meteor. Soc.,* **102,** 103–122.

——, 1978: Baroclinic instability and frontogenesis. *Rotating Fluids in Geophysics,* P. H. Roberts and A. M. Soward, Eds., Academic Press, 171–203.

——, 1982: The mathematical theory of frontogenesis. *Annu. Rev. Fluid Mech.,* **14,** 131–151.

——, 1990: Theory of extratropical cyclones. *Extratropical Cyclones: The Erik Palmén Memorial Volume,* C. W. Newton and E. O. Holopainen, Eds., Amer. Meteor. Soc., 64–80.

——, 1999: Sutcliffe and his development theory. *The Life Cycles of Extratropical Cyclones,* M. A. Shapiro and S. Gronås, Eds., Amer. Meteor. Soc., 81–86.

——, and F. P. Bretherton, 1972: Atmospheric frontogenesis models: Mathematical formulation and solutions. *J. Atmos. Sci.,* **29,** 11–37.

——, and I. Draghici, 1977: The forcing of ageostrophic motion according to the semigeostrophic equations and in an isentropic coordinate model. *J. Atmos. Sci.,* **34,** 1859–1867.

——, and N. V. West, 1979: Baroclinic waves and frontogenesis. Part II: Uniform potential vorticity jet flows. *J. Atmos. Sci.,* **36,** 1663–1680.

——, and M. A. Pedder, 1980: The diagnosis of middle latitude synoptic development. *Quart. J. Roy. Meteor. Soc.,* **106,** 707–719.

——, and P. Berrisford, 1988: A potential vorticity perspective of the storm of 15–16 October 1987. *Weather,* **43,** 122–129.

Keyser, D., 1999: On the representation and diagnosis of frontal circulations in two and three dimensions. *The Life Cycles of Extratropical Cyclones,* M. A. Shapiro and S. Gronås, Eds., Amer. Meteor. Soc., 239–264.

——, and M. J. Pecnick, 1985: A two-dimensional primitive equation model of frontogenesis forced by confluence and horizontal shear. *J. Atmos. Sci.,* **42,** 1259–1282.

Koshyk, J. N., and H.-R. Cho, 1992: Dynamics of a mature front in a uniform potential vorticity semigeostrophic model. *J. Atmos. Sci.,* **49,** 497–510.

McWilliams, J. C., and P. R. Gent, 1980: Intermediate models of planetary circulations in the atmosphere and ocean. *J. Atmos. Sci.,* **37,** 1657–1678.

Moore, G. W. K., 1993: The development of tropopause folds in two-dimensional models of frontogenesis. *J. Atmos. Sci.,* **50,** 2321–2334.

Newton, C. W., and E. O. Holopainen, Ed., 1990: *Extratropical Cyclones: The Erik Palmen Memorial Volume,* Amer. Meteor. Soc., 262 pp.

Ooyama, K., 1966: On the stability of the baroclinic circular vortex: A sufficient criterion for instability. *J. Atmos. Sci.,* **23,** 43–53.

Reed, R. J., 1955: A study of a characteristic type of upper-level frontogenesis. *J. Meteor.,* **12,** 226–237.

——, 1990: Advances in knowledge and understanding of extratropical cyclones during the past quarter century: An overview. *Extratropical Cyclones: The Erik Palmén Memorial Volume,* C. W. Newton and E. O. Holopainen, Eds., Amer. Meteor. Soc., 27–45.

——, and F. Sanders, 1953: An investigation of the development of a mid-tropospheric frontal zone and its associated vorticity field. *J. Meteor.,* **10,** 338–349.

——, and E. F. Danielsen, 1959: Fronts in the vicinity of the tropopause. *Arch. Meteor. Geophys. Bioklimatol.,* **11A,** 1–17.

Sanders, F., 1955: Investigation of the structure and dynamics of an intense surface frontal zone, *J. Meteor.,* **12,** 542–552.

Sawyers, J. S., 1956: The vertical circulation at meteorological fronts and its relation to frontogenesis. *Proc. Roy. Soc. London.,* **A234,** 346–362.

Shapiro, M. A., 1981: Frontogenesis and geostrophically forced secondary circulations in the vicinity of jet stream-frontal zone systems. *J. Atmos. Sci.,* **38,** 954–973.

——, and S. Gronås, Eds., 1999: *The Life Cycles of Extratropical Cyclones.* Amer. Meteor. Soc., 359 pp.

——, and Coauthors, 1999: A planetary-scale to mesoscale perspective of the life cycles of extratropical cyclones: The bridge between theory and observations. *The Life Cycles of Extratropical Cyclones,* M. A. Shapiro and S. Gronås, Eds., Amer. Meteor. Soc., 139–185.

Simmons, A. J., 1999: Numerical simulations of cyclone life cycles. *The Life Cycles of Extratropical Cyclones,* M. A. Shapiro and S. Gronås, Eds., Amer. Meteor. Soc., 123–137.

Snyder, C., W. C. Skamarock, and R. Rotunno, 1991: A comparison of primitive-equation and semigeostrophic simulations of baroclinic waves. *J. Atmos. Sci.,* **48,** 2179–2194.

——, ——, and ——, 1993: Frontal dynamics near and following frontal collapse. *J. Atmos. Sci.,* **50,** 3194–3211.

Staley, D. O., 1960: Evaluation of potential-vorticity changes near the tropopause and the related vertical motions, vertical advection of vorticity, and transfer of radioactive debris from stratosphere to troposphere. *J. Meteor.,* **17,** 591–620.

Stone, P. H., 1966: Frontogenesis by horizontal wind deformation fields. *J. Atmos. Sci.,* **23,** 455–465.

Sutcliffe, R. C., 1938: A remark on divergent winds. *Meteor. Mag. London.,* **73,** 44–45.

——, 1939: Cyclonic and anticyclonic development. *Quart. J. Roy. Meteor. Soc.,* **65,** 519–524.

——, 1952: Principles of synoptic weather forecasting. *Quart. J. Roy. Meteor. Soc.,* **78,** 291–320.

Thorpe, A. J., 1999: Dynamics of mesoscale structure associated with extratropical cyclones. *The Life Cycles of Extratropical Cyclones,* M. A. Shapiro and S. Gronås, Eds., Amer. Meteor. Soc., 285–296.

——, and K. A. Emanuel, 1985: Frontogenesis in the presence of small stability to slantwise convection. *J. Atmos. Sci.,* **42,** 1809–1824.

Williams, R. T., 1967: Atmospheric frontogenesis: A numerical experiment. *J. Atmos. Sci.,* **24,** 627–641.

——, 1968: A note on quasi-geostrophic frontogenesis. *J. Atmos. Sci.,* **25,** 1157–1159.

——, and J. Plotkin, 1968: Quasi-geostropic frontogenesis. *J. Atmos. Sci.,* **25,** 201–206.

Chapter 4

Polar Lows

ERIK A. RASMUSSEN

Geophysical Department, University of Copenhagen, Copenhagen, Denmark

"In treating the subject of extratropical cyclogenesis most textbooks of meteorology deal almost exclusively with the Norwegian cyclone model in which the cyclone originates as a wave perturbation on a polar front separating tropical and polar air masses. Yet it is not uncommon, especially over the oceans in winter, for cyclones to form in polar air streams behind or poleward of the polar front."—(Reed 1979)

1. Introduction

Systematic research of polar lows began in the 1960s with Harrold and Browning (1969). Almost from the very beginning, Richard Reed (RR) contributed to the research into this "new" phenomenon through a number of significant papers. A perusal of RR's papers on this subject shows that he, in several cases, was the first to point out significant features of these systems, introduced new ideas, and pointed out problems related to the formation and dynamics of polar lows. Reed in this way had a big influence in introducing the relatively unknown concept of "polar lows" into the common meteorological vocabulary. On the other hand, he recently and quite ironically, has introduced some question marks regarding the concept, such as shown by the title of an invited paper: "Is there such a thing as a polar low?" (10th Cyclone Workshop, 21–27 September 1997, Val Morin, Quebec, Canada). In an attempt to answer that question the history of polar low research is briefly reviewed and illustrated mainly by the 12 RR papers listed in the reference section.

The papers can be divided into three groups. The first comprising the papers from 1979 until 1987 are studies of polar lows for which baroclinic processes were central. Three papers published from 1987 and 1989 are general reviews including proposals for definitions and classifications of polar lows. Finally the last three from 1995 to 2001 are essentially numerical studies, based on the use of advanced mesoscale models, and aiming to identify and evaluate different physical processes of potential importance for polar lows.

2. The beginning: The first "comma-cloud paper"

a. The beginning

Over the years there has been, and still is, some confusion concerning the terms "polar low" and "comma cloud" as well as the other names that have been used to describe small-scale disturbances in polar air masses. These "other names" include Arctic instability low, polar air depression, mesoscale cyclone or vortex, Arctic hurricane, and polar airstream cyclone.

From the published literature it appears that up until about 1970 it was only Norwegian and British meteorologists who were aware of these systems that, whatever their names, have been responsible for the loss of many small vessels over the centuries. Besides causing problems at sea the effects of these small-scale systems were also felt during the winter months in coastal areas where the weather could deteriorate very rapidly with winds increasing to gale force and with heavy snow. While meteorologists were aware of the existence of these lows it was nearly impossible to forecast them.

One of the earliest references to what nowadays are known as polar lows was made by the Norwegian meteorologist Peter Dannevig (1954) in a book for pilots. On a schematic weather chart (see Fig. 4.1) Dannevig indicated where such disturbances, denoted therein in as "Arctic instability lows," typically should be expected within a cold air outbreak around Norway. The term Arctic instability low, the name used for polar lows in Norway up to the 1980s, reflects the fact that the lows seemed to develop due to thermal instability within cold air masses flowing over a warm sea. Dannevig in his work considered the possible mechanisms behind the formation of these systems and concluded that the lows could develop in the same way as tropical cyclones.

Since the 1960s British meteorologists have also taken a keen interest in mesoscale weather systems in polar airstreams since such systems could lead to significant snowfall across the British Isles and especially in Scotland. Until the 1960s such systems generally were called cold air depressions (Meteorological Office 1962). The earliest known case studies of "polar lows" were published in the British magazine *Weather* in the 1960s and

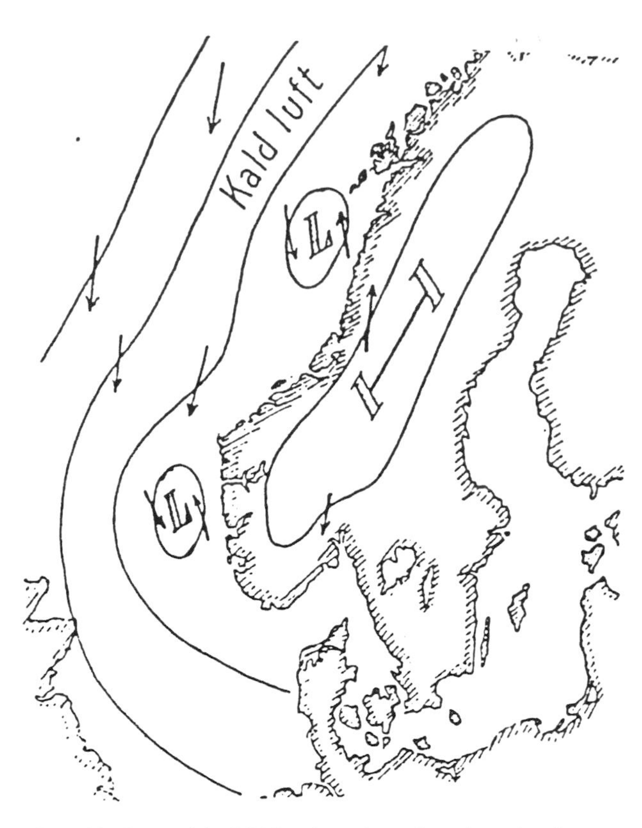

FIG. 4.1. Dannevig's (1954) schematic surface chart showing two polar lows (called instability lows by Dannevig) within a northerly outbreak of polar air near the Norwegian coast.

1970s, where Harley (1960) probably first used the term polar low. In the *Handbook of Weather Forecasting* (Meteorological Office 1964) the term polar low was taken to refer to "fairly small-scale cyclones or troughs embedded in a deep cold current which had recently left northerly latitudes."

German meteorologist were also aware of the polar lows, termed by them "polartief" (Scherhag and Klauser 1962).

Reed in his 1979 paper and in two following papers written together with Blier discussed the occurence, structure, and development of the systems now generally known as "comma clouds." In 1979 there was still some confusion regarding the meaning of the term polar low and even, among quite a number of meteorologists, a dispute over the very existence of such a phenomenon. The uncertainty about the new subject was reflected by RR in the introduction to his 1979 paper in which he wrote, "The purpose of the present paper is to stimulate discussion of the phenomenon in question, which, except for studies done in Britain and to a lesser degree in Scandinavia, is a neglected subject in the meteorological literature." Reed related and more or less identified the systems studied by him, that is, disturbances in polar airstreams over the Pacific, to the polar lows or "polar troughs" as described in the British meteorological literature. Figure 1 in RR's 1979 paper, reproduced here as Fig. 4.2, shows a schematic diagram of a polar low and polar trough as originally shown in the British publication *A Course in Elementary Meteorology* (Meteorological Office 1962).

In England and Norway polar lows or thermal instability lows traditionally had been considered as a kind of thermal instability phenomenon associated with deep convection (for a history of polar lows see Rasmussen and Turner (2003). Harrold and Browning (1969), taking issue with this explanation, argued that the fact that the precipitation associated with two polar lows passing the British Isles was produced by stable upgliding motion instead of convection showed that polar lows essentially were low-level baroclinic disturbances owing their small horizontal size to the shallow depth of the baroclinic zone.

In parallel with the studies of comma clouds by RR, research had started in Scandinavia focusing on the more northerly polar lows, deep within the cold air masses away from the main baroclinic zone (Økland 1977; Rasmussen 1977, 1979). These authors challenged the point of view of Harrold and Browning and other British meteorologists by supporting the old idea that some polar lows were basically driven by deep convection and in important ways were akin to tropical cyclones. Økland and Rasmussen, using simple models and utilizing the conditional instability of the second kind (CISK) theory introduced by Charney and Eliasen (1964) and Ooyama (1964) found evidence that polar lows, even over the cold polar seas, could develop through the action of surface fluxes and convection in the same way as tropical cyclones. Satellite observations were not available for these early studies, but conventional data, from the surface as well as upper-air observations, supported the point of view that small-scale synoptic circulations such as shown in Fig. 4.3 were driven by deep convection. The idea was reluctantly received by many meteorologists but gradually won

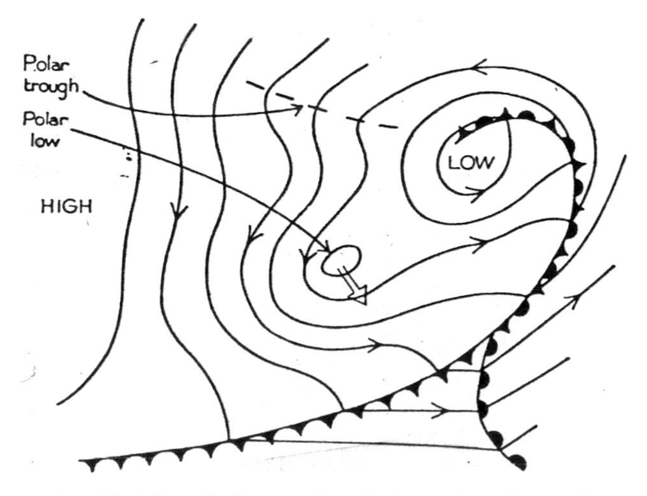

FIG. 4.2. Schematic diagram of a polar low and a polar trough (Meteorological Office 1962; from Reed 1979).

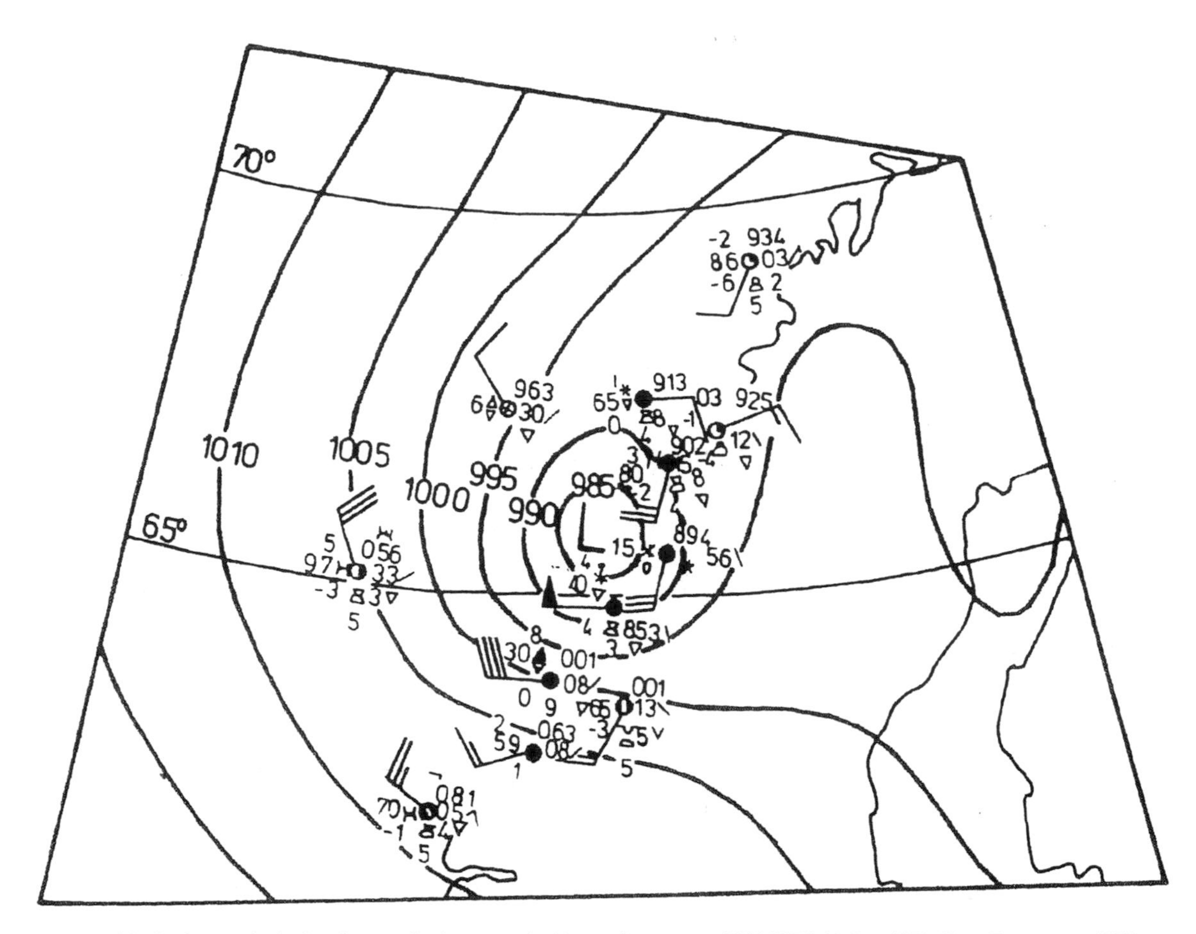

FIG. 4.3. Surface analysis showing a polar low near the Norwegian coast at 0000 UTC 13 Oct 1971 (from Rasmussen 1979).

acceptance, especially after satellite images such as shown in Fig. 4.4 became available. The satellite image shows a polar low near North Cape on 27 February 1987 (see Nordeng and Rasmussen 1992). The polar low shows a striking similarity with a small tropical cyclone including a well-defined eye and a spiral structure. This image and others demonstrated better than words that some polar lows, and especially polar lows at high latitudes, were unique systems in their own right. As such, they should not be considered merely as small-scale extratropical cyclones more or less akin to the large-scale cyclones forming on the polar front. On the other hand, polar lows that could best be described as small-scale versions of the traditional extratropical cyclones were found as well. An example is illustrated in Fig. 4.5. The satellite image shows a baroclinic polar low in its occluding stage over the eastern Baltic on 29 March 1985 (from Rasmussen and Aakjær 1992). The southern part of the cloud system associated with the low is merging with clouds belonging to the polar front situated farther south.

b. The first comma-cloud paper

In his 1979 paper RR drew, based on inspection of several comma cloud developments in the Pacific, a number of important conclusions concerning the nature and structure of comma clouds, all of which were more or less confirmed by later studies by RR himself and other authors. According to RR the cyclones in polar air masses are generally of small dimension. They form most often over the oceans in winter, originating within regions of low-level heating and enhanced convection and acquiring a comma-shaped cloud pattern as they mature. They are associated with well-developed baroclinicity throughout the troposphere and are located on the poleward side of the jet stream in a region marked by strong cyclonic wind shear and by conditional instability through a substantial depth of the troposphere. A surface pressure trough is found under the trailing edge of the comma tail and, in the more intense cases, a surface low pressure center under the comma head.

Like Anderson et al. (1969) RR noted that wave cyclones were initiated when comma clouds approached sufficiently close to the polar front. Such waves appear to occlude rapidly, jumping from an open wave to a fully occluded stage without going through the intermediate stages of development described by the Norwegian frontal model. In forecast practice this process was named instant occlusion.

FIG. 4.4. Infrared satellite image at 0418 UTC 27 Feb 1987 showing a polar low with spiral structure near North Cape. [Image courtesy of the National Environmental Research Council (NERC) Satellite Receiving Station, University of Dundee.]

Reed concluded that the small lows he studied were essentially baroclinic phenomena owing their small horizontal scale to the effect of small static stabilities at low levels in reducing the wavelength of maximum instability. Unlike the phenomena studied by Harrold and Browning, they were invariably associated with baroclinicity of appreciable depth and the presence of an upper-tropospheric jets stream. He found little or no evidence that the polar lows were a "pure CISK phenomenon" or that barotropic instability played any significant role in their formation.

Reed in his studies considered comma clouds over the Pacific. However, comma clouds are often observed in other regions as well, as illustrated by the well-developed comma cloud over the North Atlantic near Iceland shown in Fig. 4.6. In this case an intense small-scale low, that is, a polar low, developed beneath the head of the comma cloud.

Reed's 1979 paper marked the starting point of a debate over whether comma clouds were "true" polar lows, the latter being defined as the systems observed far to the north, away from the main baroclinic zone, near the ice edge and generally associated with a spiraliform cloud signature. As a consequence investigators tended to divide themselves into two schools of thought, one emphasizing baroclinic effects and the other championing CISK or other heating mechanisms (see Reed 1987). This division reflected the fact that the small-scale cold-air disturbances may appear in many

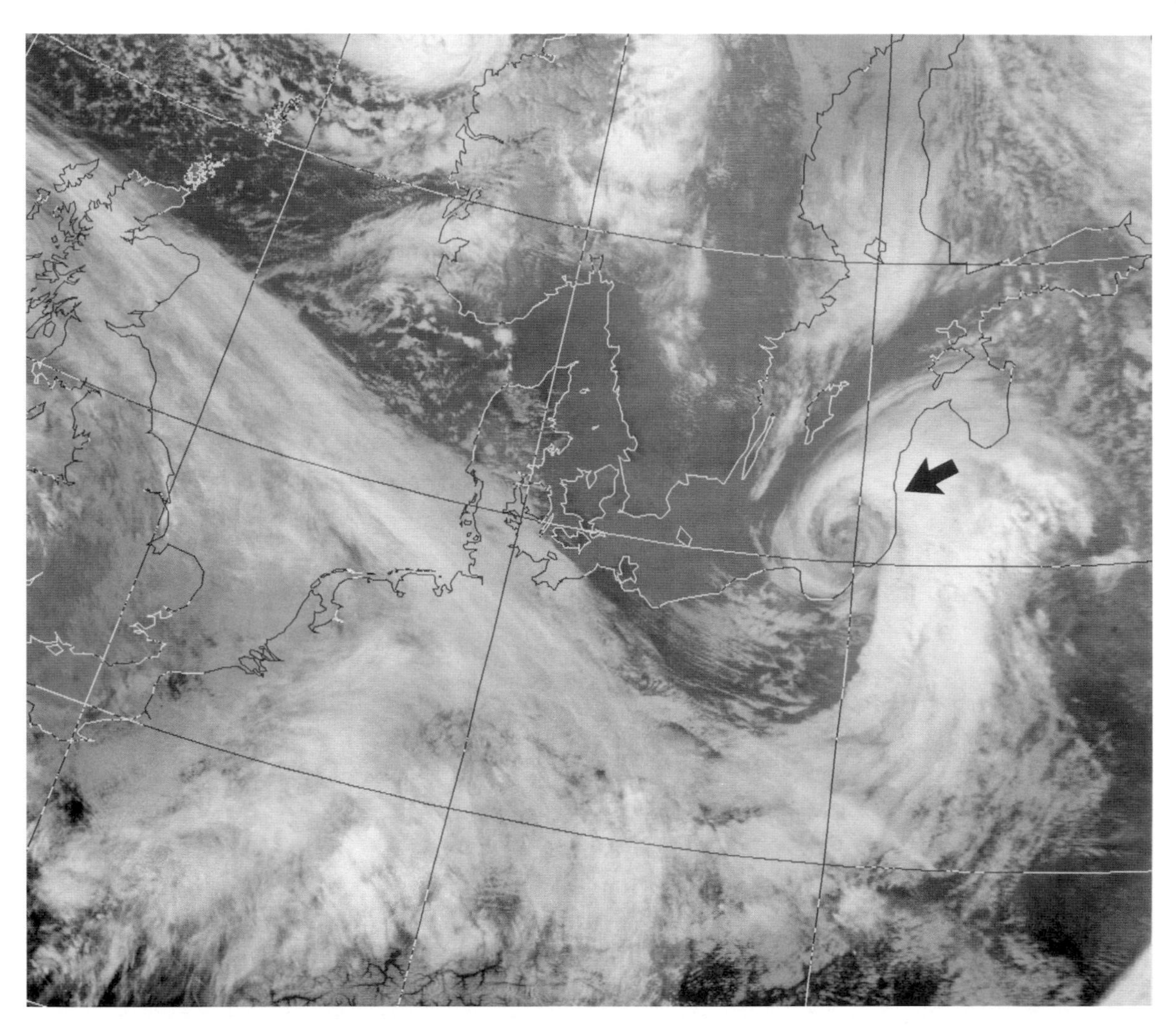

FIG. 4.5. Infrared satellite image at 1128 UTC 29 Mar 1985 showing a baroclinic polar low (marked with arrow) in its occluding stage over the eastern Baltic (from Rasmussen and Aakjær 1992). (Image courtesy of the NERC Satellite Receiving Station, University of Dundee.)

forms and that the different scientists studying these phenomena tended to focus on only one type of low pressure system, typically comma clouds or the so-called spiral polar lows, the latter also being denoted "real" or "true" polar lows by some workers. In Fig. 4.7 are shown, close to each other, representatives examples of a spiral polar low (southeast of Iceland) and a comma cloud over the northwestern part of the British Isles. The image illustrates the striking difference between the spiraliform system that formed deep within the cold air mass, and the comma cloud that formed farther south on the northern flank of the polar front, the latter indicated by the band of clouds seen at the bottom part of the figure.

In his 1979 paper RR did not distinguish between different types of cold-air disturbances but in general concluded that, "the small lows that develop in polar airstreams are essentially baroclinic phenomena." He did not, however, at that time use the term "comma cloud" to denote these systems as such. Instead he used the term polar low as a generic term for the different kinds of polar air disturbances or, alternatively, terms such as "small cyclones" or "small lows."

The studies by RR and many others have clearly documented the connection with regions of upper-level, positive vorticity advection (PVA) and the occurence of comma clouds, one of the major cloud forms associated with polar lows. The second major cloud form is the "spiraliform" signature, characterized by one or more spiral bands of convective clouds around a circulation center, an example of which is shown in Fig. 4.4. (Other examples are shown in Figs. 4.7 and 4.9.)

Spiraliform cloud systems are characteristic of disturbances forming in northerly airstreams at high latitudes deep within cold air masses and they often develop near the central region of upper-level cold-core vortices. In such cases the upper-level PVA will be small and the formation of polar lows will be due

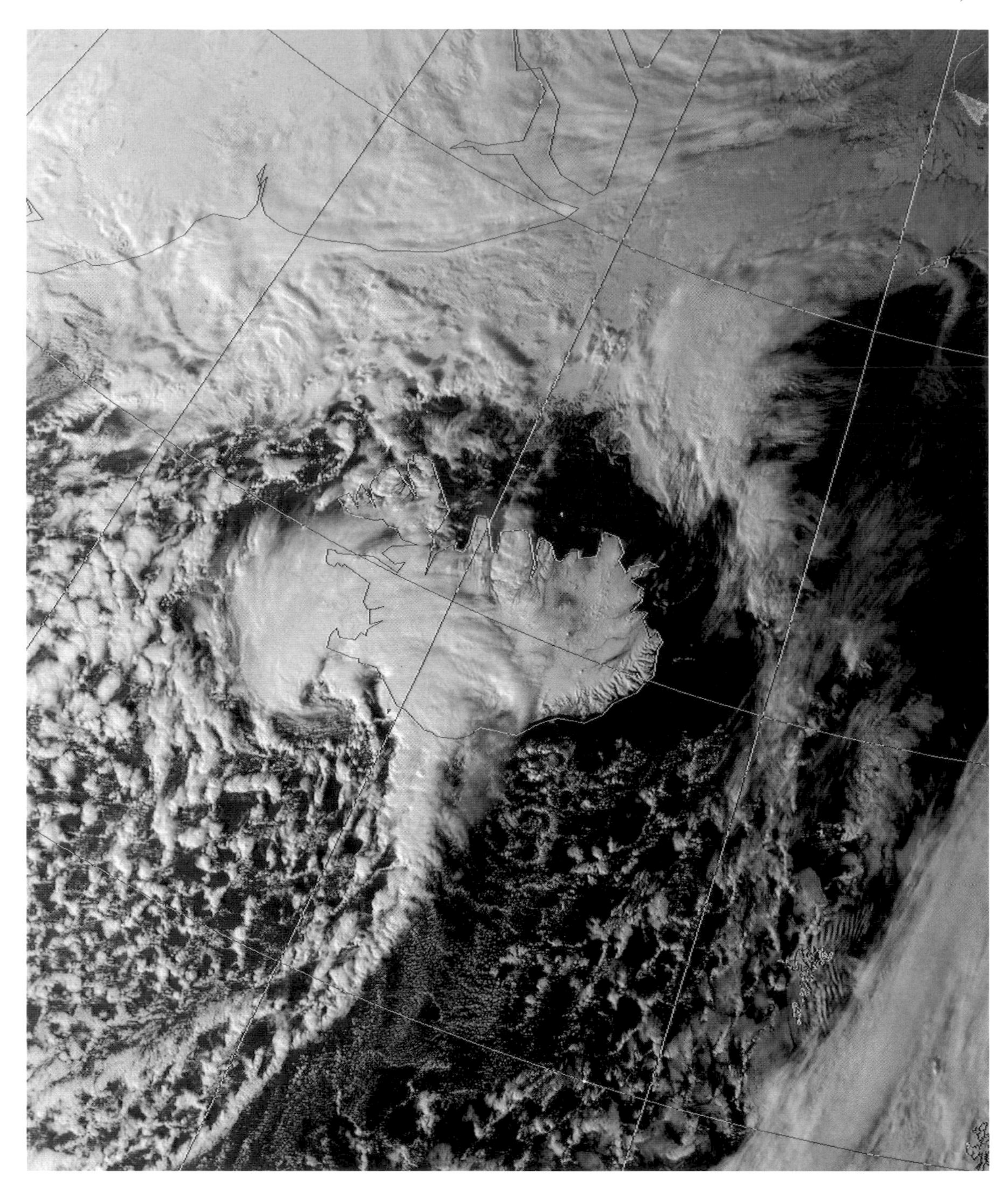

FIG. 4.6. Visible satellite image at 1528 UTC 8 Mar 1988 showing a comma cloud over the North Atlantic near Iceland. (Image courtesy of the NERC Satellite Receiving Station, University of Dundee.)

mainly to thermal instability caused by the combination of relatively warm low-level temperatures over a "warm" sea surface, and by the low upper-level temperatures near the center of the upper-level systems. The cold-core upper-level vortices are characteristic features of the polar regions and have their origin within the central region of the circumpolar (planetary) vortex from whence they occasionally drift southward.

Many comma clouds are associated with only weak surface circulations, and this, together with the fact that they typically are found at relatively southerly positions within the westerlies, led in the 1980s to a still ongoing discussion over whether they should be considered to be polar lows or not. It was soon realized, however, that polar air disturbances can appear in many forms so that one could talk about a "polar low spectrum" with

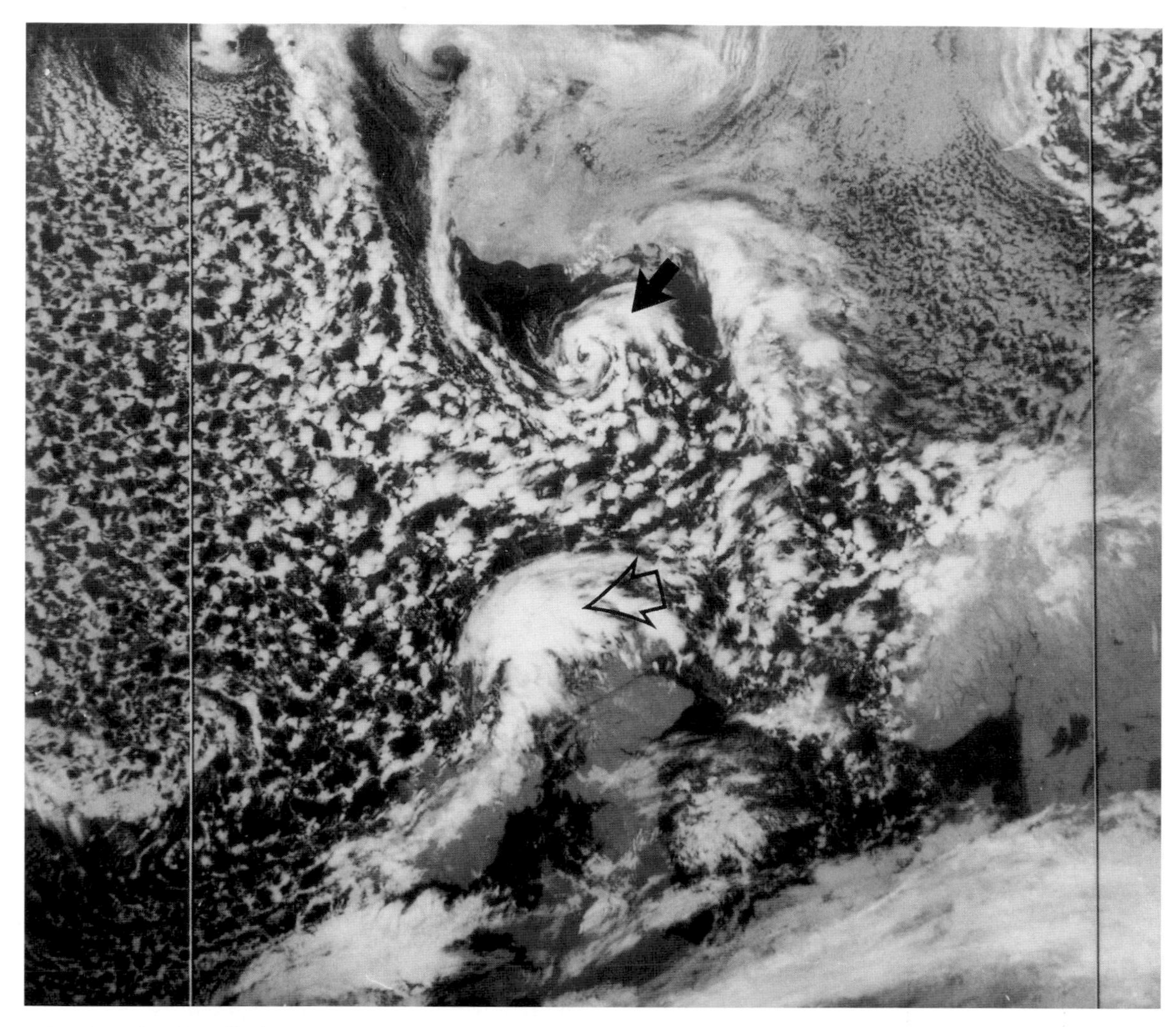

FIG. 4.7. Infrared satellite image at 1932 UTC 25 Nov 1978 showing a spiralformed polar low (indicated by black arrow) southeast of Iceland, and a comma cloud (indicated by open arrow) over Scotland. (Image courtesy of the NERC Satellite Receiving Station, University of Dundee.)

baroclinic systems such as comma clouds at one end of the spectrum, and convective systems, in some way driven by thermal instability, at the other end. This compromise was reflected in the practical definitions that for a number of years have been generally used. According to a recent, and rather typical definition (Rasmussen and Turner 2003), "A polar low is a small, but fairly intense maritime cyclone that forms poleward of the main baroclinic zone (the polar front or other major baroclinic zones). The horizontal scale of the polar low is approximately between 200 and 1000 kilometers and surface wind generally near or above gale force." This definition, as well as a number of earlier ones, are fairly general and does not distinguish between comma clouds and convective systems.

As mentioned in the beginning of this section terms other than polar low have been used to denote cold-air disturbances. This, together with the fact that the term polar low sometimes is used for systems that do not strictly fulfill the requirements set up in the definition above (mostly because of the wind force requirement), occasionally have lead to some confusion. In this connection it should be noted that polar low research during the 1980s was almost exlusively focused upon Northern Hemisphere systems. During the late 1980s and 1990s, however, increasing interest was focused on Southern Hemisphere systems. While Southern Hemisphere small-scale lows at fairly *low* latitudes resemble their Northern Hemisphere counterparts, higher-latitude systems, those near the Antarctic coast, differ in import ways from the Northern Hemisphere systems. Most significantly, most of these vortices near the Antarctic ice edges are much weaker and of smaller scale compared to the Northern Hemisphere polar lows. In order to avoid confusion between the weak Southern Hemisphere systems and the generally more vigorous Northern Hemisphere polar lows it soon became customary to denote the Southern Hemisphere systems by terms such as mesocyclone, Antarctic mesocyclone,

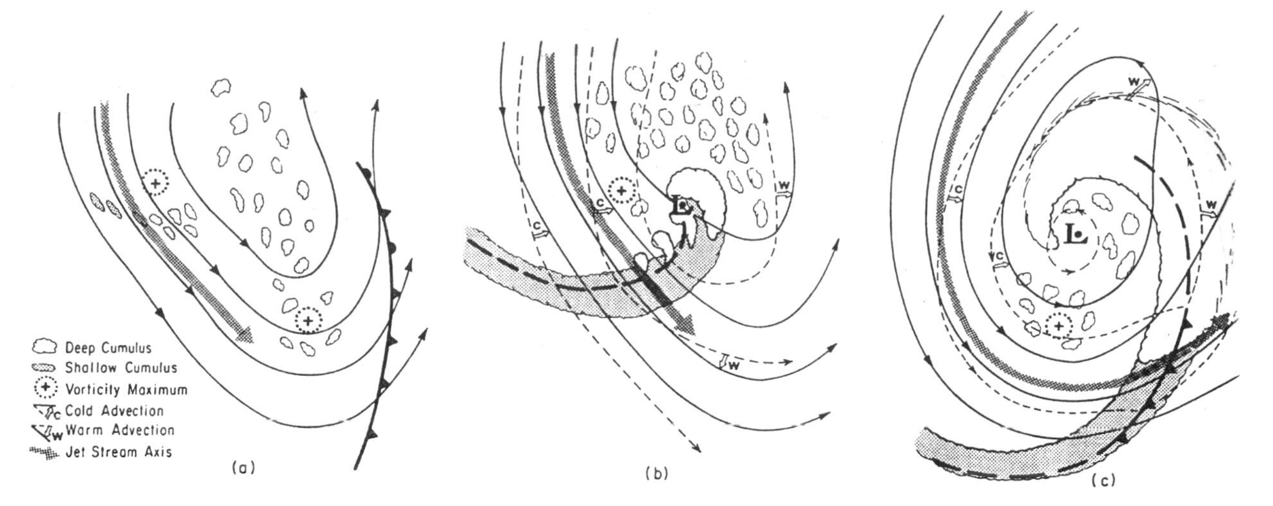

FIG. 4.8. Schematic diagram of comma cloud developement. (a) incipient stage, (b) intensifying stage, and (c) mature stage. Solid lines are 500-hPa contours; dashed lines are surface isobars (from Reed and Blier 1986a).

mesoscale vortex, etc. [for a more detailed discussion see Rasmussen and Turner (2003)].

Because of the rather striking similarities between some high-latitude polar lows and tropical systems (hurricanes), more exotic names such as "Arctic hurricanes" ocassionally have been applied to especially strong polar lows (e.g., Emanuel and Rotunno 1989; Businger and Baik 1991). Reed (1992), in a comment to Businger and Baik's study, raised objections to the use of the term Arctic hurricanes arguing that in no cases had the surface wind exceeded the 33 m s^{-1} wind speed required of a hurricane.[1] Also, referring to the fact that similar systems have been observed over the Mediterranean (e.g., Rasmussen and Zick 1987), he noted that these systems are not peculiar to the polar regions.

3. Further studies on comma clouds

Mullen (1979, 1982, 1983) in a series of papers extended RR's work in significant ways, describing the large-scale environment within which comma clouds developed over the wintertime North Pacific behind or poleward of major frontal bands. Based on composites for 22 cases, Mullen concluded that the oceanic polar air cyclones were baroclinic instability phenomena whose small scales related to the low values of the Richardson number near the surface and whose large upper-level amplitude related to the effect of latent heat release on baroclinic development.

In two further papers written together with Blier (Reed and Blier 1986a, b), the two authors extended Reed's 1979 study through a detailed investigation of two comma cloud developments over the eastern Pacific in, respectively, March and November 1982. Noting the similarity between the two pronounced developments and earlier studies, the authors summarized their findings

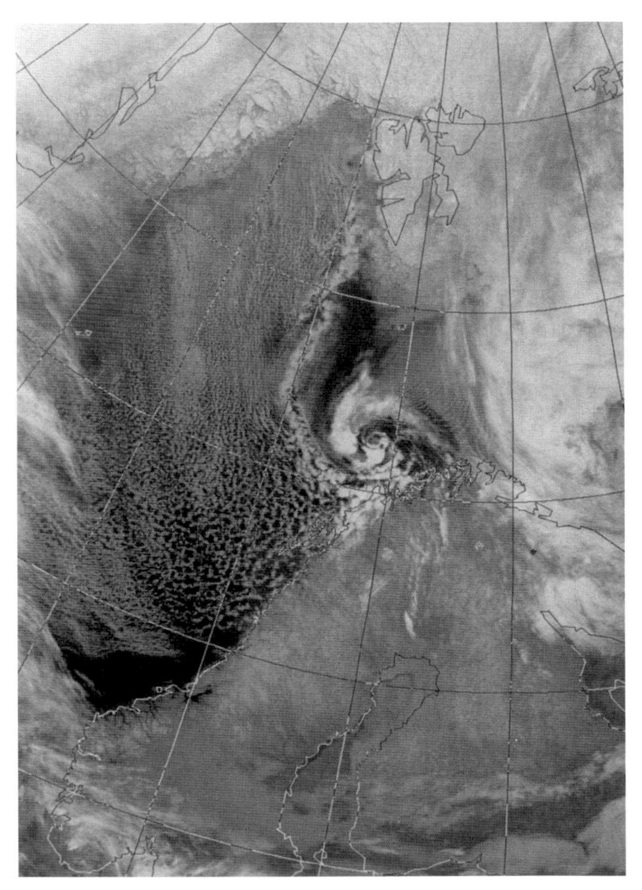

FIG. 4.9. Infrared satellite image at 1756 UTC 26 Mar 1981 showing a polar low with spiral structure over the Norwegian Sea west of North Cape. (Image courtesy of the NERC Satellite Receiving Station, University of Dundee.)

[1] Concerning the maximum surface mean wind speed associated with polar lows there have been a few reports of winds exceeding the 33 m s^{-1} threshold for hurricane force (see Rasmussen and Turner 2003).

via the conceptual model of a comma cloud development in the form of the schematic diagram shown in Fig. 4.8.

Using the quasigeostrophic omega equation to make qualitative deductions, Reed and Blier concluded that baroclinic instability as expressed by the differential vorticity advection term in the equation, was, no doubt, of importance in initiating and maintaining the disturbances, and that the development was enhanced by larger than normal lapse rates over a warm ocean. Sensible heating from the ocean, while not an immediate cause of the deepening of the systems, nevertheless played an important role by decreasing the stability of the environment. Also the moisture fluxes at the surface were important for enhancing the latent heat release.

Reed and Blier pointed out the need for an "initiating" and/or "organizing element" for the formation of the comma clouds and found that this ingredient in general was the advance of a preexisting short-wave trough, or of a secondary vorticity center. They found no fundamental differences for southward- and eastward-moving systems. They did note, however, that northerly flows may have an important enhancing effect on clomma cloud developments because the systems in this case are steered southward over increasingly warmer waters and as such are more strongly invigorated by deep convection. A similar effect has been noted for polar low developments over the Norwegian Sea.

Following the oil crisis in the 1970s, oil drilling companies started to search for oil at sea at higher latitudes along the Norwegian continental shelf. The sudden occurence of polar lows and the associated severe weather presented a serious threat to these operations. To improve timely forecasts of the polar lows the oil companies sponsored the three-year Norwegian Polar Lows Project from 1982 to 1985. The project spurred the interest in polar lows dramatically.

In a paper presented at The International Conference on Polar Lows in Oslo [20–23 May 1986 (Reed 1986)] RR noted that "some controversy has surrounded the use of the term polar low." Doing this, he referred to the fact that comma clouds, as well as other kinds of small-scale cyclones that are different from the comma systems, *all* were called polar lows. To overcome this difficulty[2] he suggested a rather general definition, defining polar lows as "small synoptic or subsynoptic cyclones that (1) form in cold air masses poleward of major fronts or jet streams, (2) lack pronounced frontal structure throughout all, or nearly all, of their lifetimes, and (3) appear on satellite images as comma or spiral shaped cloud masses that display convective features at an early stage in the cyclonic development," a definition that can be applied to both the comma cloud type of polar lows studied by Reed as well as for the more convective, high-latitude systems often associated with a spiral of clouds such as shown is Figs. 4.9 and 4.4). However, as noted by Bier (1996) and others "any cyclonic system developing under the influence of positive vorticity advection will tend to have a comma-shaped cloud mass" for which reason it may sometimes be difficult to decide whether a cloud field should be described as a "comma cloud" or a "spiral."

Although RR in general ascribed polar low development to baroclinic instability, he noted that there was observational evidence that "supports the idea that the growth of baroclinically-induced disturbances is much enhanced by CISK" and that numerical experiments had shown that "in general both baroclinicity and CISK are needed to account for the full intensity of polar lows." Reed specifically noted that his definition did "not require the existence of baroclinicity" and "that convective activity, and hence presumably conditional instability, were essential features of polar lows."

Reed (1986) gave three examples of different types of polar lows: the first being the November 1982 comma cloud development over the eastern Pacific discussed by Reed and Blier (1986b). The second case was a so-called reverse shear baroclinic development along a shallow baroclinic zone seaward of and parallel to the boundary of the pack ice along the northeast Greenland coast. In reverse shear situations the direction of the thermal wind is opposite to the surface wind and to the direction of the movement of the disturbances (Duncan 1978). Several studies (e.g., Grønås et al. 1987) have documented that polar lows over the Nordic Seas often develop in reversed shear flows. Figure 4.10 shows a wave train including two well-developed polar lows over the sea northeast of Iceland (a third less well-developed wave was located along the northeast coast of Iceland) that formed within a reverse shear flow over the northern Norwegian Sea and the Greenland Sea. The orientation of the low-level flow within the cold air mass west of the lows was clearly indicated by the cloud streets. The movement of the disturbances was along the surface flow, that is, in a southwesterly direction opposite to the thermal wind. Note that the warm-advection and therefore the cloud mass in such cases will trail the low pressure center rather than precede it.

In their 1987 paper Reed and Duncan investigated a case in which a wave train consisting of no less than four polar lows developed over the sea nearly parallel to the ice edge along the northeast Greenland coast. The disturbances formed in the shallow baroclinic zone of small static stability, which extended outward from the boundary of the Arctic ice pack in a situation where the strongest flow was near the surface and the 500-mb flow was nearly stagnant. Reed and Duncan concluded that baroclinic instability could explain the initial growth of the disturbances and the observed wavelength of 500–600 km, but that some other mechanism, presumably deep convection, was required in order to account for the

[2] At the same meeting Kerry Emanuel introduced the idea of a "polar low spectrum" reflecting the understanding that polar lows can appear in many forms due to the variety of effective forcing mechanisms.

FIG. 4.10. Infrared satellite image at 0840 UTC 20 Mar 1981 showing a wave train of polar lows over the Barents–Norwegian Sea. (Image courtesy of the NERC Satellite Receiving Station, University of Dundee.)

observed speed of travel and for the extremely fast growth rate found with these systems.

The third example considered by RR (1986) combined shallow and deep baroclinicty. In this case, which was also discussed by Businger (1985), it was demonstrated that migratory, upper-level vorticity maxima can play a role in the development of polar lows over the Norwegian Sea. Reed in his discussion of this case somewhat ridiculed the idea that the Norwegian Sea is "a region sometimes thought to be populated by a class of polar lows quite unlike the comma cloud." Doing this, he argued that apart from a different *orientation,* the upper-level features were similar for a comma cloud development on November 1982 (Reed and Blier 1986b) over the Pacific and a development near Bear Island south of Spitzbergen on 21–22 November 1983 studied by Businger (1985).

In a paper presented in September 1987 at a seminar at the European Centre for Medium-Range Weather Forecasts in Reading, United Kingdom, RR (1987) in a little-known work reviewed the observed features of polar lows, the theoretical ideas concerning their origin, and the results of some numerical experiments aimed at simulating their development. In this connection he altered his previous definition given at the Oslo meeting, now defining a polar low as "any type of small synoptic or subsynoptic cyclone, of an essentially non-frontal nature, that forms in a cold air mass poleward of major jet streams or frontal zones and whose main cloud mass is largely of

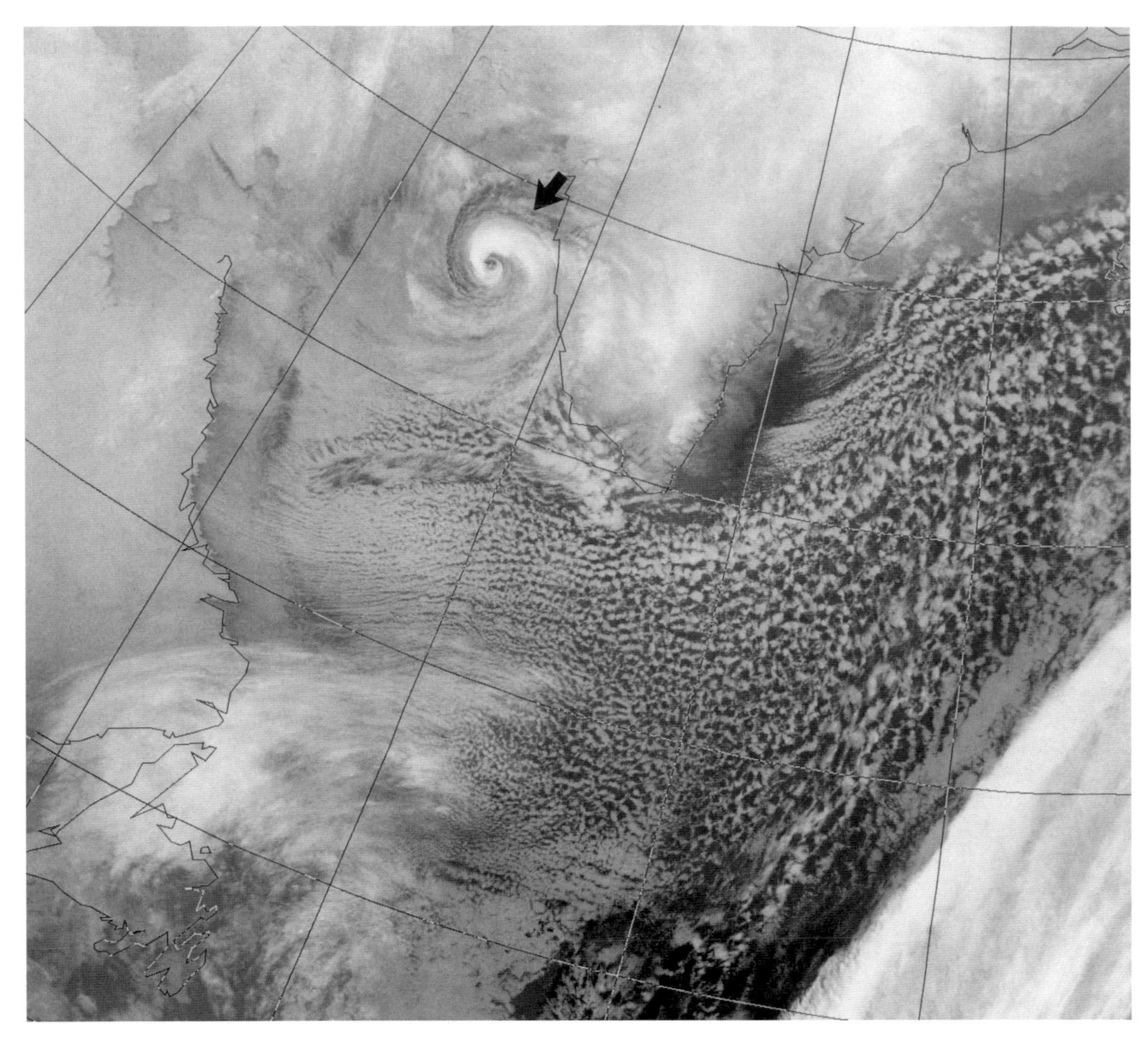

FIG. 4.11. Infrared satellite image at 1540 UTC 18 Jan 1989 showing a polar low (indicated by the arrow) of the cold-low type. Part of an extensive cloud band associated with the polar front is seen on the lower-right corner of the image. (Image courtesy of the NERC Satellite Receiving Station, University of Dundee.)

convective origin. The cloud mass in question may appear on satellite images as a small comma-shaped system or alternatively as a spiral or circular pattern, sometimes with an eye-like center."

As an important part of this work RR proposed a classification scheme based on a combination of synoptic features and physical considerations including upper-and lower-level baroclinic systems, combined baroclinic systems, and what RR called the inner occlusion type. The first of the these types, the "upper-level, short-wave/jet streak type" was identical to the comma cloud type discussed earlier. The second type considered by Reed was the "reverse shear type" commonly found over the seas to the west and north of Norway and also briefly discussed above. In the "combined type" two distinct flow patterns were involved, one characterized by upper-level PVA, and another by shallow baroclinicity. When an upper-level short wave traverses the marginal ice zone, it is possible for a system to develop that combines both features. According to RR "cases presented by Rasmussen (1985) and Businger (1985) *are perhaps* [italics added] good examples of such mixed developments." The two words in italics show how carefully RR presents his arguments. For many years it was taken almost for granted that the two cases mentioned by Reed, and a number of similar cases as well, where a development was observed to take place in connection with the passage of an upper-level trough across an ice edge, involved high-level as well as low-level baroclinic instability. More recent work, however (i.e., Albright et al. 1995; see also Rasmussen and Turner 2003), has cast doubt on this.

The final type considered in the classification was the "inner occlusion type" defined by RR as the small comma or spiral-shaped cloud patterns of convective character that sometimes flare up within the inner cores of old occlusions or cold lows without any obvious association

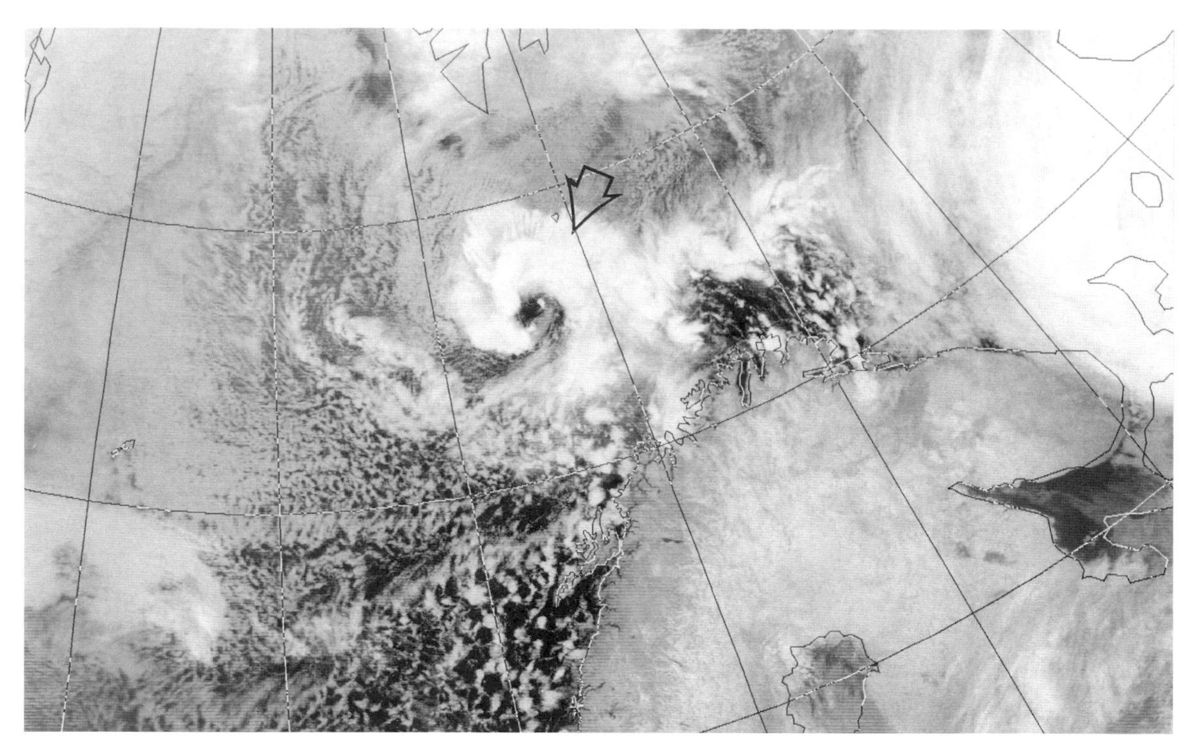

FIG. 4.12. Infrared satellite image at 0250 UTC 13 Dec 1982 showing a cloud spiral (indicated by the open arrow) associated with an upper-level vortex (PV anomaly) over the sea south of Svalbard. Clouds associated with a wave on the polar front are seen farther east over the eastern part of the Barents Sea. (Image courtesy of the NERC Satellite Receiving Station, University of Dundee.)

with upper-level short waves or low-level baroclinic features. An example of such a system over the David Strait is shown in Fig. 4.11.

In extreme cases the inner occlusion type of polar lows may, according to RR, be associated with small, hurricane-like vortices embedded within a larger polar low or a synoptic-scale low. As an example of this RR refers to the 13–14 December 1982 development that took place when an upper-level cold-core low, on a southerly track, entered the sea around Svalbard and Bear Island (see Rasmussen 1985; Rasmussen et al. 1992; Rasmussen and Turner 2003). As the upper-level low moved over the ice-free sea, the cloud spiral shown in Fig. 4.12 developed. In spite of the impressive cloud spiral, none of the surface observations within the region indicated the formation of any significant surface circulation at that time. However, as the low-level stable layer below the upper-level circulation was gradually eroded away over the warm sea surface, the upper-level vortex penetrated downward. As a result of this the surface circulation increased, creating a "parent circulation" in the form of a polar low of moderate strength. Later on during 13 December observations from weather ship *AMI*, located close to the center of the polar low, showed that a hurricane-like core had formed within the "parent circulation." As the hurricane-like core formed, most likely due to the outbreak of deep convection, the cloud field completely changed its structure from the spiral seen in Fig. 4.12 into the complex cloud field seen in Fig. 4.13. The center of the surface low was situated near the small cloud cluster indicated by the open arrow. The polar low as seen in the surface maps is illustrated in Fig. 4.14.

In the conclusions of his review paper Reed (1987), summarized briefly some of the main points concerning the state of knowledge of polar lows in the mid-1980s after the Norwegian Polar Lows Project. From the observational work carried out during this project it had become clear that a variety of systems fell under the general name of polar lows. Work was still needed, however, to develop a meaningful classification of the systems and, according to RR, to resolve the issue of whether the small, intense vortices that are sometimes termed "real" polar lows should be regarded as a distinct phenomenon or as a subgroup embedded within larger, subsynoptic systems. Regarding the theory, RR mentioned that theoretical investigators had tended to divide into two schools of thought, one emphasizing baroclinic effects and the other championing CISK or some other heating mechanism, but that lately there had been widespread recognition that both mechanisms were important. Finally in his conclusion RR pointed out the promising role for future studies using advanced mesoscale models. As will be demonstrated in the following, RR himself played an important role in utilizing these models, and through this work contributed in important ways to the further understanding of polar lows.

In two almost identical review papers written together with Businger (Businger and Reed 1989a,b) the classi-

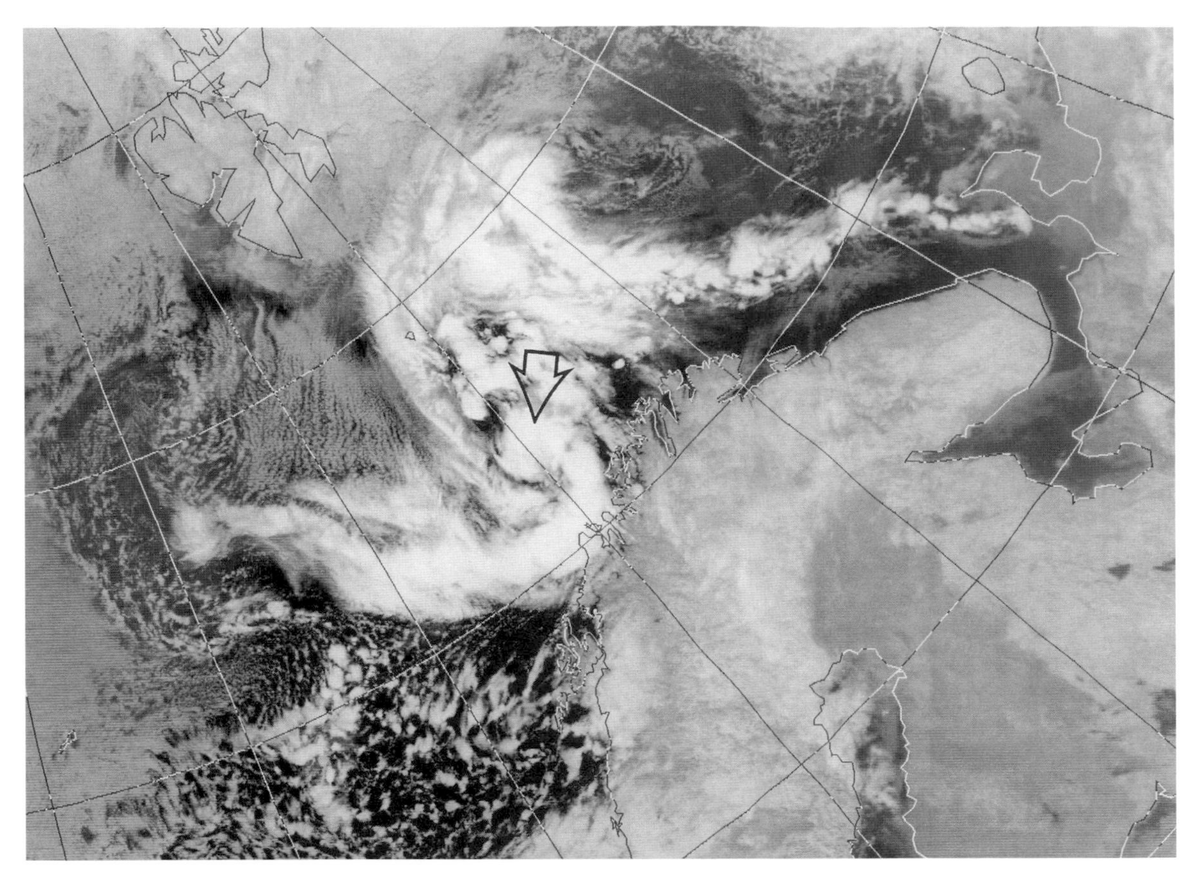

FIG. 4.13. Infrared satellite image at 0419 UTC 14 Dec 1982 showing a complex cloud field. A small cloud cluster associated with a polar low northwest of North Cape is indicated by the open arrow. The cloud band emanating from the cloud cluster, i.e., from the polar low, is associated with an Arctic front heading an outbreak of a shallow Arctic air mass within the strong circulation around the low. (Image courtesy of the NERC Satellite Receiving Station, University of Dundee.)

fication scheme proposed by Reed in his 1987 paper was slightly changed. In their joint paper Businger and Reed differentiated between three elementary types of polar-low development based on associated distinctive synoptic patterns. The types are also physically distinct in the degree and distribution of baroclinicity, static stability, and surface fluxes of latent and sensible heat. The three types are (i) the short-wave/jet streak type, characterized by a secondary vorticity maximum and PVA aloft, deep, moderate baroclinicity, and modest surface fluxes; (ii) the Arctic front type, associated with ice boundaries and characterized by shallow baroclinicity, and strong surface fluxes; and (iii) the cold-low type, characterized by weak baroclinicity, strong surface fluxes, and deep convection. In addition combinations of the above types may exist.

Although the classification by Businger and Reed has been widely used, some problems still remain as pointed out by Grønås and Kvamstø (1995) in connection with a study of polar low developments over the Norwegian Sea during Arctic air outbreaks. In the cases studied by Grønås and Kvamstø an Arctic inversion at the top of the convective boundary layer was formed as cold air from the ice was advected out over the warm ocean. Grønås and Kvamstø claimed that, "In the Norwegian and Barents Seas, the northerly flow and the Arctic inversion seem to be necessary conditions for polar lows to develop." In connection with this they suggested a modification to the definition of the *Arctic front polar low* class by Businger and Reed, arguing that these lows "do not form at the leading edge of the Arctic front" (Grønås and Kvamstø 1995, p. 811). As an alternative to the term Arctic front polar low, Grønås and Kvamstø proposed the name *Arctic outbreak lows*. Concerning the development of these lows, they added that "it seems evident that a mobile upper disturbance is also active," and that "The presence of an upper disturbance, using the PV concept, should therefore be included in the definition."

The classification by Businger and Reed comprises three main types of polar lows. In a paper presented at the 75th anniversary of the Bergen School, Grønås and Kvamstø (1994) suggested a further simplification to the Businger–Reed classification, defining only two main types. The first type in the Grønås–Kvamstø classification is the same as the first type in the Businger–Reed classification, that is, the comma clouds. However,

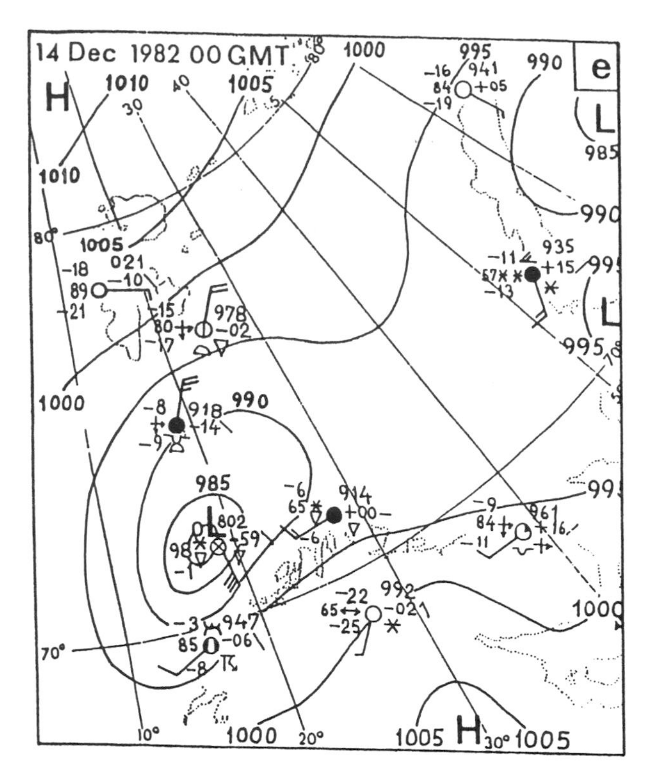

FIG. 4.14. Surface map at 0000 UTC 14 Dec 1982 showing a polar low over the Norwegian Sea.

according to Grønås and Kvamstø, "European meteorologists do normally not call them polar lows, but secondary extratropical cyclones." The second type considered by Grønås and Kvamstø was, as discussed in the previous paragraph, an extension of the Businger–Reed Arctic front type, including contributions from upper-air disturbances and deep convection. Concerning the third group in the Businger–Reed classification, that is, the cold-low type, Grønås and Kvamstø simply (and somewhat surprisingly) excluded this class, claiming that "it is infrequent in northern waters."

Classification schemes such as those suggested by Businger and Reed and later by Grønås and Kvamstø are attractive because of their simplicity. On the other hand they may be *too simplified*. Rasmussen (see Rasmussen and Turner 2003) attempted to improve the classification presented by Reed and Businger. As a basis for this study he selected the group of gale-producing polar lows that occurred around Norway in the period 1978–82 as found by Wilhelmsen (1985). Except for a single case for which satellite data were not available it was possible to place each of the cases within one of the seven categories or types as shown in Table 4.1. Like Businger and Reed's classification, the scheme defined in Table 1 was partly based on distinctive synoptic patterns associated with the occurence of the polar lows and partly on physical considerations. When considering the numbers of cases within the different groups in the table, it should be remembered that it is based on the occurrence of high-latitude polar lows as found around Norway. The categories as such form a pragmatic classification valid for parts of the Nordic Seas alone, and may in principle not be valid for other regions. It should be especially noted that the "Wilhelmsen file" does not include any examples of polar lows forming along occlusions. Such developments, however, are also quite often observed over the Nordic Seas and may be classified as "reverse shear systems" (Bond and Shapiro 1991) or, alternatively, added as an extra group.

Polar lows within the largest group, the cold-low type, an example of which is shown in Fig. 4.11, are more or less identical with the cold-low type considered by Businger and Reed. The cold-low type of polar lows is characterized by little or vanishing PVA aloft, but with very low upper-air temperatures. This, through a lowering of the static stability, promotes deep convection, which in turn may lead to the development of a polar low.

In addition, a large number of reverse shear systems were found. Polar lows in the form of "normal" baroclinic, forward shear systems were rare, with only two cases being found. Only one comma cloud was found, and a dubious one at that, reflecting the fact that most of the developments took place at high latitudes far away from the polar front.

Developments along boundary layer fronts (BLFs, group 3), a type related to the Arctic front type in the Businger–Reed classification, were quite frequently found. These kinds of shallow fronts are often observed near the west coast of Svalbard in situations with a northerly flow. An example of a polar low development of this type is shown in Fig. 4.15. The numerous small-scale vortices along the BLF were most likely caused by barotropic shear instability and such vortices seldom become significant. Only in the case of upper-level forcing, generally in form of a short-wave trough, may a vortex develop into a polar low as seen in the figure.

Developments within synoptic-scale troughs behind large-scale cyclones constitute another large group ("trough systems"). During winter a number of large-scale cyclones follow a primary cyclone track over the northern part of the Norwegian Sea and the Barents Sea. In this situation polar lows like the one shown in Fig. 4.9 may form in troughs characterized by large values of low-level vorticity trailing the main pressure low center.

TABLE 4.1. The number of polar lows as found by Wilhelmsen in the Nordic Sea for the period 1978–82 and their distribution within seven different categories.

Group	Type of polar low	No. of cases
1	Cold-low type	9
2	Reverse shear systems	8
3	Boundary layer fronts	5
4	Trough systems	5
5	Baroclinic waves, foreward shear	2
6	Orographic polar lows	2
7	Comma clouds	1

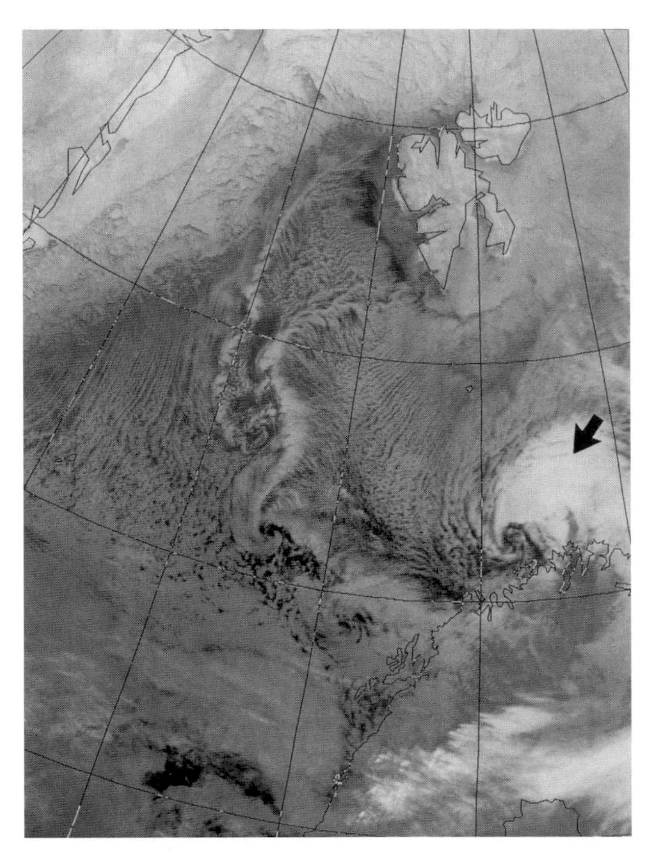

FIG. 4.15. Infrared satellite image at 1833 UTC 2 Mar 1981 showing a polar low (indicated by an arrow) near North Cape, formed at the end of a BLF seen farther to the west. Several minor vortices are seen along the front, which after its formation near the ice edge along the west coast of Svalbard drifted out over the open sea. (Image courtesy of the NERC. Satellite Receiving Station, University of Dundee.)

Virtually all of these developments are triggered by short-wave troughs aloft.

Reed in his 1987 paper stressed the role of advanced mesoscale models for future studies of polar lows and in 1995 he coauthored two important polar low papers (Albright et al. 1995; Bresch et al. 1997) both utilizing advanced numerical mesoscale models. In the first of these papers (Albright et al. 1995) a convective-type polar low development in December 1988 was chosen as the subject. The highly successful study showed how a polar low formed over an ice-free region over the eastern part of the Hudson Bay as an amplifying upper-level cold trough advanced into the region. The model depicted the polar low as a small, relatively shallow, system embedded within a larger cold low, resembling a miniature hurricane in structure but lacking hurricane force winds. Sensitivity experiments revealed that fluxes of heat and moisture from the region of open water and the associated condensation heating in deep organized convection were essential to the development. The feedback between the surface fluxes and the wind speed enhanced the intensification, a finding that, as pointed out by Albright et al. "lends strong support to the hypothesis of Emanuel and Rotunno (1989) that air–sea interaction instability (ASII) is a major factor in at least some types of polar low developments."

The results highlighted the central role of deep, organized convection for the type of polar low studied by Albright, et al. The authors also considered the role of baroclinic influence, either in the form of energy release from a low-level baroclinically unstable Arctic front or of forced lifting by an upper-level, short-wave trough. The authors found it questionable to characterize the zone of temperature contrast along the border of the open water as a front that related to the problem of baroclinic instability. In the same way they found that upper-level baroclinic forcing had at most a small effect on the surface development. As the key factor for the development, they instead pointed toward the general cooling aloft accompanying the advance of the upper-level cold low. The cooling, coupled with low-level warming and moistening by the surface fluxes, resulted in a drastic destabilization of the layer between the surface and 550 hPa. The destabilization gave rise to substantial condensational heating in deep mesoscale convection, and, from the point of view of potential vorticity diagnostics, it allowed a large midlevel, positive PV anomaly to induce cyclonic flow all the way to the surface.

The authors concluded that "condensation heating in deep convection was the overwhelming cause of intensification, and the growth of the convection was a consequence of the concurrent warming and moistening of the lower layers by the surface fluxes and the simultaneous cooling of the layers above by the advance of the upper-level low." Also, in their conclusions, they pointed toward the absence, or near absence, of baroclinic forcing as a triggering and organizing mechanism.

The type of mechanism described by Albright et al. was not only of interest in connection with polar lows forming in this particular region but also for polar low developments in other regions as well. For many years the prevalent point of view had been that many polar low developments near the ice edges were "a two-stage process." According to this belief evolution of a typical polar low started as a kind of type-B baroclinic development involving an interaction between a short-wave upper-level trough and an ice-edge-generated low-level baroclinic zone. Subsequently, when the baroclinically induced low-level circulation had achieved a certain strength, convection might take over as the primary mechanism for further, rapid development (Rasmussen 1985; Emanuel and Rotunno 1989).

However, evidence for a first-stage baroclinic development was small or lacking for a number of polar low developments for which the key factor most likely was the same as in the Hudson Bay case, that is, the general cooling aloft (these remarks should *not* be understood to mean that baroclinic processes in general are not important for polar low developments at high latitudes; cf., for example, the role of reverse shear instability).

As shown above, RR has played a central role for roughly 20 years in the study of the "new" phenomenon of the polar low, and through this has helped to broaden the understanding of these systems among the wider meteorological community. It is quite ironic therefore that RR quite recently presented a paper with the following title: "Is there such a thing as a polar low?" (Invited paper, 10th Cyclone Workshop, 21–27 September 1997, Val Morin, Quebec, Canada; Richard Reed in collaboration with Mark D. Albright and James F. Bresch). Part of the background for this, according to RR, "deliberately provocative title" was some experiments of a polar low development on 7 March 1977 discussed in a paper by Bresch et al. (1997). A series of fine-mesh simulations (20 km) were performed using the fifth-generation Pennsylvania State University–National Center for Atmospheric Research Mesoscale Model (MM5) to simulate this polar low, which formed near the ice edge in a moderately baroclinic cold airstream behind a predecessor low. The development showed many features similar to those observed in the Norwegian-Barents Sea such as "the sudden appearance of an eyelike feature in satellite imagery," and the "characteristic dip and rise in pressure and fall in temperature" seen at observing stations as the low passed by.

Sensitivity experiments showed that low-level heating by surface fluxes of sensible heat and midlevel heating from latent heating in clouds were equally effective in promoting the development while baroclinic instability, like in the Hudson Bay case, probably played a minor role.

A prominent feature of the simulation was the presence of several narrow bands (shear lines) of extremely large low-level vorticity. Actually the simulated polar low formed from a wave on one of those bands, which were a significant source of low-level heating and vorticity and provided a low-level PV anomaly with which a mobile, upper-level PV anomaly could interact. That polar lows may develop along shear lines through a mechanism involving barotropic instability has also been directly observed on several occasions from satellite imagery (see Rasmussen and Turner 2003). As noted by Bresh et al. (1997), "The dependence of the ultimate low development on the timing, location, and intensity of the essentially unpredictable bands highlights the difficulty of making precise predictions of polar lows, even in situations where their formation seems assured."

Another of the principal results of this study was that the modeled low in some ways resembled an extratropical cyclone rather than a hurricane, with pronounced *asymmetries* in wind, temperature, humidity, and PV fields. However, the fact that polar lows are not complete "symmetric systems" is also known from other studies of polar lows and a consequence of the fact that the vortices typically develop within a more or less strong basic current and within a more or less pronounced low-level baroclinic environment (Rasmussen and Turner 2003).

From a PV perspective the development was fairly simple. Once over an open sea with a reduced static stability in the lower atmosphere the downward penetration from the upper-level PV anomaly (upper-level trough) was sufficient to induce a low-level thermal anomaly. Through mutual interaction and phase-locking with the upper disturbance a significant low-level circulation was formed, and a low-level PV anomaly was created by condensation heating in organized convection. The PV so produced then added its contribution to the low-level circulation, further intensifying the low. The fact that these ideas commonly applied to midlatitude cyclogenesis are also applicable to the development of polar lows, plus the above-mentioned asymmetric features, suggest, according to Bresch et al. (referring to polar lows), "that the latter can be regarded in many instances as small, primarily Arctic counterparts of the larger middle-latitude systems" and "Inasmuch as typical polar lows possess many physical characteristics in common with other extratropical cyclones, they are perhaps better viewed as part of a broad spectrum of ocean cyclones that differ in such respects as degree of baroclinicity, strength of upper-level forcing, tropospheric stability, amount of latent heat release, and the magnitude and arrangement of the surface fluxes, *rather than as a truly distinct phenomenon*" (italics added).

In 2001 RR (together with a number of coauthors) published a paper on a tropical-like cyclone in the Mediterranean Sea (Reed et al. 2001). Because the kinds of small cyclones studied by Reed and coworkers in this paper form in polar air masses that are heated and moistened by the warm sea, and because they are situated within a large-scale, upper-level cold trough or cutoff low, they have also been likened to polar lows (Rasmussen and Zick 1987).

The storm first formed over the warm waters between Sicily and Libya, and, as it developed, assumed a hurricane-like appearance in satellite imagery. Ships near the vortex center reported near-hurricane-force winds. The attempts to simulate the storm development reported in the paper were "met with mixed success." The authors pointed out that "the predicted track departed substantially from the observed and the contraction of the storm to mesoscale dimensions was missed." However, as noted by the authors, "the case in point occurred in 1982, when satellite data were less plentiful than at present, and when gross errors in North African soundings were evident," and that "enhanced use of satellite data holds the key to future success in predicting tropical-like cyclones over the Mediterranean Sea, since these data offer the only practical means of defining initial fields with the required resolution."

The polar low is a comparatively "new" phenomenon, hardly being known within the general meteorological community 30 years ago. Up through the 1980s many meteorologist were still very skeptical as to their existence, Kellogg and Twitchell (1986) vividly described the situation facing meteorologists around

1980 concerning the meteorological phenomenon known as polar lows: "The history of meteorology is replete with instances of some phenomenon in the atmosphere that defies an adequate description. We know that something exists, sometimes with disastrous consequences to people and their possessions, but its origins and evolution and characteristics are only vaguely understood. Furthermore, it may even be hard for meteorologists to agree what to call it." However, partly due to the increasing amounts of satellite imagery that became available from the 1970s to the present showing well-organized small-scale systems like the one seen in Fig. 4.4 as well as in other figures in this paper, the skepticism gradually disappeared. The polar low as seen in Fig. 4.4 and "normal" extratropical cyclones clearly have different horizontal scales. However, they are both balanced systems and the differences in scale may partly be explained by the small value of the Rossby radius of deformation NH/f associated with the polar low. In polar maritime air masses with a reduced static stability and a low tropopause, both N and H will be small. In addition, due to the high latitude, f will be large; all of which contributes to a small horizontal scale. In addition to differences in the horizontal scale, some of the polar lows within the "polar low spectrum," that is, the important group of "cold lows" ("real polar lows"), have a mechanism of development very different from that of the large synoptic-scale extratropical cyclones. While it may be reasonable to view in particular the comma clouds as "part of a broad spectrum of ocean cyclones" as suggested by Bresch et al. rather than as "a truly distinct phenomenon," the same procedure seems questionable for polar lows such as seen, for example, in Figs. 4.4, 4.9, and 4.11, the latter one having its their roots in processes within the core region of the circumpolar vortex. Interestingly in this way the debate about the nature of polar lows has returned to the point where it started when RR published his first paper on cyclogenesis in polar air streams in 1979, that is, a discussion about the differences between the high-latitude "real polar lows" characterized by a cloud field of predominately spiral structure, and the baroclinic comma clouds found close to the polar front, and whether any of these should be considered as "truly distinct phenomena" or not.

As documented in this brief review RR has, over a period of more than 20 years, contributed in a most significant way to the study of this "new" phenomenon: the polar low. Meteorologists have often lamented the gap between synoptic and dynamic meteorology. Reed has in his work combined the best from the art of synoptic meteorology with a deep insight into dynamic meteorology. While he for example, in the 1980s, in a skillful way utilized the quasigeostrophic omega equation to elucidate the processes involved in the development of the baroclinic comma clouds, he also later on successfully applied the potential vorticity concept to illustrate and explain the basic processes involved in the formation of the more convective high-latitude polar lows. There are still many gaps in our understanding of polar lows, but thanks to the dedicated work of RR and others within this field, we now know much more about this important weather phenomenon than when RR published his first paper on comma clouds back in 1979.

In this brief review the discussion has been focused on RR's contributions to the understanding of the polar low. Many other scientists, however, have contributed to this area of meteorological research as is evident by a large number of publications in the international literature. For a comprehensive, up-to-date review of polar low research over the last 30 years the reader is referred to Rasmussen and Turner (2003).

REFERENCES

Albright, M. D., R. J. Reed, and D. W. Ovens, 1995: Origin and structure of a numerically simulated polar low over Hudson Bay, *Tellus,* **47A,** 834–848.

Anderson, R. K., J. P. Ashman, F. Bittner, G. R. Farr, E. W. Fergusson, V. J. Oliver, and A. H. Smith, 1969: Application of meteorological satellite data in analysis and forecasting. ESSA Tech. Rep., Washington, DC.

Bier, W., 1996: A numerical modeling investigation of a case of polar airstream cyclogenesis over the Gulf of Alaska. *Mon. Wea. Rev.,* **124,** 2703–2725.

Bond, N. A., and M. A. Shapiro, 1991: Polar lows over the Gulf of Alaska in conditions of reverse shear. *Mon. Wea. Rev.,* **119,** 551–572.

Bresch, J. F., R. J. Reed, and M. D. Albright, 1997: A polar-low development over the Bering Sea: Analysis, numerical simulation, and sensitivity experiments. *Mon. Wea. Rev.,* **125,** 3109–3130.

Businger, S., 1985: The synoptic climatology of polar-low outbreaks, *Tellus,* **37A,** 419–432.

——, and R. J. Reed, 1989a: Cyclogenesis in cold air masses. *Wea. Forecasting,* **4,** 133–156.

——, and ——, 1989b: Polar lows. *Polar and Arctic Lows*, P. F. Twitchell, E. A. Rasmussen, and K. L. Davidson, Eds., Deepak, 3–45.

——, and J. J. Baik, 1991: An Arctic hurricane over the Bering Sea. *Mon. Wea. Rev.,* **119,** 2293–2322.

Charney, J., and E. Eliasen, 1964: On the growth of the hurricane depression. *J. Atmos. Sci.,* **21,** 68–75.

Dannevig, P., 1954: *Meteorologi for Flygere.* Aschehough and Co., 224 pp.

Duncan, C. N., 1978: Baroclinic instability in a reversed shear flow. *Meteor. Mag.,* **107,** 17–23.

Emanuel, K. A., and R. Rotunno, 1989: Polar lows as Arctic huricanes. *Tellus,* **41A,** 1–17.

Grønås, S., and N. G. Kvamstø, 1994: Synoptic conditions for Arctic front polar lows. *The Life Cycles of Extratropical Cyclones*, S. Grønås and M. A. Shapiro, Eds., Amer. Meteor. Soc., 89–95.

——, and ——, 1995: Numerical simulations of the synoptic conditions and development of Arctic outbreak polar lows. *Tellus,* **47,** 797–814.

——, A. Foss, and M. Lystad, 1987: Numerical simulations of polar lows in the Norwegian Sea. *Tellus,* **39,** 334–353.

Harley, D. G., 1960: Frontal contour analysis of a "polar" low. *Meteor. Mag.,* **89,** 141–147.

Harrold, T. W., and K. A. Browning, 1969: The polar low as a baroclinic disturbance. *Quart. J. Roy. Meteor. Soc.,* **95,** 710–723.

Kellogg, W. W., and P. F. Twitchell, 1986: Summary of the workshop on arctic lows 9–10 May 1985, Boulder, Colorado. *Bull. Amer. Meteor. Soc.,* **67,** 186–193.

Meteorological Office, 1962: *A Course in Elementary Meteorology.* HMSO.

——, 1964: *The Handbook of Weather Forecasting.* Meteorological Office.

Mullen, S. L., 1979: An investigation of small synoptic cyclones in polar air streams. *Mon. Wea. Rev.,* **107,** 1636–1647.

——, 1982: Cyclone development in polar airstreams over the wintertime continent. *Mon. Wea. Rev.,* **110,** 1664–1676.

——, 1983: Explosive cyclogenesis associated with cyclones in polar air streams. *Mon. Wea. Rev.,* **111,** 1537–1553.

Nordeng, T. E., and E. A. Rasmussen, 1992: A most beautiful polar low—A case study of a polar low development in the Bear Island region. *Tellus,* **44,** 81–99.

Økland, H., 1977: On the intensification of small-scale cyclones formed in very cold air masses heated over the ocean. Institute Report Series, No. **26,** Institutt for Geofysik, Universitet Oslo, 25 pp.

Ooyama, K., 1964: A dynamical model for the study of tropical cyclone development. *Geofis. Int.,* **4,** 187–198.

Rasmussen, E., 1977: The polar low as a CISK phenomena. Institute for Theoretical Meteorology Rep. 6, University of Copenhagen, 77 pp.

——, 1979: The polar low as an extratropical CISK disturbance. *Quart. J. Roy. Meteor. Soc.,* **105,** 531–549.

——, 1985: A case study of a polar low development over the Barents Sea. *Tellus,* **37A,** 407–418.

——, and C. Zick, 1987: A subsynoptic vortex over the Mediterranean with some resemblance to polar lows. *Tellus,* **39A,** 408–425.

——, and P. Aakjær, 1992: Two polar lows affecting Denmark. *Weather,* **47,** 326–338.

——, and J. Turner, Eds., 2003: *Polar Lows.* Cambridge University Press, 612 pp.

——, T. S. Pedersen, L. T. Pedersen, and J. Turner, 1992: Polar lows and arctic instability lows in the Bear Island region. *Tellus,* **44A,** 133–154.

Reed, R. J., 1979: Cyclogenesis in polar air streams. *Mon. Wea. Rev.,* **107,** 38–52.

——, 1986: Baroclinic instability as a mechanism for polar low formation. *Proc. Int. Conf. on Polar Lows,* 141–150.

——, 1987: Polar lows. *Proc. Seminar on the Nature and Prediction of Extra Tropical Weather Systems,* Reading, United Kingdom, European Centre for Medium-Range Weather Forecasts, 213–236.

——, 1992: Comments on "An Arctic hurricane over the Bering Sea." *Mon. Wea. Rev.,* **120,** 2713.

——, and W. Blier 1986a: A case study of comma cloud development in the eastern Pacific. *Mon. Wea. Rev.,* **114,** 1681–1695.

——, 1986b: A further study of comma cloud development in the eastern Pacific. *Mon. Wea. Rev.,* **114,** 1696–1708.

——, and C. N. Duncan, 1987: Baroclinic instability as a mechanism for the serial development of polar lows. A case study. *Tellus,* **39A,** 376–384.

——, Y.-H. Kuo, M. D. Albright, K. Gao, Y.-R. Guo, and W. Huang, 2001: Analysis and modelling of a tropical-like cyclone in the Mediterranean Sea. *Meteor. Atmos. Phys., 76,* 183–202.

Scherhag, R., and L. Klauser, 1962: Grundlagen der Wettervorhersage. *Meteorologisches Taschenbuch,* F. Baur, Ed., Akademische Verlagsgesellschaft Geest & Portig.

Wilhelmsen, K., 1985: Climatological study of gale-producing polar lows near Norway. *Tellus,* **37A,** 451–459.

Chapter 5

The Role of the Quasi-Biennial Oscillation in Stratospheric Dehydration

James R. Holton

University of Washington, Seattle, Washington

"Perhaps a simple explanation will soon be found, and what now seems an intriguing mystery will be relegated to the category of a meteorological freak. Or perhaps the phenomenon will prove to have a greater significance than we now might envisage, either because of some intrinsic property it possesses or because of its effect on other related areas of research." —(Reed 1967)

1. Introduction

Among Professor Richard Reed's many contributions to the understanding of the meteorology of the middle atmosphere, his discovery of the equatorial quasi-biennial oscillation (QBO) has perhaps had the most enduring influence. In a paper presented at the 40th anniversary meeting of the American Meteorological Society, January 1960, he showed that rawinsonde data from Canton Island (2.8°S) revealed "alternate bands of easterly and westerly winds which originate above 30 km and which move downward through the stratosphere at a speed of about 1 km per month." He pointed out that the alternating wind regimes "appear at intervals of roughly 13 months, 26 months being required for a complete cycle." This work appeared in print in Reed et al. (1961). Although several other workers noted the occurrence of this strange wind oscillation at about the same time, unlike Reed they seem not to have appreciated its dynamical significance, and the challenges that the momentum budget and periodicity of this strange oscillation presented for theoreticians.

The story of the development of the theory of the QBO has been related by Lindzen (1987), and all aspects of the QBO have very recently been the subject of a major review (Baldwin et al. 2001), so it is unnecessary to revisit the history here. Various aspects of the QBO have been a major source of research for the past 40 years, and remain so to this day. A search in the Inspec database using the keyword "QBO" reveals more than 400 entries since 1980.

Shortly after Reed's discovery of the QBO several workers reported oscillations of a similar period in column ozone (Funk and Garnham 1962) Again, however, it was Reed who first sought a dynamical explanation. His simple analytical model given in Reed (1964) and Reed (1967) showed that the observed QBO in ozone in the equatorial region could be approximately explained by the advection of the time mean ozone profile by the vertical circulation associated with the QBO. The equatorial vertical velocity driven by the QBO is approximately out of phase with the vertical shear of the zonal wind (and hence with the temperature). Thus the QBO component of the vertical velocity consists of subsidence in the westerly shear phase and ascent in the easterly shear phase. Reed (1967) showed that since the ozone mixing ratio increases rapidly with height in the lower tropical stratosphere, and the chemical timescale is long, advection by the QBO-driven vertical circulation produces an ozone maximum a few months after maximum westerly shear, and a minimum a few months after maximum easterly shear. Following the early work of Reed, there have been a large number of observational and modeling studies related to the QBO in ozone. More than 1/4 of the Inspec references to the QBO relate to the ozone issue.

Investigation of the QBO signal in other trace constituents, which must rely on satellite observations, has a more limited history. The work of Trepte and Hitchman (1992) is an important contribution to this subject. They analyzed the volcanic aerosol distributions in the tropical stratosphere during westerly and easterly QBO shear maximum near the 30-hPa level. Their analysis clearly showed that the distribution of the aerosol in the equatorial stratosphere was strongly influenced by the secondary meridional circulation of the QBO. During the westerly shear phase, equatorial aerosol amounts were depressed above 30 hPa owing to the downward phase of the QBO-related vertical motion, while during the easterly shear phase, equatorial aerosol amounts were enhanced above 30 hPa owing to the upward phase of the QBO-related vertical motion.

The nearly 10 years of methane and water vapor observations from the Halogen Occultation Experiment (HALOE) on the *Upper Atmosphere Research Satellite* (*UARS*) have provided an excellent opportunity for study of the QBO in two vertically stratified long-lived stratospheric constituents (Dunkerton 2001; Randel et al. 1998). These studies, and related modeling studies (e.g., Gray 2000), have revealed that meridional and vertical transport by the secondary meridional circulation, and meridional mixing by planetary waves, all contribute to the total QBO signal in long-lived tracer fields. The focus of these studies has mainly been on the middle and upper stratosphere.

The QBO signal in equatorial temperature extends, however, not only into the lower stratosphere, but at least to the tropopause near the 100-hPa level. The amplitude of the QBO temperature oscillation at 100 hPa is about 1 K (Reed 1965). The QBO temperature oscillation is maintained in the presence of radiative damping by adiabatic heating and cooling associated with vertical motion; thus, if the QBO temperature oscillation extends down to the tropopause, the secondary vertical circulation of the QBO must also extend to the tropopause. Stratospheric water vapor can then be influenced by the QBO in at least three ways. First, the tropopause temperature modulation will modulate the saturation mixing ratio and, hence, the freeze-drying effect believed responsible for stratospheric dehydration (Brewer 1949). Second, the vertical motion associated with the QBO will modulate the rate at which water vapor is transported across the tropopause and upward into the equatorial stratosphere. Third, the zonal wind of the QBO will influence the rate at which air passes through the tropical "cold trap" region of the west Pacific (where tropopause temperatures are coldest) and, hence, the rate at which dehydration occurs (Holton and Gettelman 2001).

Owing to the sensitive dependence of the saturation mixing ratio on temperature, the QBO modulation of tropopause temperature should provide a QBO signal in water vapor even in the absence of a QBO in the meridional circulation field. The impact of the QBO on stratospheric dehydration should depend strongly on the relative phase of the QBO temperature oscillation at the tropopause, and the phase of the annual cycle of the coldest tropopause temperatures. In this paper a modified version of the model of Holton and Gettelman (2001) is used to explore the role of the QBO in stratospheric dehydration for specified QBO oscillations in temperature and vertical velocity.

2. The water vapor model

The model is a simple two-dimensional (longitude–height) transport model for the equatorial tropopause region. In the version of the model used here, basic-state annually varying fields are specified for the temperature distribution $T(x,z,t)$, and for the zonal mean horizontal and vertical wind components, $u(z,t)$ and $w(z,t)$. These basic state fields are perturbed by specified QBO wind and temperature oscillations based upon the analytic model of Reed (1967). For simplicity the zonal wind is assumed to consist of a constant easterly component $u_0 = -5\ \mathrm{m\ s^{-1}}$ plus a downward-propagating biennial component whose amplitude increases monotonically from 0 at 14 km to 15 m s^{-1} at 20 km, and whose phase propagates downward as a sinusoid with vertical wavenumber $m = 0.26\ \mathrm{km^{-1}}$. Thus,

$$u(z,t)) = u_0 + u_B(z)\cos\left(\frac{\pi t}{360} + mz + \lambda\right), \qquad (1)$$

where z and t are in kilometers and days, respectively, and λ gives the phase relative to the annual cycle. For simplicity the annual period is 360 days, and the QBO is assumed to have a period of 720 days.

The QBO temperature field is proportional to the vertical shear of the zonal wind. Thus,

$$T_B(z,t) = A\,du/dz, \qquad (2)$$

where, based on Reed's model, $A = 500$ K s. In order to include the important effects of the annual variation of temperature in the west Pacific cold trap, a temperature distribution based on Seidel et al. (2001) is specified consisting of a time mean standard profile $T_0(z)$, an annually varying component with amplitude $T_A(z)$, a quasi-biennially varying component $T_B(z,t)$, and a longitudinally dependent annually varying perturbation centered at $x = 0$, $z = 16$ km. Thus the zonally averaged temperature distribution is given by

$$T_M(z,t)) = T_0(z) + T_A(z)\cos\left(\frac{\pi t}{180}\right) + T_B(z,t), \qquad (3)$$

while the total temperature field is given by

$$T(x,z,t)) = T_M(z,t)$$

$$+\, T_C(z)\exp\left[-\left(\frac{z - z_{trop}}{d}\right)^2\right]\exp\left[-\left(\frac{x}{L}\right)^2\right]$$

$$(1 + \cos(\pi t/180))/2, \qquad (4)$$

where $T_c = 4$ K, $d = 1$ km, $L = 2500$ km, and t is the time in days. The temperature distribution given by (4) at $x = 0$ and $z = 16.5$ km has an annual range of 6.6 K and a QBO range of 0.4 K. The annual range decreases monotonically to zero at 20 km, while the QBO range increases to 5.5 K at 20 km.

The vertical velocity is specified as

$$w(z,t) = w_0 + w_A \cos\left(\frac{\pi t}{180}\right) - BT_B. \quad (5)$$

Here $w_0 = +0.225$ mm s^{-1}, $w_A = +0.075$ mm s^{-1}, and $B = 3.5 \times 10^{-5}$ K^{-1}. The first two terms on the right side of (5) give the annually varying Brewer–Dobson upwelling (Rosenlof 1995), while the last term is the QBO component based on Reed (1967).

Water vapor and cloud ice are assumed to satisfy the following relations:

$$\frac{Dq_v}{Dt} = -\alpha(q_v - \tilde{q}) + K\frac{\partial^2 q_v}{\partial z^2} + E - C \text{ and} \quad (6)$$

$$\frac{Dq_i}{Dt} = -\alpha q_i + K\frac{\partial^2 q_i}{\partial z^2} - E + C. \quad (7)$$

Here, $\tilde{q}$ (= 6 ppmv) designates the mean subtropical lower-stratospheric water vapor mixing ratio (Randel et al. 2001); E and C stand for evaporation and condensation, respectively. Here, $E = q_i/\tau_E$ and C is zero unless $q_v > q_s$, in which case $C = (q_v - q_s)/\delta t$. The timescale for E is $\tau_E = 1$ day and the timescale for C is $\delta t = 1$ h.

TABLE 5.1. Model configuration for the water vapor mixing ratio simulations. The control case has only an annual cycle of temperature and vertical velocity. Cases 1–4 include QBO components of various fields. The Xs show which QBO fields are included in each case. See text for details.

Case	T_{qbo}	w_{qbo}	U_{qbo}
Control			
1	X	X	X
2	X		
3		X	X
4			X

The rate coefficient for mixing with the subtropical air, α, and the vertical diffusion coefficient, K, are based on the estimates of Mote et al. (1998). Here α is assumed to decrease linearly from 3.1×10^{-7} s^{-1} at 14 km to 2.0 ×

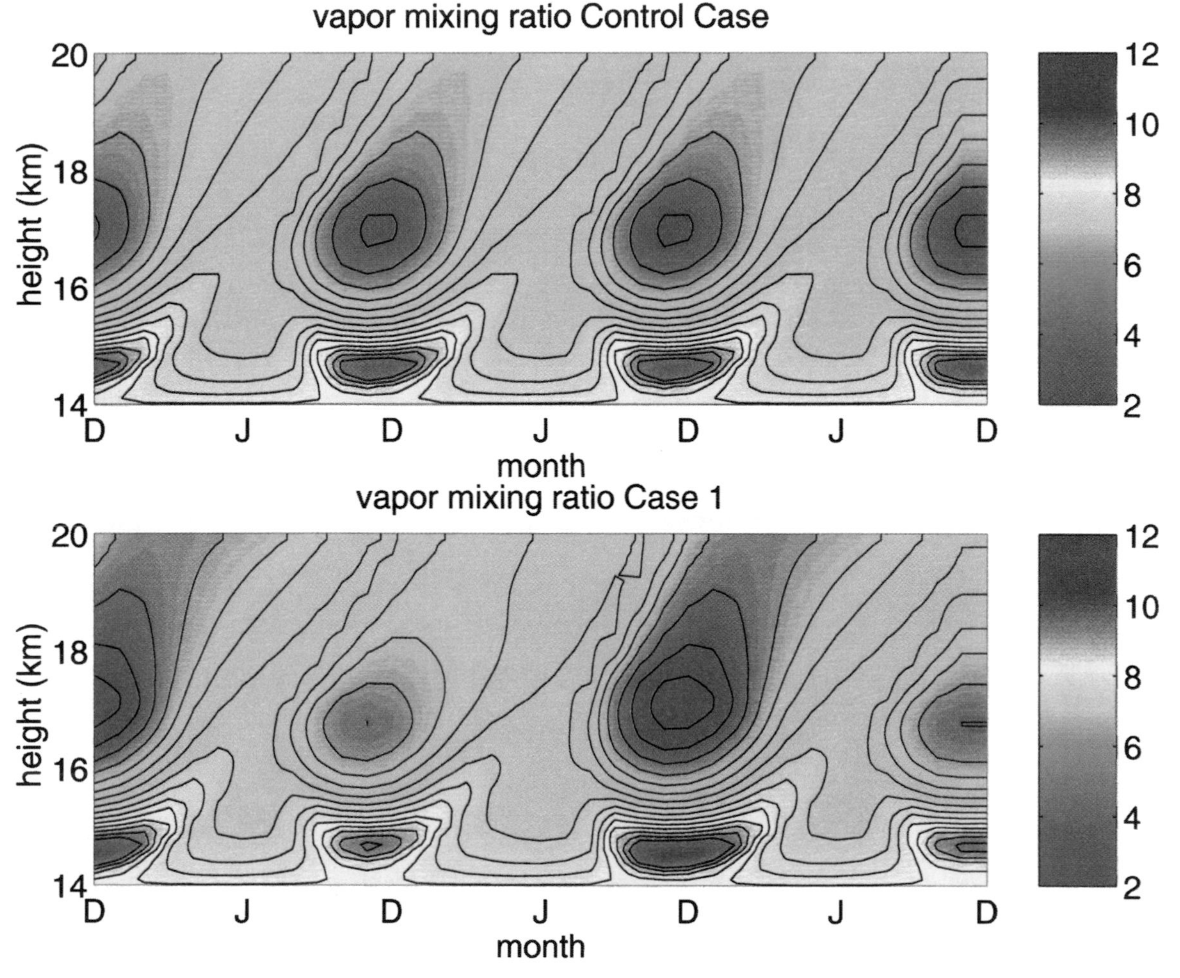

FIG. 5.1. Time–height sections of water vapor mixing ratio in ppmv for (top) the control case with only the annual cycle included and (bottom) case 1 with QBO perturbations in the temperature, vertical velocity, and zonal wind. Contour interval is 0.5 ppmv.

10^{-8} s^{-1} at 20 km, and K is set to the constant value of 0.003 m^2 s^{-1}. Sedimentation of ice particles is represented by adding a fall speed of 6 mm s^{-1} to the vertical velocity used in (2) for computing the advection of cloud ice. This fall speed is consistent with a 5-μm particle radius (Boehm et al. 1999).

Equations (6) and (7) are solved for the time evolution of q_v and q_i using a semi-Lagrangian algorithm with biquadratic interpolation (Durran 1999). Humidity in the region of the tropical tropopause is poorly constrained by observations. For simplicity the model is here initialized at 40% relative humidity and is maintained at that value

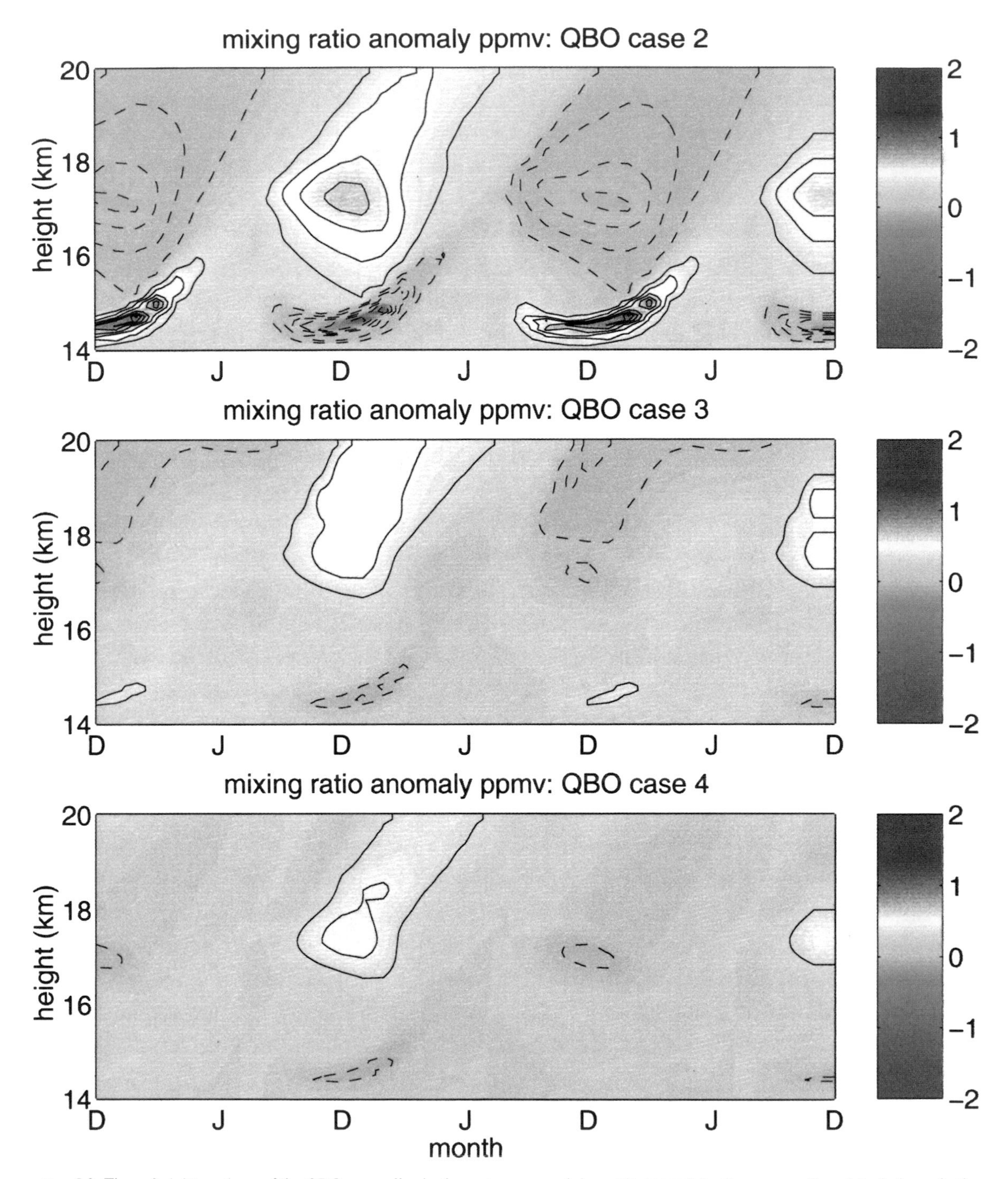

FIG. 5.2. Time–height sections of the QBO anomalies in the water vapor mixing ratio (ppmv) for three cases. Case 2 includes only the temperature QBO, case 3 includes only the zonal wind and vertical velocity QBOs, and case 4 includes only the zonal wind QBO. Contours are plotted at an interval of 0.2 ppmv. Zero contours are omitted.

at the lower boundary (14 km). The domain is assumed to have a zonal scale of 18 000 km, with periodic boundary conditions. This represents the approximate distance for recirculation of air transported through the tropopause-level subtropical anticyclone over the western Pacific (Pfister et al. 2001).

3. Results

The model described above was run for a number of differing specifications of the QBO temperature and wind anomalies. In all cases described here the model was run for 48 months. In order to eliminate transient effects of initialization, only the last 36 months are shown in the diagrams. Cases displayed in the figures are indicated in Table 5.1. A time–height section of the water vapor mixing ratio in the control case is shown in the top panel of Fig. 5.1. There is a strong annual cycle, with a minimum that appears at the tropopause in Northern Hemisphere winter and gradually propagates upward owing to advection by the Brewer–Dobson circulation. The lower panel in Fig. 5.1 shows a similar time–height section for QBO case 1. For this case the maximum QBO signal in temperature at the tropopause occurs near the winter solstices, leading to a strong QBO signal at the same season. As in the case of the annual cycle, the imprint on stratospheric water vapor produced by the QBO temperature oscillation at the tropopause is carried upward by the Brewer–Dobson circulation, leading to a QBO component of the tropical "tape recorder" signal (Zhou 2000). Thus, although the QBO in wind and temperature propagates downward in time, the QBO in water vapor propagates upward!

In both the runs shown in Fig. 5.1 maxima in the water vapor mixing ratio occur below the 15-km level in association with the winter minima in the stratosphere. These maxima are caused by the evaporation of ice particles that fall from the cirrus clouds produced by the freeze-drying process in the cold trap.

As noted above, the QBO in water vapor is influenced not only by temperature but by the horizontal and vertical velocity components of the QBO as well. This is most easily seen by plotting the anomaly in water vapor mixing ratio obtained by subtracting the control case from the cases shown in Table 5.1. The upper panel of Fig. 5.2 shows the QBO water vapor mixing ratio anomaly for case 2. This case retains the temperature oscillation of the QBO but omits the QBO contributions to the zonal wind and vertical velocity. The middle panel in Fig. 5.2 shows the anomaly for case 3 in which the QBO temperature signal is omitted but both vertical and horizontal velocity components of the QBO are included. Finally, the lower panel in Fig. 5.2 shows the result for case 4, in which only the zonal wind component of the QBO is included.

All of the anomaly fields of Fig. 5.2 clearly show upward-propagating water vapor maxima and minima with a 2-yr period, and with maxima and minima at the tropopause level coinciding with the Northern Hemisphere solstice. Figure 5.2 confirms that the temperature perturbation near the tropopause is the primary cause of the QBO in the water vapor mixing ratio. The meridional circulation does, however, make a substantial contribution above 17 km (middle panel of Fig. 5.2), which is in phase with the contribution from the temperature anomaly. The increased upward motion during the cold phase of the QBO enhances the upward propagation of dehydrated air; the decreased upward motion during the warm phase of the QBO suppresses the upward propagation of dehydrated air.

Finally, as mentioned earlier, because the QBO oscillation in the zonal wind modulates the rate at which air parcels pass through the west Pacific cold trap, even in the absence of QBO signals in temperature and vertical velocity there remains a small QBO signal in water vapor (lower panel of Fig. 5.2). Although the amplitude is much smaller than the signals associated with temperature and vertical velocity, it is clear that, at least in this model, the zonal wind oscillation associated with the QBO has an influence on stratospheric dehydration associated with its influence on horizontal advection through the cold trap. This signal is also in phase with that produced by the temperature oscillation. Maximum dehydration occurs in the easterly phase when the zonal wind near the tropopause is strong, while less dehydration occurs during the westerly phase when the zonal wind near the tropopause is weak.

Model runs were also carried out in which the QBO temperature anomaly was specified to be a maximum at the tropopause during the equinoctial season. These (not shown) confirmed that the strongest dehydrating effect occurs when the negative phase of the QBO temperature oscillation at the tropopause coincides with the time of tropopause temperature minimum in the annual cycle (i.e., northern winter).

4. Concluding remarks

Through its modulation of both wave and mean-flow transport the QBO of the equatorial stratosphere influences the distribution of many constituents. In addition the QBO in temperature influences constituent distributions by modulating temperature-dependent reaction rates, a factor important for understanding the stratospheric ozone distribution. Finally, as discussed above, the QBO in the near-tropopause temperature introduces a significant quasi-biennial component to the water vapor distribution, which may vary in intensity from one cycle to the next depending on the phase of the QBO relative to the annual cycle.

Owing to the role of water vapor in the climate system, and the critical role of polar stratospheric clouds in ozone depletion, it is important that all processes influencing the distribution and variability of water in the tropical tropopause layer and in the lower stratosphere be well understood. The association of low stratospheric water vapor mixing ratios with cold tropopause temperatures is consistent with observations, and is hardly new. However, one aspect of the present results seems not to have been remarked upon previously. Namely, in the layer below the cold winter tropopause, there is a distinct maximum in mixing ratio. As shown previously by Holton and Gettelman (2001) the model produces substantial cirrus cloud in the region of the west Pacific cold trap during northern winter. The maximum in water vapor mixing ratio in the upper troposphere below the cold trap is produced by the evaporation of ice particles that have fallen out of the cirrus clouds (produced in the model wherever the mixing ratio exceeds 110% of ice saturation). Unfortunately there are insufficient observations of the seasonal variations of water vapor in this region to determine whether the model-produced upper-troposphere signal consisting of annual and QBO cycles out of phase with the cycles above the tropopause is found in nature.

Acknowledgments. I wish to thank Professor Richard Reed for many productive discussions of the QBO and many other topics over the past 35 years. This work was sponsored by NASA Contract NAS 5 97046 (HIRDLS).

REFERENCES

Baldwin, M. P., and Coauthors, 2001: The quasi-biennial oscillation. *Rev. Geophys.,* **38,** 179–229.

Boehm, M. T., J. Verlinde, and T. P. Ackerman, 1999: On the maintenance of high tropical cirrus. *J. Geophys. Res.,* **104,** 24 423–24 433.

Brewer, A. M., 1949: Evidence for a world circulation provided by the measurements of helium and water vapor distribution in the stratosphere. *Quart. J. Roy. Meteor. Soc.,* **75,** 351–363.

Dunkerton, T. J., 2001: Quasi-biennial and subbiennial variations of stratospheric trace constituents derived from HALOE observations. *J. Atmos. Sci.,* **58,** 7–25.

Durran, D. R., 1999: *Numerical Methods for Wave Equations in Geophysical Fluid Dynamics.* Springer, 465 pp.

Funk, J. F., and G. L. Garnham, 1962: Australian ozone observations and a suggested 24-month cycle. *Tellus,* **14,** 378–382.

Gray, L. J., 2000: A model study of the influence of the quasi-biennial oscillation on trace gas distributions in the middle and upper stratosphere. *J. Geophys. Res.,* **105,** 4539–4551.

Holton, J. R., and A. Gettelman, 2001: Horizontal transport and the dehydration of the stratosphere. *Geophys. Res. Lett.,* **28,** 2799–2802.

Lindzen, R. S., 1987: On the development of the theory of the QBO. *Bull. Amer. Meteor. Soc.,* **68,** 329–337.

Mote, P. W., T. J. Dunkerton, M. E. McIntyre, E. A. Ray, P. H. Haynes, and J. M. I. Russell, 1998: Vertical velocity, vertical diffusion, and dilution by midlatitude air in the tropical lower stratosphere. *J. Geophys. Res.,* **103,** 8651–8666.

Pfister, L., and Coauthors, 2001: Aircraft observations of thin cirrus clouds near the tropical tropopause. *J. Geophys. Res.,* **106,** 9765–9786.

Randel, W. J., F. Wu, J. M. Russell III, A. Roche, and J. W. Waters, 1998: Seasonal cycles and QBO variations in stratospheric CH_4 and H_2O observed in *UARS* HALOE data. *J. Atmos. Sci.,* **55,** 163–185.

——, ——, A. Gettelman, J. M. Russell III, J. M. Zawodny, and S. J. Oltmans, 2001: Seasonal variation of water vapor in the lower stratosphere observed in Halogen Occultation Experiment data. *J. Geophys. Res.,* **106,** 14 313–14 326.

Reed, R. J., 1964: A tentative model of the 26-month oscillation in tropical latitudes. *Quart. J. Roy. Meteor. Soc.,* **90,** 441–465.

——, 1965: The present status of the 26-month oscillation. *Bull. Amer. Meteor. Soc.,* **46,** 374–387.

——, 1967: The structure and dynamics of the 26-month oscillation. *Proc. Int. Symp. on Dynamics of Large-Scale Processes in the Atmosphere,* Moscow, Russia, Russian Academy of Science, 376–387.

——, W. J. Campbell, L. A. Rasmussen, and D. G. Rogers, 1961: Evidence of a downward-propagating, annual wind reversal in the equatorial stratosphere. *J. Geophys. Res.,* **66,** 813–817.

Rosenlof, K. H., 1995: The seasonal cycle of the residual mean meridional circulation in the stratosphere. *J. Geophys. Res.,* **100,** 5173–5191.

Seidel, D. J., R. J. Ross, and J. K. Angell, 2001: Climatological characteristics of the tropical tropopause as revealed by radiosondes. *J. Geophys. Res.,* **106,** 7857–7878.

Trepte, C. R., and M. H. Hitchman, 1992: Tropical stratospheric circulation deduced from satellite aerosol data. *Nature,* **355,** 626–628.

Zhou, X. L., 2000: The tropical cold point tropopause and stratospheric water vapor. Ph.D. thesis, State University of New York at Stony Brook, 121 pp.

Chapter 6

Richard J. Reed and Atmospheric Tides

RICHARD S. LINDZEN

Program in Atmospheres, Oceans, and Climate, Massachusetts Institute of Technology, Cambridge, Massachusetts

"This study has confirmed earlier findings of the existence of pronounced diurnal tidal motions near the stratopause and has provided first clues concerning the worldwide pattern of the wind fluctuations."—(Reed 1966a)

1. Introduction

In many respects, atmospheric tides are one of Dick's more minor interests. I think I am correct that Dick became interested in this problem as a result of consulting activities at the White Sands Missile Range during the early 1960s. However, with his usual combination of insight and originality of viewpoint, Dick's contribution to this area was major. Luck was also hardly irrelevant to his work here, but Dick, as usual, was astute in both recognizing and exploiting luck. His contributions were characteristically useful to theoreticians—a hallmark of Dick's work. Dick's efforts in this area consist of five relatively brief papers from the period 1965–69 (Beyers et al. 1966; Reed et al. 1966a,b; Reed 1967; Reed et al. 1969). There is no indication of prior or subsequent interest. Within these papers, Dick and coworkers simply Fourier analyzed (in time) meteorological rocketsonde data for horizontal wind and temperature as a function of altitude over the range covered by the rockets (approximately 40–60 km). Data binned according to hour as well as data from campaigns of relatively frequent soundings over periods of about 2 days in length were used. Although today the needed mathematical apparatus to conduct such analyses is literally built into word processors, there was no particular novelty or challenge in the analyses 35 years ago either. What proved important were the care exercised in the analyses as well as the specific findings and the fact that we were given vertical profiles of amplitude and phase, both of which sometimes varied significantly in ways that systematically depended on location and period (diurnal or semidiurnal). Before turning to the details, it will be useful to describe the state of atmospheric tidal studies at the time of Dick's contributions. However, before doing even this, I feel it personally necessary to point out the important impact of Dick's work on my own efforts.

In August 1964, I completed my thesis at Harvard. The thesis dealt with the interactions of dynamics, chemistry, and radiation in the stratosphere, and among the topics looked at were the quasi-biennial oscillation and sudden warmings. Harvard's small program had little emphasis on data, and I was delighted when Dick Reed invited me to spend some time at the University of Washington. This happened to be when Dick was most involved with the tidal observations for the middle atmosphere. Although I had not worked on tides before, I was familiar with the work of classmates at Harvard who were concerned with inertial oscillations (where the frequency equaled the local Coriolis frequency) in the ocean, and were analyzing the problem using the equatorial β plane (Merl Hendershott and Robert Blandford). Dennis Moore was also using this methodology to study equatorial waves. Not surprisingly, Dick Reed pressed me to look at the problem of mesospheric tides, which I began to do using the methodology my friends had exposed me to. At the time, I had almost no familiarity with the general problem of atmospheric tides. However, in continuing work on this problem while a North Atlantic Treaty Organization (NATO) postdoctoral fellow at the University of Oslo with Arnt Eliassen who, though generous and supportive, rarely interfered in ones choice of topics to work on, I found it necessary to look at the problem on a sphere with the traditional methodology of classical tidal theory (i.e., the use of Hough functions). Although Hough functions were frequently mentioned in the literature, I had to go to Hough's original papers to get the computational details (Hough 1897). It was, incidentally, another postdoctoral at the University of Oslo, Larry Larsen, who introduced me to the topic of gravity waves and critical levels, which was to play a major part in my subsequent work on the quasi-biennial oscillation.

2. Status of atmospheric tides prior to 1964

In fact, in the mid-1960's, our understanding of atmospheric tides was undergoing a rapid change after almost a century of concentration on surface data (primarily for pressure) and a focus on a particular theory: namely, the resonance theory suggested by Lord Kelvin (1882) and brought to a state of remarkable development by Pekeris (1937). This theory was effectively canonized in the text by Wilkes (1949), and this was the standard reference used even by the White Sands group (Miers 1965). Details of the history are given in Lindzen (1990) and in Chapman and Lindzen (1970) and need not be repeated here, though they reveal an excellent example of how progress occurs within an observational (as opposed to a laboratory) science. The problem Lord Kelvin focused on was the fact that the surface pressure field was dominated by the solar semidiurnal component as opposed to the diurnal. This would have been expected for gravitational forcing, but gravitational forcing was dominated by lunar periods. The dominance of solar periods suggested thermal forcing, which was predominantly diurnal. Kelvin suggested that the atmosphere might have a resonance that would favor the semidiurnal component. Efforts over the next 50–60 years by Horace Lamb and G. I. Taylor among others suggested that the atmosphere's primary resonance was not sufficiently close to the main semidiurnal mode, but that it might be possible for the atmosphere to have multiple resonances. Pekeris (1937) in a tour de force showed that an atmosphere where temperature rose to a peak at what we subsequently referred to as the stratopause and diminished sharply above would indeed have a suitable second resonance, and simultaneous measurements by a variety of indirect means confirmed that the atmosphere qualitatively had the requisite structure (Martyn and Pulley 1936). However, in subsequent years, in situ measurements of temperature showed quantitative differences from what Pekeris required that were sufficient to remove the required resonance, and in 1963, Butler and Small showed that direct forcing by insolation absorbed by ozone and water vapor were sufficient to produce the observed semidiurnal surface pressure oscillation. They further suggested that since the eigenmodes for diurnal oscillations were confined to within 30° of the equator and were oscillatory with latitude, little diurnal forcing would project on them. That there was a potential problem with this picture was suggested by meteor radar observations of the layer from 80 to 100 km, which showed significant diurnal oscillations at 53°N though semidiurnal amplitudes were larger (Greenhow and Neufeld 1961); at 35°S, the diurnal oscillations were larger (Elford 1959). There were also analyses of radiosonde data (Harris et al. 1962); these exploited the availability of stations that for a while had four soundings per day and, thus, permitted analysis for tidal periods. The early radio meteor work showed the presence of strong tidal oscillations in wind with the then-surprising presence of diurnal as well as semidiurnal components. However, the early radio meteorological data consisted simply of slab measurements for the layer at 80–100 km, and offered no information concerning vertical structure [this was later remedied by the use of high vertical resolution continuous wave (cw) radar results reviewed in Glass and Spizzichino (1974)]. The radiosonde data were plagued by low signal to noise as well as irregular results. At least part of the reason was later noted by Wallace and Hartranft (1969): namely, the diurnal tide below 30 km was significantly regional rather than migrating. The possibility of this problem was also noted by Reed et al. (1966b).

3. Advent of meteorological rocket data

At this point the early analysis of rocketsonde data began suggesting very strong meridional winds at levels near 50 km (Miers and Beyers 1964). If these winds were characteristic of zonal means (and given the sparsity of data, there was a strong temptation to assume this), then they implied huge Coriolis torques. As it turned out, rocket soundings were preferentially made near noon local time. A series of rocketsondes over a 2-day period showed that there was a large 24-h oscillation in meridional wind with a very small daily mean (Miers 1965). Of course, it might also have been the case that the rocketsondes were sampling a planetary wave rather than the mean flow, and, in some respects, that is what a tide is. In any event, it appears that the problem was solved by recognition that noon soundings were preferentially sounding one phase of a diurnal oscillation. The fact that the bias was uniform at all stations, however, led Reed et al. (1966b) to insightfully suggest that the tides were primarily migrating with the sun. That this was, in fact, the case, was a piece of extraordinary luck since it implied that tidal results from any station were likely to be representative of that whole latitude circle.

Although the presence of strong tidal oscillations in the stratosphere was recognized by both Miers (1965) at Whites Sands in New Mexico and Lenhard (1963) at Eglin Air Force Base in Florida, Dick was quick to explore the availability of relevant data from several other stations (Reed et al. 1966a,b), binning data according to hour groups in a reasonable manner in order to infer tidal behavior over a wide range of latitudes. Early on, he and colleagues recognized that the temperature data was problematic—having much larger diurnal amplitudes than implied by radiative calculation, and also lacking the structure that characterized the wind oscillations (Beyers et al. 1966). He also observed that the presence of strong monsoonal zonal winds made the analysis of the zonal wind component of the tide (given the limited data) difficult. He, therefore, focused on the meridional wind component for which the mean was indistinguishable from zero (Reed et al. 1966a). Although having credible analyses for both wind components and temperature would have provided important tests for theory, less than credible data would simply serve to confuse matters. Dick also intuitively recognized that the extraordinary and robust variation of the meridional component with both latitude and altitude would alone provide a rather stringent test of any theory.

Given that tidal theory until the 1960s, had focused almost exclusively on the semidiurnal tide (because of its dominance in the well-observed surface pressure field), the discovery of dominant diurnal oscillations in stratospheric winds was itself novel. The initial rocket data showed little evidence of any semidiurnal oscillation at all. This can be seen in Fig. 6.1., from Beyers et al. (1966). Moreover, the traditional approach to tides paid little attention to phase variations with height since the main semidiurnal mode had almost infinite vertical wavelength. Indeed, the whole notion that tides might consist of vertically propagating waves was foreign to the literature of that time. The rocket data, for the first time, made such notions unavoidable. This is readily seen in Fig. 6.2., from Beyers et al. (1966). The presence of a downward phase progression characteristic of the upward vertical propagation is quite clear.

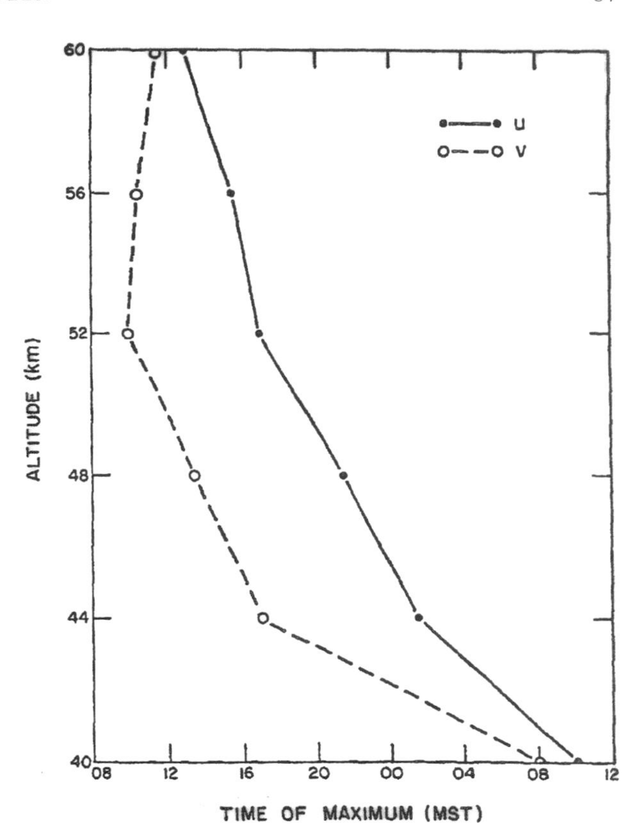

FIG. 6.2. Time of maximum of the diurnal zonal and meridional wind components as determined from harmonic analysis. Time is local standard. [From Beyers et al. (1966).]

4. Evolution of tidal picture between 1965 and 1969

As already noted, Dick encouraged me to look at stratospheric tides during my stay at the University of Washington from August 1964 until March 1965. Relatively ignorant of atmospheric tidal theory, I began looking at the problem by attempting to calculate the response to Leovy's (1964) heating results using an equatorial β plane, and quickly ran into the problem noted by Butler and Small (1963), though they did not seem to realize that it was a problem: namely, that the eigenfunctions (in my calculations Hermite polynomials) were restricted largely to latitudes equatorward of 30°. I got no further at the University of Washington [though the work was used later in Lindzen (1967b)], and in March of 1965, I sailed for Oslo to take up a NATO postdoctoral fellowship. In Oslo, I decided to bite the bullet and look at the tidal problem on a sphere. Siebert's (1961) review article was a little short on details, so I went to Hough's paper, and proceeded to follow him in looking for solutions to Laplace's tidal equation in the form of spherical harmonic expansions. Rather than use continued fractions, I decided to obtain the eigenfunctions from truncated determinants. The computing situation at the University of Oslo was such that I decided to evaluate the determinants on a new desk calculator that Eliassen had just acquired. After a couple of weeks of painful calculations, I discovered that there were both positive and negative equivalent depths. Nothing in the literature that I was aware of indicated the existence of these negative "equivalent depths," but the arithmetic seemed correct, and the associated eigenfunctions had the virtue of covering latitudes poleward of 30°. I quickly

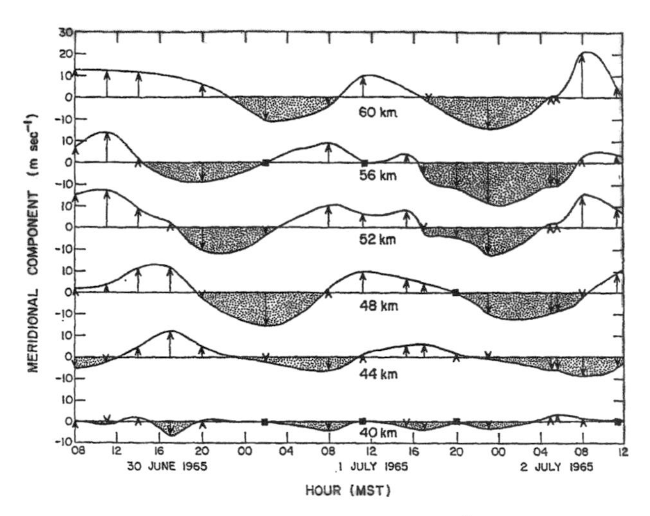

FIG. 6.1. Meridional wind components in m s^{-1} averaged over 4-km layers centered at 40, 44, 48, 52, 56, and 60 km over White Sands. Positive values indicate a south to north flow. [From Beyers et al. (1966).]

wrote up the results (Lindzen 1966), and Bernhard Haurwitz, one of the reviewers, noted that Susumo Kato had simultaneously found the negative equivalent depths in connection with studies of tides in the ionosphere (Kato 1966). I then used the Hough functions with both negative and positive equivalent depths to calculate the atmospheric response to heating due to absorption by both water vapor and ozone, finding that in contrast to the semidiurnal tide, the former was of greater importance at low latitudes. The results pretty much replicated the data analyzed by Dick for 30°N, but they also indicated profound qualitative differences between high latitudes and lower latitudes due to the respective importance of Hough modes with negative and positive equivalent depths (Lindzen 1967c). This was contrary to Dick's assumption that the diurnal tide was a smooth global system (Reed et al. 1966a). I also mentioned to Dick that I was calculating the semidiurnal tide in the stratosphere [these results appeared in Lindzen (1968)].

Dick, presumably to get ahead of the theory again, analyzed the rocket data for the semidiurnal tide (Reed 1967), and discovered that although the semidiurnal wind oscillation was much smaller than the diurnal oscillation below 50 km, it grew markedly above this level and became comparable with the diurnal. He also noted that although Pekeris's theory predicted a 180° phase shift near 30 km for the semidiurnal tide, such a phase shift was not observed until about 45 km. As it turned out, the amplitudes were reasonably close between theory and observation, but the theory continued to predict a 180° phase shift near 30 km even with thermal forcing largely due to ozone absorption. Green (1965), in fact, explained that this was due to the dominance of forcing above 30 km. Some years later (Lindzen 1978) I found that this discrepancy could be removed if the forcing by the semidiurnal component of latent heat release associated with precipitation were included. The existence of such a component was confirmed by Hamilton (1981), but its origin remains to be explained. Reed et al. (1969) carefully composited data from the Meteorological Rocket Network (MRN) to analyze for the diurnal stratospheric tide at a number of latitudes, and confirmed the marked differences in behavior between high and low latitudes. They noted that the relatively minor discrepancies between theory and observation could be corrected by adjusting the relative importance of forcing by water vapor and by ozone, and they argued that since the forcing by water vapor was well established, there might be some point in reassessing the forcing by ozone. Typical of the surprises that geophysics provides, it is now the forcing by water vapor that is being contested (Cess et al. 1995; Arking 1996). It is suggested there may be significant unaccounted for absorption by either clouds or water vapor. In a recent paper (Braswell and Lindzen 1998) it has even been argued that the so-called anomalous absorption could improve the theoretical description of the diurnal tide.

5. Reflections

In rereading Dick Reed's tidal papers from the 1960s, I was reminded of the excitement of the period. Data above radiosonde levels were previously so rare that one inevitably generalized its importance. Any rocket sounding was initially taken as typical of the whole earth. As more data became available and strong seasonal and latitudinal variations were recognized, data were taken as characteristic of season and latitude. Thus, the winter–summer difference was taken as a surrogate for differences between the Northern and Southern Hemispheres (Murgatroyd 1957; Batten 1961) The situation was, in many respects, similar to the current situation in planetary exploration. With the advent of the Meteorological Rocket Network (and the related development of small and relatively cheap rockets), there was hope of in situ coverage within the middle atmosphere. Tides were an obvious application for such data, and Reed was quick to sense this. Unfortunately, the MRN basically fizzled out by the early 1970s. In part, this was due to the emphasis on remote sensing from space. Unfortunately, vertical resolution of space data has been no match for what the MRN might have provided. (A similar situation exists in the troposphere as radiosonde stations diminish.) Also, meteorological rockets held out the possibility of flexible campaigns with high temporal resolution—a possibility exploited for the exploration of tides.

With tides, as with easterly waves and sudden warmings, the contribution of Reed and his collaborators was not discovery. Rather it was intelligent and discerning analysis. In particular, they were able to determine what was significant in the data, and what was needed to guide and check theory. This, in part, required the proper identification of "red herrings." One of these red herrings was, as I have already noted, the measurement of temperature on the rockets using thermistor beads. These measurements showed huge daily variations in temperature in the neighborhood of 50 km with ranges on the order of 15°C. The White Sands team under Willis Webb were obsessed with these oscillations, insisting on an accuracy within 1°C for the beads (Beyers and Miers 1965) and arguing for validity for years (Thiele and Beyers 1967).[1] The problem with this magnitude was that it was much larger than was expected from radiative forcing due to ozone absorption (Leovy 1964). Beyers et al. (1966) showed that it was also unlikely that dynamics would account for the difference; quite the contrary, they showed that at these levels, dynamics would diminish the radiative range. While Beyers et al. (1966) diplomatically allowed that the discrepancy was unresolved, they proceeded to focus on horizontal wind oscillations, and even here they properly noted that signal to noise was much better for the

[1] I had, myself, ventured into this issue by noting that the observed temperature oscillations recorded by the thermistor beads as a result of their swinging in and out of the sunlight were much larger than the claimed specifications for the device permitted (Lindzen 1967a).

meridional component, which they therefore concentrated on, thus providing a credible basis for assessing theory. As it turned out, classical tidal theory plausibly replicated observed meridional wind oscillations as functions of both latitude and altitude using known radiative forcing (Lindzen 1967c; Reed et al. 1969), though as Reed et al. (1969) noted, somewhat different balances of forcing by water vapor and ozone absorption could improve the detailed fit. This brings one back to the personal perspective. Not only was it at Reed's urging that I turned my theoretical attention to the problem of atmospheric tides, but the continuing interaction between us continued to advance both theory and observational analysis. There was great pleasure for me, as a then-young scientist, to have a colleague who was both eager to check my results in the real world and to provide new results demanding explanation. Such an interaction between theory and data at the personal level was rare then, and in today's world of extensively planned large programs, is rarer still. Thirty-five years later, I remain grateful to Dick Reed for this experience.

Acknowledgments. The preparation of this paper was supported by Grant DE-FG02-93ER61673 from the Department of Energy.

REFERENCES

Arking, A., 1996: Absorption of solar energy in the atmosphere: Discrepancy between model and observations. *Science,* **273,** 779–782.

Batten, E. S., 1961: Wind systems in the mesosphere and lower ionosphere. *J. Meteor.,* **18,** 283–291.

Beyers, N. J., and B. T. Miers, 1965: Diurnal temperature changes in the atmosphere between 30 and 60 km over White Sands Missile Range. *J. Atmos. Sci.,* **22,** 262–266.

——, and R. J. Reed, 1966: Diurnal tidal motions near the stratopause during 48 hours at White Sands Missile Range. *J. Atmos. Sci.,* **23,** 325–333.

Braswell, D., and R. S. Lindzen, 1998: Anomalous solar absorption and the diurnal atmospheric tide. *Geophys. Res. Lett.,* **25,** 1293–1296.

Butler, S. T., and K. A. Small, 1963: The excitation of atmospheric oscillations. *Proc. Roy. Soc. London,* **A274,** 91–121.

Cess, R. D., and Coauthors, 1995: Absorption of solar radiation by clouds: Observations versus models. *Science,* **273,** 496–499.

Chapman, S., and R. S. Lindzen, 1970: *Atmospheric Tides.* D. Reidel, 200 pp.

Elford, W. G., 1959: A study of winds between 80 and 100 km in medium latitudes. *Planet. Space Sci.,* **1,** 94–101.

Glass, M., and A. Spizzichino, 1974: Waves in the lower thermosphere: Recent experimental investigation. *J. Atmos. Terr. Phys.,* **36,** 1825–1839.

Green, J. S. A., 1965: Atmospheric tidal oscillations: An analysis of the mechanics. *Proc. Roy. Soc. London,* **A288,** 564–574.

Greenhow, J. S., and E. L. Neufeld, 1961: Winds in the upper atmosphere. *Quart. J. Roy. Meteor. Soc.,* **87,** 472–489.

Hamilton, K., 1981: A note on the observed diurnal and semidiurnal rainfall variations. *J. Geophys. Res.,* **86,** 12 122–12 126.

Harris, M. F., F. G. Finger, and S. Teweles, 1962: Diurnal variations of wind, pressure and temperature in the troposphere and stratosphere over the Azores. *J. Atmos. Sci.,* **19,** 136–149.

Hough, S. S., 1897: On the application of harmonic analysis to the dynamical theory of tides. Part 1: On Laplace's "Oscillations of the first species," and on the dynamics of ocean currents. *Philos. Trans. Roy. Soc. London,* **A189,** 201–257.

Kato, S., 1966: Diurnal atmospheric oscillation. 1. Eigenvalues and Hough functions. *J. Geophys. Res.,* **71,** 3201–3209.

Kelvin, L., [W. Thompson], 1882: On the thermodynamic acceleration of the earth's rotation. *Proc. Roy. Soc. Edinburgh,* **11,** 396–405.

Lenhard, R. W., 1963: Variation of hourly winds at 35–65 kilometers during one day at Eglin Air Force Base, Florida. *J. Geophys. Res.,* **68,** 227–234.

Leovy, C., 1964: Radiative equilibrium in the mesosphere. *J. Atmos. Sci.,* **21,** 238–248.

Lindzen, R. S., 1966: On the theory of the diurnal tide. *Mon. Wea. Rev.,* **94,** 295–301.

——, 1967a: On the consistency of thermistor measurements of upper air temperatures. *J. Atmos. Sci.,* **24,** 317–318.

——, 1967b: Planetary waves on beta planes. *Mon. Wea. Rev.,* **95,** 441–451.

——, 1967c: Thermally driven diurnal tide in the atmosphere. *Quart. J. Roy. Meteor. Soc.,* **93,** 18–42.

——, 1968: The application of classical atmospheric tidal theory. *Proc. Roy. Soc. London,* **A303,** 299–316.

——, 1978: Effect of daily variations of cumulonimbus activity on the atmospheric semidiurnal tide. *Mon. Wea. Rev.,* **106,** 526–533.

——, 1990: *Dynamics in Atmospheric Physics.* Cambridge University Press, 310 pp.

Martyn, D. F., and O. O. Pulley, 1936: The temperature and constituents of the upper atmosphere. *Proc. Roy. Soc. London,* **A154,** 455–486.

Miers, B. T., 1965: Wind oscillations between 30 and 60 km over White Sands Missile Range, New Mexico. *J. Atmos. Sci.,* **22,** 382–387.

——, and N. J. Beyers, 1964: Rocketsonde wind and temperature measurements between 30 and 70 km for selected stations. *J. Appl. Meteor.,* **3,** 16–26.

Murgatroyd, R. J., 1957: Winds and temperatures between 20 km and 100 km — A review. *Quart. J. Roy. Meteor. Soc.,* **83,** 417–458.

Pekeris, C. L., 1937: Atmospheric oscillations. *Proc. Roy. Soc. London,* **A158,** 650–671.

Reed, R. J., D. J. McKenzie, and J. C., and Vyverberg, 1966a: Diurnal tidal motions between 30 and 60 kilometers in summer. *J. Atmos. Sci.,* **23,** 416–423.

——, ——, and ——, 1966b: Further evidence of enhanced diurnal tidal motions near the stratopause. *J. Atmos. Sci.,* **23,** 247–251.

——, 1967: Semidiurnal tidal motions between 30 and 60 km. *J. Atmos. Sci.,* **24,** 315–317.

——, M. J. Oard, and M. Siemisnski, 1969: A comparison of observed and theoretical diurnal tidal motions between 30 and 60 kilometers. *Mon. Wea. Rev.,* **97,** 456–459.

Siebert, M., 1961: Atmospheric tides. *Advances in Geophysics,* Vol. 7, Academic Press, 105–182.

Thiele, O. W., and N. J. Beyers, 1967: Upper atmosphere pressure measurements with thermal conductivity gages. *J. Atmos. Sci.,* **24,** 551–557.

Wallace, J. M., and F. R. Hartranft, 1969: Diurnal wind variations; surface to 30 km. *Mon. Wea. Rev.,* **97,** 446–455.

Wilkes, M. V., 1949: *Oscillations of the Earth's Atmosphere.* Cambridge University Press, 72 pp.

Chapter 7

Characteristics of African Easterly Waves

ROBERT W. BURPEE

Cooperative Institute of Marine and Atmospheric Studies, Miami, Florida

"The energetics of the [easterly] waves in the Atlantic ITCZ are seen to be quite different from those of the waves in the Pacific ITCZ. In the Atlantic ITCZ the strongest upward motions occur at relatively low levels in the trough where the temperature anomaly is negative. Thus upward motion is correlated with low temperatures and eddy kinetic energy is converted to eddy available potential energy. In the Pacific ITCZ on the other hand, the largest rising motions occur in the upper part of the trough where the temperature anomaly is positive, and the energy conversion is in the opposite direction." —(Reed 1978)

1. Introduction

African easterly waves [AEWs, the term now preferred by R. Reed (2001, personal communication)] are synoptic-scale disturbances that form over tropical northern Africa and propagate westward, in a nearly unbroken progression, across Africa and the Atlantic Ocean during the summer months. Piersig (1936) was the first to describe a type of cyclones on historical surface weather charts that included the eastern tropical Atlantic and must have been the surface part of an unobserved AEW. He reported that these cyclones moved westward in the trade wind belt and occurred most frequently in August and September near the West African coast between 5° and 20°N. In view of the upstream location of these cyclones from the North Atlantic breeding grounds of hurricanes and their seasonal similarity to hurricanes, he speculated that the disturbances could be the precursors of some hurricanes. Subsequently, Hubert (1939) tracked a disturbance from near 5°W over Africa westward to the West African coast. He pointed out that if he extrapolated the disturbance westward to the central Atlantic, at a similar speed, it would have arrived in the general area, at about the same time, where the 1938 New England hurricane was first detected.

Regula (1936) noted that squall lines along the coast of West Africa were frequently accompanied by some unidentified system that produced 24–48 h of surface pressure falls before squall occurrence and 24–48 h of pressure rises afterward. Schove (1946) noted that Nigerian "thundery" weather and line squalls often occur in troughs in the easterlies.

Erickson (1963) and Arnold (1966) described images from early weather satellites that revealed the frequent summertime occurrence of westward-propagating, synoptic-scale disturbances. These disturbances, likely AEWs, were accompanied by convective cloudiness as they crossed tropical western Africa and the eastern North Atlantic. Erickson argued that one of the strongest waves possessed features of an incipient hurricane (Debbie of 1961) near the coast.

Carlson (1969a) analyzed daily streamline maps with observations from the conventional West African network at two levels, ~600 and 3000 m (~700 hPa) for 3 1/2 months during the summer of 1968. He showed that wavelike wind disturbances (AEWs) could be tracked westward nearly continuously in the wind field from about 10°E. The AEWs had an average period of 3.2 days and a wavelength of ~2100 km. In 1967, Carlson (1969b) tracked four of these disturbances westward from central tropical Africa to the Atlantic and showed that the three strongest became hurricanes. He (1969a) speculated that the formation of AEWs might be related to convective processes over elevated terrain.

Using visible satellite images, Frank (1969) tracked "inverted v" cloud patterns, associated with AEWs, across the Atlantic. Frank (1970) showed that during the summer from 1967 to 1969 the average number of AEWs passing Dakar, Senegal, and the average 200-hPa zonal wind component in 10-day periods at Dakar have similar distributions. This finding led him to suggest that the AEWs are related to the subtropical easterly jet in the upper troposphere of West Africa.

Carlson and Prospero (1972) pointed out that the prolonged summertime heating of air passing over the

Sahara creates a dry-adiabatic, mixed layer that extends to altitudes of 5000–6500 m and contains Saharan dust particles. As northern parts of AEWs pass over the Sahara, they entrain the dusty air. Carlson and Prospero showed that the Saharan air layer (SAL) becomes elevated near the West African coast as it is undercut by surface marine air from the north. This leaves the SAL and its dust near 600–800 hPa where AEWs transport the SAL westward across the Atlantic and the dust frequently reaches the Caribbean Sea.

Burpee (1972) searched for the typical region of origin of the AEWs and a mechanism that could explain their formation and growth. Applying spectral analysis methods to the meridional wind data from several widely spaced upper-air African stations north of the equator (Fig. 7.1), he identified the region of usual wave formation as the area between Khartoum, Sudan (32°E), and Fort-Lamy (now N' Djamena), Chad (15°E). He decided that this area is the wave source region on the basis of spectral calculations of random time series of the observations that identified a statistically significant spectral peak in the range of 3.1–5.7 days in the meridional winds at 700 hPa at Fort-Lamy. In addition, upper-air stations farther to the west have significant spectral peaks at 3.1–5.7 days throughout an increasingly deep layer up to the West African coast, but he did not find significant peaks to the east of Fort-Lamy.

On a meridional cross section along 35°E near 10°–15°N, Burpee (1972) documented the presence of a midtropospheric, maximum in the easterly wind component that becomes an easterly jet in a meridional cross section at 5°E. Burpee (1971) determined that the absolute vorticity of the zonal winds on cross sections at 5° and 35°E (Figs. 7.2d and 7.2e) increases monotonically from the equator toward the north except on the north side of the midtropospheric jet where the shear vorticity is anticyclonic and negative. The presence of the negative relative shear vorticity causes the absolute vorticity to reach a maximum. Kuo (1949) demonstrated that this is a necessary condition for barotropic instability of the flow. This condition exists in the AEW growth region, from May through October, based on cross sections that included observations averaged for several years.

Burpee (1972) subsequently established that the horizontal and vertical shears of the mean jet satisfy the Charney–Stern (1962) criterion for the instability of an internal baroclinic jet, also from May through October. He found that the horizontal and vertical shears of the long-term mean zonal wind appear to contribute equally to the wave growth. These findings led to the accepted theory that AEWs have their origins in the instability of the midtropospheric easterly jet.

On the south side of the easterly jet, Burpee (1972) noted that in the lower-troposphere AEWs advect warm

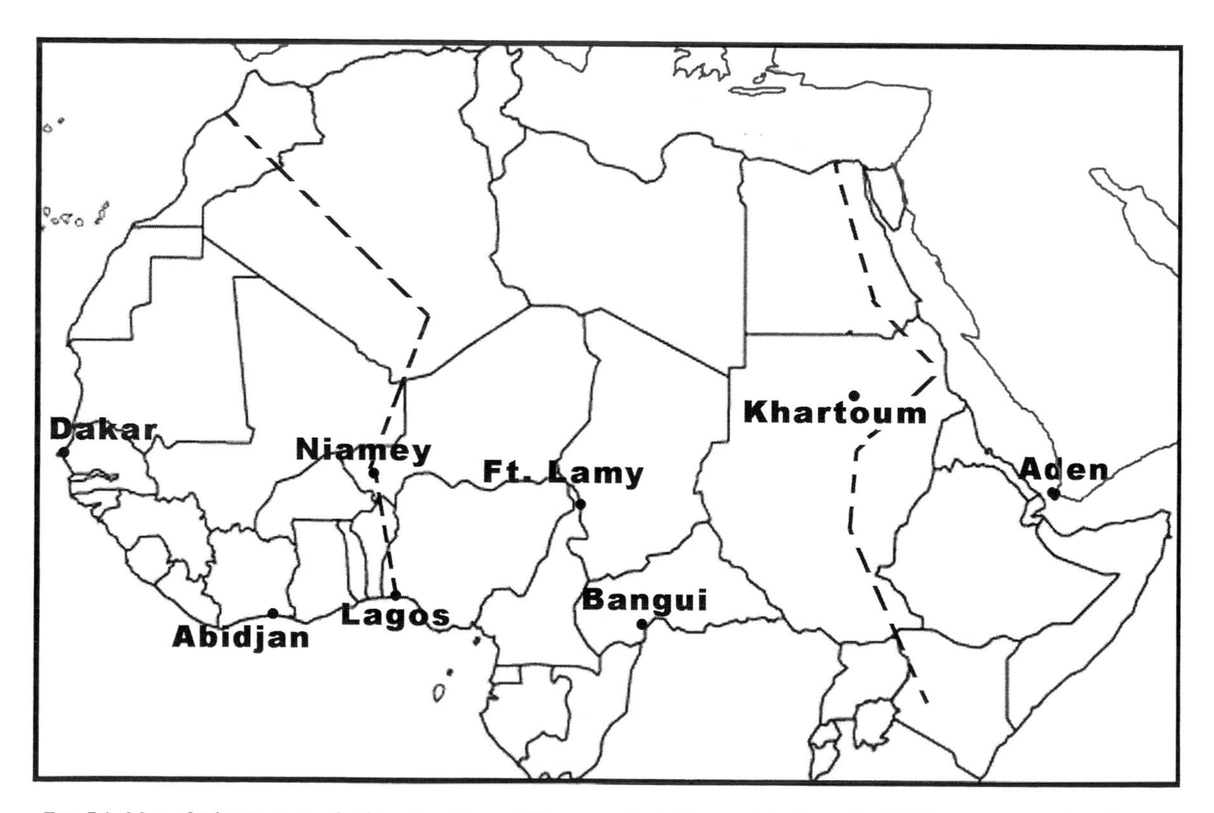

FIG. 7.1. Map of relevant parts of Africa. Locations of the upper-air stations used in Burpes's (1972) power spectral analyses are identified. The dashed lines join the stations providing the zonal wind data in the meridional cross sections of Figs. 7.2d,e. Reproduced from Burpee (1972) and modified slightly.

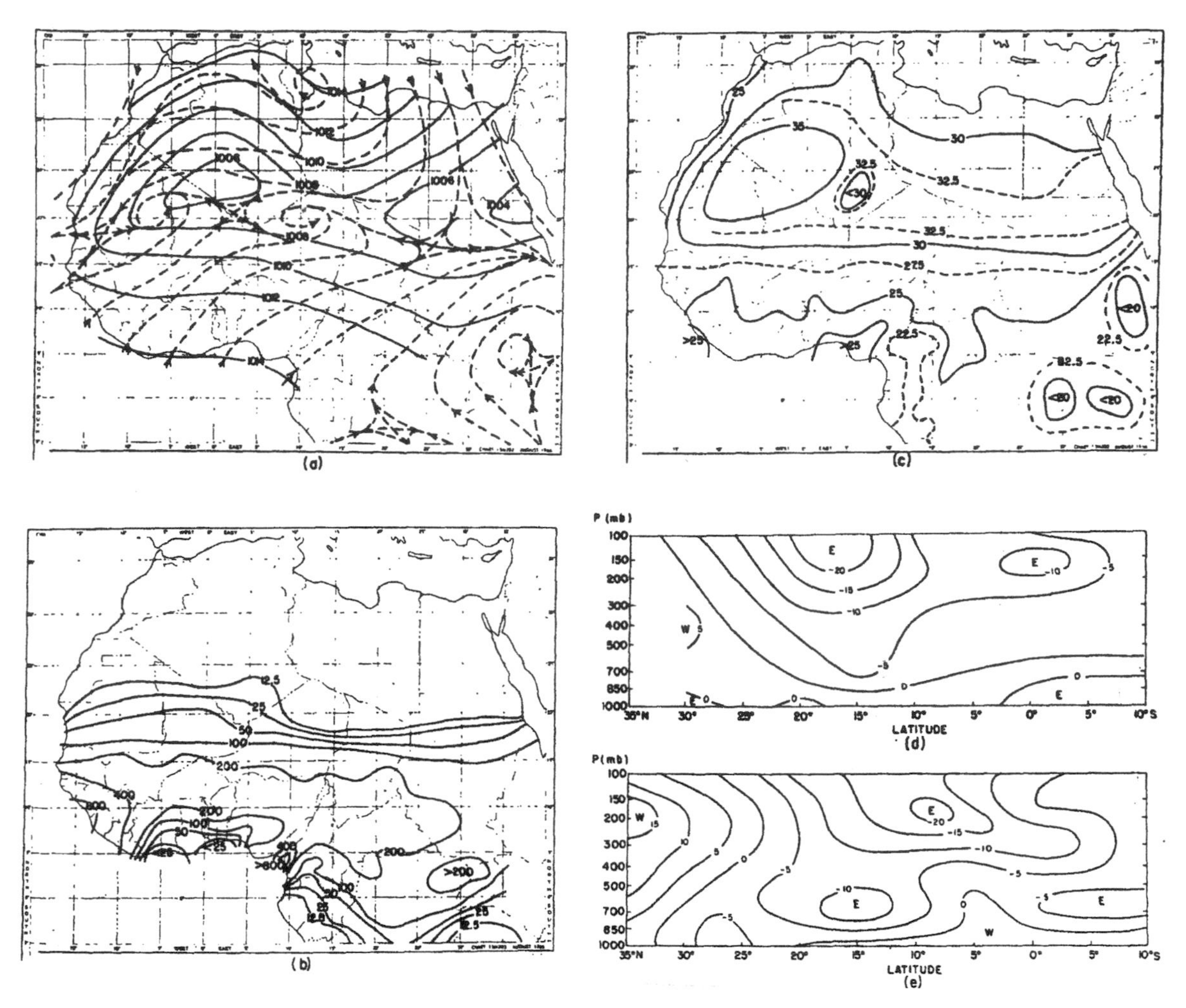

FIG. 7.2. Monthly mean maps for Aug: (a) surface pressure (hPa) and gradient-level streamlines shown, respectively, by solid and dashed lines; (b) rainfall (mm); (c) daily average surface temperature (°C); and (d),(e), meridional cross sections of zonal wind (m s^{-1}) at 35° and 5°E, respectively. Reproduced from Burpee (1972).

air southward, down the low-level temperature gradient. In the middle troposphere, the AEWs transport easterly momentum southward, also down the easterly wind gradient. Burpee (1974) documented that the AEWs advect relative humidity northward, nearly out of phase with the temperature advection. Carlson (1969a) indicated that the maximum cloudiness is near the wave trough. South of 12.5°N. Burpee (1974) showed that the highest frequencies of cloudiness and precipitation occur near the 700-hPa wave trough. Farther to the north, he found that the maximum thunderstorm and rainfall frequencies are typically to the east of the trough and in the region with higher humidity.

Carlson (1969a) and Burpee (1974) described the presence of two separate surface centers of circulation, related to the AEWs, in western Africa. One, located near 20°N on the southern edge of the Sahara, is shallow and largely cloud free and has properties similar to a thermal low (Carlson 1969a). The other occurs near the average position of the rainfall maximum at about 10°N. Carlson speculated that this vortex is of convective origin.

2. Mean August weather conditions in tropical northern Africa

Having demonstrated that AEWs originate, grow, and mature in the latitudinal band between the equator and 20°N, Burpee (1972) presented the average conditions in this region and surrounding areas for August, near the time that the AEWs attain maximum amplitude. Figures 7.2a–c include daily averaged surface temperature, surface pressure, and total rainfall that are 10-yr means as well as mean gradient-level winds [the latter are reproduced from Atkinson and Sadler (1970)] that reveal the location of the shallow southwesterly monsoon. Figures 7.2d and 7.2e show the mean zonal wind for

August, as a function of pressure and longitude at 35° and 5°E, respectively.

One unusual aspect of this continental region compared with the tropical oceans is the intense low-level temperature gradient created by the hot, dry Sahara Desert air to the north and the relatively cool, but warm, moist air from the Gulf of Guinea to the south. Figure 7.2c indicates that the average surface temperature gradient reaches 10°C in 10° of latitude and is nearly independent of longitude. Reed et al. (1977) pointed out that, except for the surface layer, this gradient extends westward over the eastern Atlantic Ocean in the easterly offshore flow, but the magnitude of the gradient weakens with increasing distance to the west of the West African coast. In response to the low-level baroclinic zone and in accord with the thermal wind equation, there is a region of easterly shear with height and a midtropospheric easterly wind maximum that occurs near 600 hPa (Fig. 7.3). The maximum becomes a midtropospheric jet in west-central Africa.

Mass (1978) pointed out that the topographic slope of the African continent, north of the equator, is fairly smooth apart from the mountains of Ethiopia that rise above a high plateau to elevations exceeding 3000 m. Thus, he noted that topography is not of primary importance in affecting the large-scale mean rainfall patterns in western and central tropical northern Africa where the lines of constant rainfall have a largely east–west orientation.

3. African easterly waves during an international field program

a. Compositing methodology—The data tabulation procedure

The GATE field program was divided into three phases with two interphase breaks that enabled most of the ships to replenish supplies and provide their crews with a shore rest in Dakar. The time interval for Phase III and its preceding interphase break began on 23 August and continued until 19 September 1974. During this time, Reed et al. (1977) tracked the AEWs westward across central and western Africa and the eastern Atlantic Ocean. The wave propagation was continuous during this time interval. They determined that the average AEW period, wavelength, and propagation speed were ~3.5 days, 2500 km, and 7–8 m s^{-1}, respectively. They noted that the wavelength was somewhat longer over Africa than in the eastern Atlantic, but that the wave propagation speed was nearly constant throughout the region. The AEWs influenced a greater vertical depth (Burpee 1972) and were stronger (Albignat and Reed 1980) in western Africa. In this region, the waves were also more convective (Payne and McGarry 1977) and had a larger effect on rainfall patterns (Reed et al. 1977).

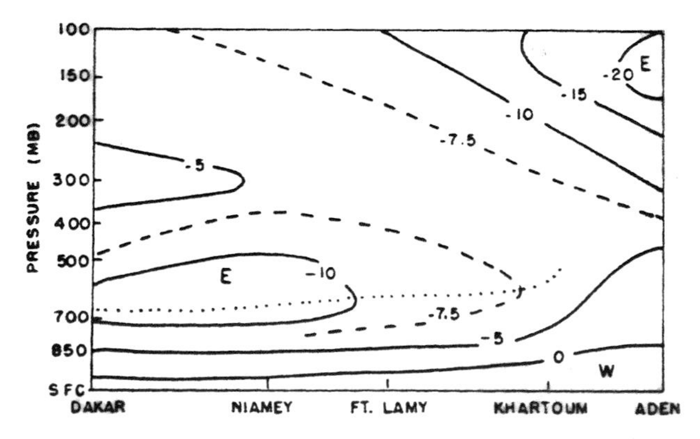

FIG. 7.3. Zonal cross section of monthly mean zonal wind (m s^{-1}) for Aug along 13°N. The dotted line indicates the position of the easterly wind maximum in the middle troposphere. Reproduced from Burpee (1972).

Reed et al. (1977) tracked the vorticity centers of eight AEWs across the region of most numerous observations, 10°E–31°W. They determined the AEW structure by compositing observations in the GATE Quick-Look Data Set relative to the 700-hPa trough position at the mean paths of the waves, at what they called the reference latitude, 11°N over the continent and 12°N over the ocean. In the mean flow, they observed the easterly jet ~5° north of the reference latitude. Reed et al. separated all available observations into eight east–west categories relative to the trough positions and seven north–south bands, 4° of latitude in width, the central one of which was centered on the reference latitude. They defined category 4 as the position of the 700-hPa trough and category 8 as the 700-hPa ridge. Categories 2 and 6 were centered on the maximum northerly and southerly wind components, respectively. The intermediate parts of the AEWs were referred to as categories 1, 3, 5, and 7. They observed relatively small differences in the wave structure over land versus the ocean so they presented average results for the entire region.

b. Horizontal and vertical structure

Figure 7.4 shows the composite perturbation streamline fields at the surface, and 850, 700, and 200 hPa. The perturbation field is calculated by subtracting the mean flow for a latitude band from the total wind at each grid point in that band. The eight winds in each latitudinal band are filtered to emphasize the signal for periods greater than 2 days and shorter than 6 days. The perturbation winds are plotted relative to wave category and their latitudinal difference from the reference or 0° latitude at 11° or 12°N. Since the average wavelength of the waves is about 2500 km, each category is about 300 km or 3° of longitude in width. The number of wind observations diminishes with height, but still, in the inner latitude band of the 200-hPa composite, there are ~50 observations per grid point.

At the lowest three levels in Fig. 7.4, cyclonic and anticyclonic centers are nearly aligned in the vertical, slope northward from the surface to 700 hPa, and tilt little

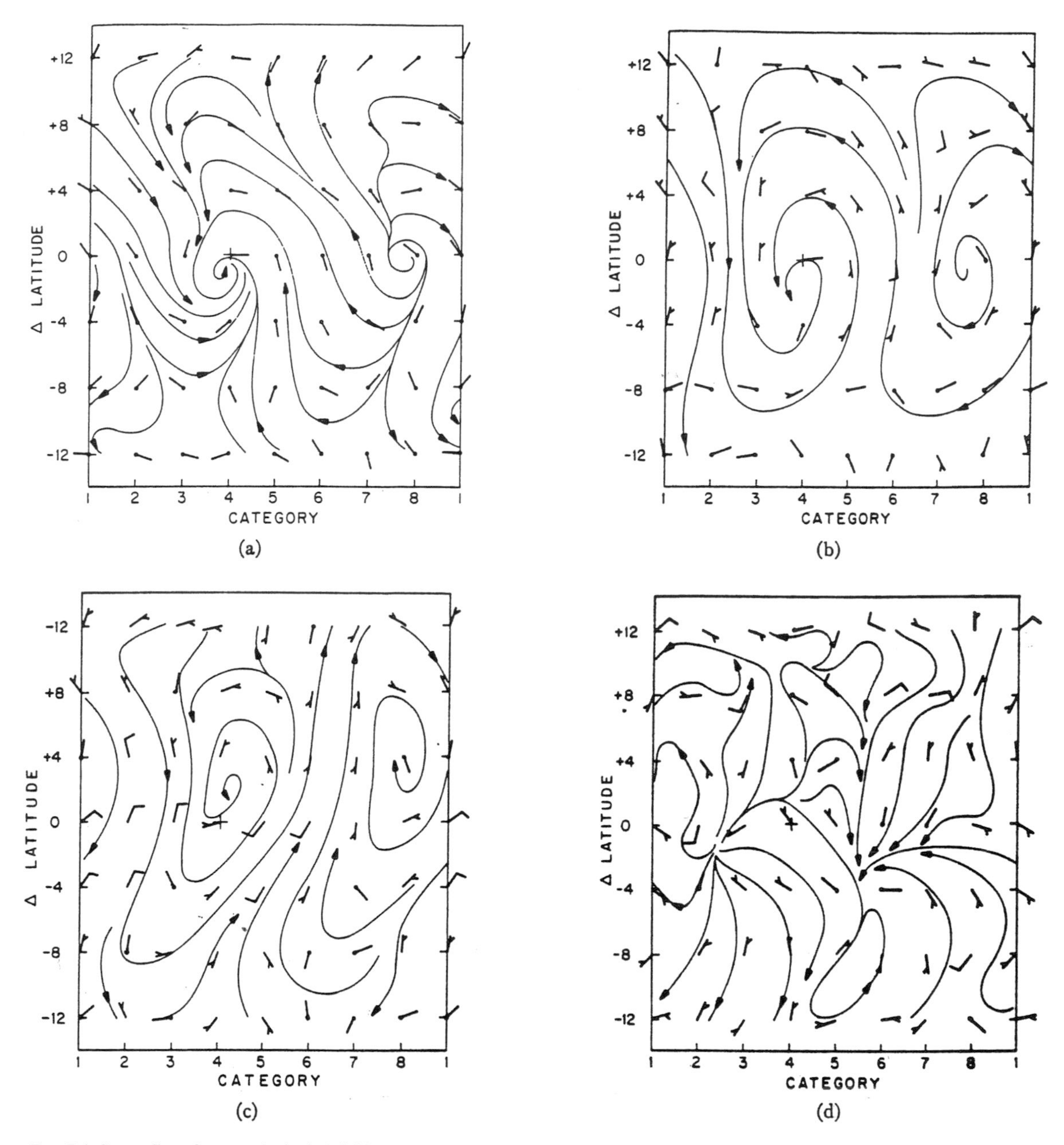

FIG. 7.4. Streamlines for perturbed wind field: (a) surface and (b) 850, (c) 700, and (d) 200 hPa. Category separation is approximately 3° longitude. Cross denotes wave center at 700 hPa. One full barb corresponds to 5 m s^{-1}, one-half barb to 2.5 m s^{-1}, and no barb to 1 m s^{-1}. Reproduced from Reed et al. (1977).

from east to west. The northward slope of the cyclonic circulation center is toward the coolest perturbation temperatures (not shown). The vorticity center (not shown), however, is nearly vertical at low levels. At the upper-tropospheric level of 200 hPa, regions of divergence and convergence are above the low-level cyclonic and anticyclonic circulation centers, respectively. At the lowest three levels, to the south of the reference latitude, the troughs and ridges tilt from southwest to northeast.

Vertical cross sections for the reference latitude are displayed in Fig. 7.5 and plotted with coordinates of pressure and wave categories. Figure 7.5 presents sections for the perturbation meridional wind and temperature, relative humidity, and vertical velocity. Figure 7.5a reveals a maximum meridional perturbation wind amplitude of ~5 m s^{-1} near 700 hPa. This amplitude is larger than the one reported by Burpee (1975) in a composite analysis for all of GATE. The general features of the meridional wind are nearly vertical below 700 hPa but tilt westward and weaken aloft. In the upper troposphere, a secondary maximum of meridional wind speed of ~2 m s^{-1} is approximately out of phase with the lower maximum. The perturbation westerly winds (not shown) are generally smaller than the meridional winds below 350 hPa and are positively correlated with the southerly winds near the reference latitude.

The temperature deviation cross section (Fig. 7.5b) reveals a three-level vertical structure with a cool core in the 700-hPa trough. In the trough region at 500–

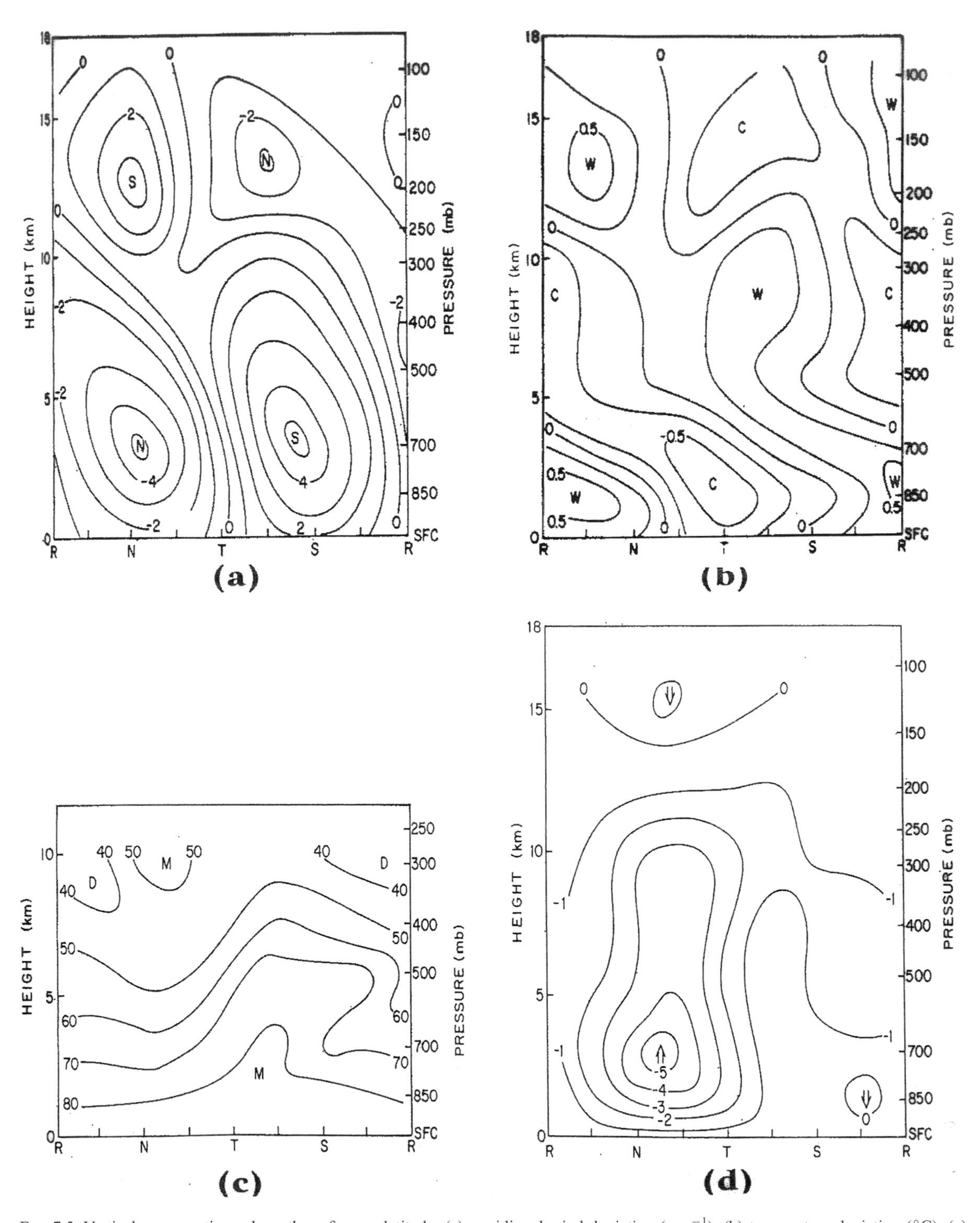

FIG. 7.5. Vertical cross sections along the reference latitude: (a) meridional wind deviation (m s^{-1}), (b) temperature deviation (°C), (c) relative humidity (%), and (d) vertical velocity (hPa h^{-1}). Letters R, N, T, and S refer to ridge, northerly wind, trough, and southerly wind sectors of the wave, respectively. Reproduced from Reed et al. (1977).

300 hPa, there is a relatively warm core and near 150 hPa a second cool area. In the vertical section of relative humidity (Fig. 7.5c), high values of moisture occur near the trough from the surface to ~400 hPa. High moisture values tend to be correlated with southerlies and low values with northerlies.

Reed et al. (1977) pointed out that the divergence pattern (not shown) is more complicated than the one Reed and Recker (1971) calculated for synoptic-scale waves in the equatorial Pacific, possibly because of the influence of the midtropospheric easterly jet over Africa. The AEW vertical velocity section (Fig. 7.5d) has a maximum value of upward motion of ~5 hPa h^{-1} (1–2 cm s^{-1}) that occurs to the west of the 700-hPa trough. At the reference latitude, subsidence is weak and present only in two small areas of the section: one near 850 hPa ahead of the ridge and the other above 150 hPa between categories N and T.

c. Large convective clusters, squall lines, and rainfall

Payne and McGarry (1977), two of Reed's students, examined the infrared (IR) brightness distribution in images of the first *Synchronous Meteorological Satellite* (*SMS-1*). They confined their study to the area from the equator to 20°N and 10°E–30°W for Phase III of GATE and the previous interphase break. They recorded the life cycles of nonsquall, large (maximum area > ~2 × 10^5 km^2) and long-lived (lifetimes > 24 h) cloud clusters by noting the wave category of cluster formation and dissipation. These clusters form most frequently just to the west of the trough and their preferred area of decay is to the east of the trough. This result agrees with their finding that large, long-lasting clusters move more slowly than the average propagation speed of AEWs.

Squall lines produce considerable rain during the summer in central and western tropical Africa. During their study, Payne and McGarry (1977) tracked 46 squall lines and noted the locations of the formation and dissipation of each squall relative to the AEW categories. They determined that category 2, the northerly wind maximum, is the favored region of squall genesis. The average squall moved at about twice the propagation speed of the AEWs, moving about one category through the waves to the west and most frequently dissipating in category 1 (Fig. 7.6). Aspliden et al. (1976) studied squall lines during the three phases of GATE. They found that, on average, squalls formed over the African continent from 1300 to 1700 LST, to the west of the 700-hPa AEW trough axis, and moved westward at ~15 m s^{-1}. Fewer squalls formed over the B-scale ship array and they tended to form between 0300 and 1000 LST in general agreement with the results of Payne and McGarry.

Payne and McGarry and Reed et al. (1977) composited, respectively, the percentage coverage of convective cloud and rainfall (Fig. 7.7) and found that both quantities were highly modulated by AEWs. The composited amounts of both quantities are about three times greater in and ahead of the wave trough than near the ridge. The locations of the maxima over Africa are in good agreement with the reference latitude vertical velocity shown previously. There is also consistency between the moisture budget and the vertical velocity computed by Thompson et al. (1979) in the GATE B-scale ship array.

d. The source region of African easterly waves

The numerous African observations recorded during GATE provide a rare opportunity to examine the otherwise largely data-void area of central and eastern tropical northern Africa. On the basis of limited data from earlier years, Burpee (1972) determined this area (15°–32°E) to be the favored formation region of AEWs. Albignat and Reed (1980), using data from 23 August to 19 September 1974 for northern tropical Africa, southern Arabia, and the GATE A/B ship array, prepared 700-hPa time series at 6-hourly intervals of the meridional wind. They computed spectral analyses from

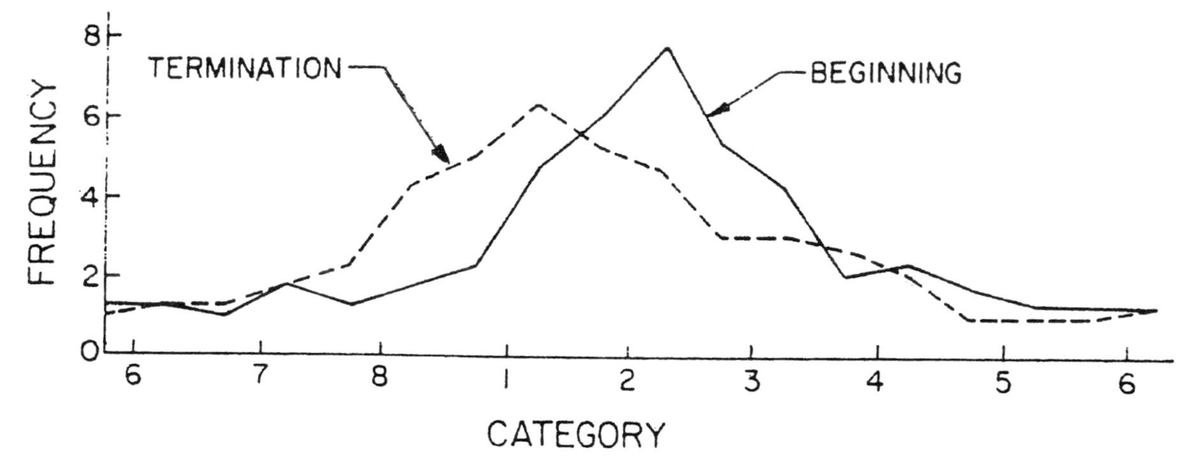

FIG. 7.6. Frequency distribution of the location of the leading edge of squall clusters vs wave phase category for 46 squall episodes at the beginning (solid line) and termination (dashed line) of each squall episode. Reprodued from Payne and McGarry (1977).

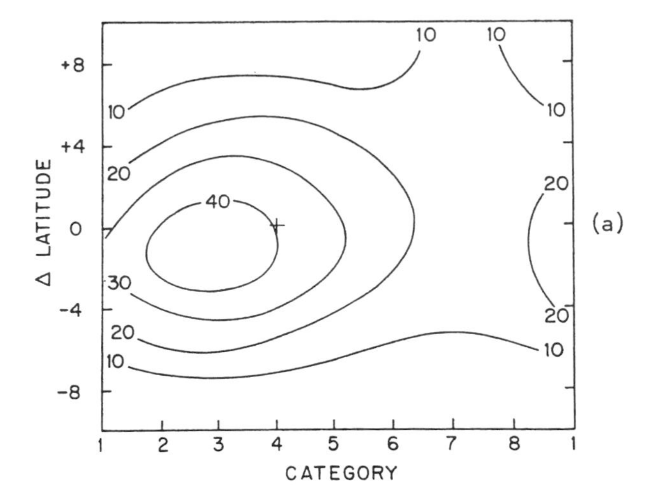

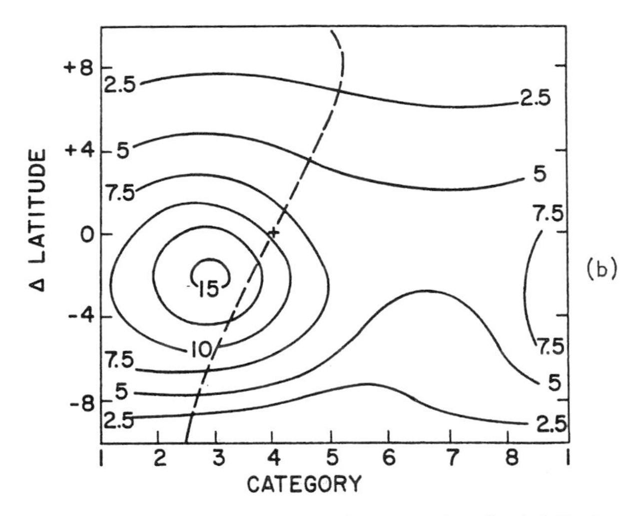

FIG. 7.7. (a) Percentage coverage by convective cloud defined as white-appearing cloud in *SMS-1* infrared images. (b) Average precipitation rate (mm day^{-1}). Dashed line depicts trough axis at 700 hPa as a function of latitude. Reproduced from Reed (1978).

the time series and plotted the spectral amplitudes in the frequency band 0.2–0.4 cycles per day (cpd), equivalent to periods between 2.5 and 5 days, at 700 hPa. Burpee (1972) had shown in his spectral study of AEWs that he initially detected the disturbances farthest to the east at the 700-hPa level of Fort-Lamy. Albignat and Reed plotted the geographical distribution of the spectral amplitudes (they did not calculate the statistical significance of their spectra) at 700 hPa (Fig. 7.8). The 28-day length of the time series at the African stations (even shorter for the ship array) is short in comparison with the observed number of AEWs (eight in western Africa, section 3a) during the period of study. Standard spectral analysis textbooks maintain that the record length of a time series in spectral analyses should be at least 10 times the periods being investigated. Nevertheless, the pattern of the spectral amplitudes computed by Albignat and Reed seems consistent with other wave features observed west of 20°E. Figure 7.8 depicts a general increase in wave amplitude to the west, beginning at longitudes west of 20°E with a general southwestward displacement of the larger amplitudes toward the West African coastline. The major increase in spectral amplitude occurs between 10°E and 0°. To the east of 20°E, meridional wind spectral amplitudes in the frequency range 0.2–0.4 cpd are either less prominent or absent. However, meridional wind spectra based on longer time series that are tested for significance are needed to corroborate the absence of any wave activity. They calculated cross correlations of the basic wave quantities and determined average AEW wavelength of 2500 km and a phase speed of ~ 9 m s^{-1}.

Albignat and Reed (1980) also show the position of the 700-hPa easterly jet axis over Africa during the 28-day period and the flux of easterly momentum, in the 0.2–0.4-cpd frequency band, away from the jet core, or equivalently, westerly momentum toward the core (Fig. 7.9). Figures 7.8 and 7.9 illustrate an important aspect of the wave dynamics with the AEWs increasing in strength in the same region where the jet loses easterly momentum in the frequency band of the waves.

Albignat and Reed conclude that the primary growth region for AEWs is between 0° and 10°E. This conclusion agrees with the origins of the wave disturbances depicted in the 850-hPa analyses for phase III by Sadler and Oda (1978). The magnitudes of the horizontal and vertical wind shears (700–850 hPa) of the zonal wind suggest an important role for the midtropospheric easterly jet in the wave growth. Albignat and Reed proposed that in the region between 0° and 10°E, wave growth is caused by a combination of two processes: the increased barotropic and baroclinic instabilities of the midtropospheric jet and the greater organization of convection that allows latent heating to energize the AEWs.

The question of the presence of AEWs in central and eastern Africa or even southern Arabia has not been answered conclusively. There is the possibility of alternate explanations including the existence of weak disturbances, undetected by the spectral analysis method or not existing in the region to the east of 10°E in August–September 1974. Certainly, if waves are present to the east of 10°E, they are too weak to be much of a factor in local weather. It is difficult to track weak disturbances across largely data-void areas and possible that this 28-day period is unrepresentative of normal conditions. This is a question that merits further inquiry.

e. *Energetics of the African easterly waves*

Norquist et al. (1977) used the meteorological fields obtained by Reed et al. (1977) during Phase III and the preceding interphase break period to diagnose energy transformations in AEWs. During this time, the AEWs were well developed and near or at their seasonal maximum amplitude. The fields are based on observa-

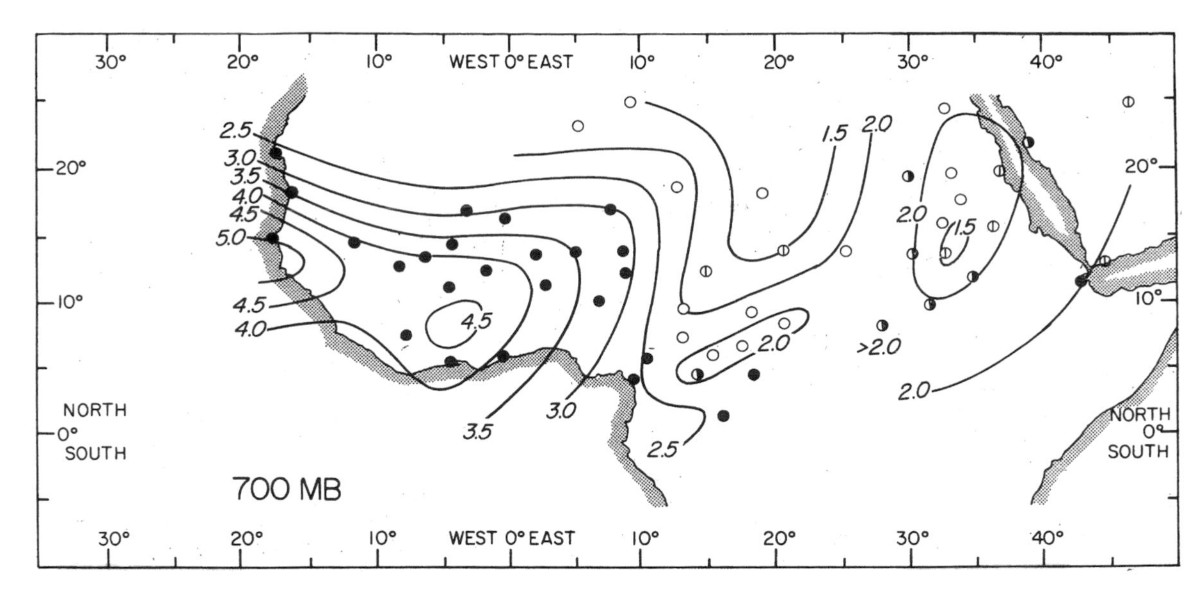

FIG. 7.8. Amplitude of meridional wind oscillation in 0.2–0.4-cpd frequency band at 700 hPa. Isopleths over land are based on a 28-day period of record, and over the ocean on a 21-day period. Solid circle signifies a pronounced spectral peak in the 0.2–0.4-cpd range, half circle a secondary peak, circle with line a weak peak, and an open circle no detectable peak. Reproduced from Albignat and Reed (1980).

tions in the region from 10°E to 31°W and 1°S to 26°N from the surface to 100 hPa with 15°W dividing the land from the ocean area. The numerical estimates of the various energy terms are displayed in the form of the Lorenz (1955) energy diagram where AZ and AE are the zonal and eddy available potential energies, KZ and KE are the zonal and eddy kinetic energies, and CA, CE, and CK are the conversion rates from AZ to AE, AE to KE, and KZ to KE.

The primary goal of the energetics analysis is to determine the processes that account for the generation and maintenance of the KE of AEWs. Adding the observations for the land and ocean areas in the combined area, they determined that the KE is about twice AE and that KE and AE are about an order of magnitude larger than the zonal energies, KZ and AZ (Fig. 7.10). Norquist et al. calculated that the KE of the AEWs is maintained almost equally by conversion from KZ to KE (the barotropic conversion) and from AE to KE (the baroclinic conversion). They estimated that in the absence of friction, the KE of the waves would double in 3 days.

Separate calculations for the land and ocean areas reveal, somewhat surprisingly, that the conversion from KZ to KE, the barotropic conversion, is larger over the

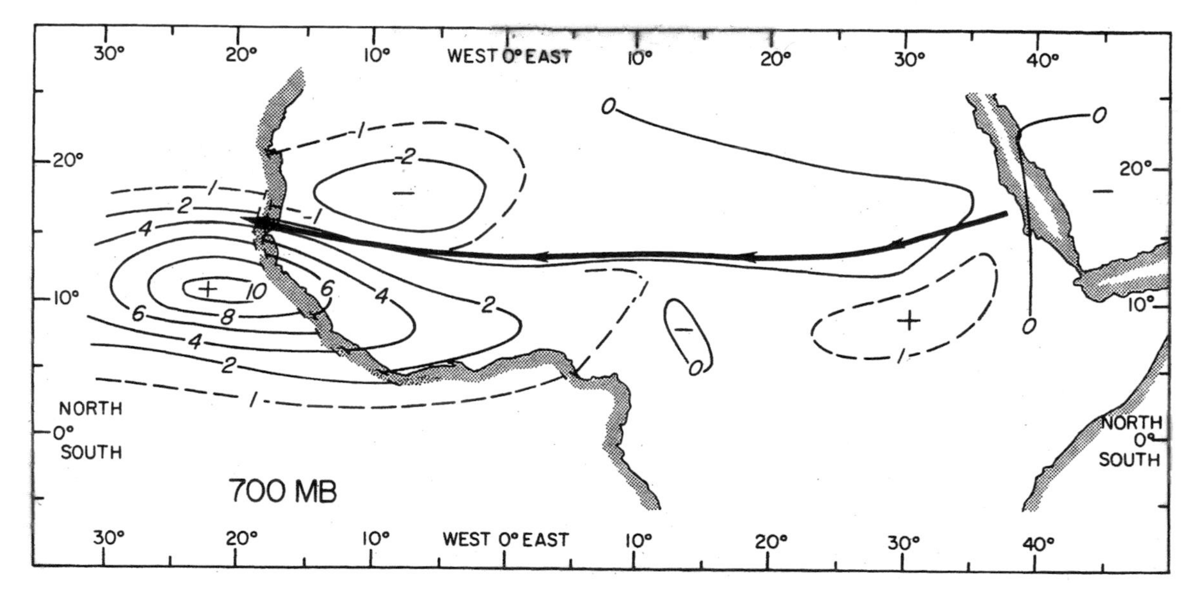

FIG. 7.9. Northward flux of zonal momentum (uv) at 700 hPa in 0.2–0.4-cpd frequency band (m^2s^{-2}). Heavy lines represent jet axis. Caption for Fig. 7.8 defines length of time series. Reproduced from Albignat and Reed (1980).

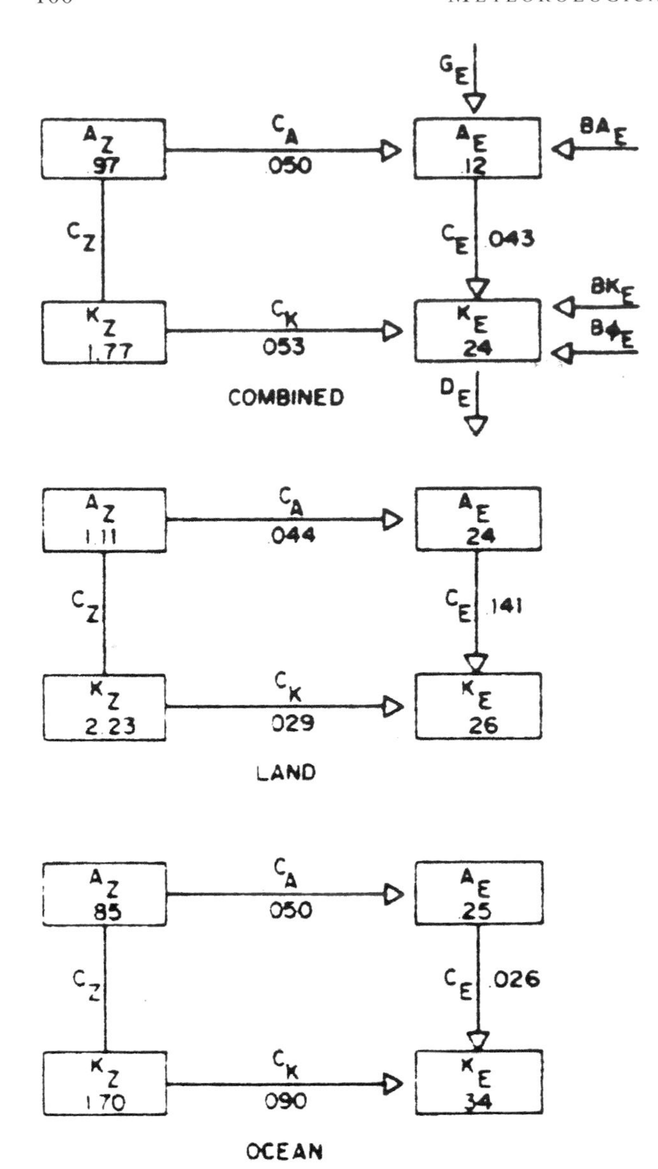

FIG. 7.10. Partitioned energies (10 J m^{-2}) and energy conversion rates (W m^{-2}) for (top) land and ocean combined, (middle) West Africa, and (bottom) the eastern Atlantic Ocean. Symbols are defined in text. Reproduced from Reed (1980).

limited ocean region than it is over the land area. Also, a surprise, the conversion from AE to KE is stronger over the land than the ocean. Norquist et al. (1977) infer from this calculation that latent heat in organized deep convection is more important in the wave growth and maintenance over West Africa than over the ocean.

One intriguing aspect of the AEWs and the midtropospheric easterly jet that requires clarification is the process that increases the jet strength across Africa (Albignat and Reed 1980) as the jet simultaneously gives up energy to the waves. R. Reed (2001, personal communication) offers an explanation that relies on his interpretation of Albignat and Reed (1980), Norquist et al. (1977), and Mouzna (1984). He argues that the midtropospheric easterly jet in summer develops as a result of the temperature difference between the hot air over the Sahara and the relatively cool air in the monsoon flow in the rainbelt that comes from the ocean to the south. The solar heating pattern produces the temperature difference and the wind flow comes into balance with the mass distribution through subtle ageostrophic motions, not unlike those in a Hadley circulation. In the presence of dissipative processes, whatever the type or scale, the jet is maintained by the restoration of the thermal gradient by the differential heating. According to the data, the jet and baroclinity increase in strength from 30°E to the Greenwich meridian and maintain their strength from there to the coast. These growth processes of the jet occur as the waves form, grow, and mature at the expense of the jet energy in the same region.

Another way to approach the problem (R. Reed 2001, personal communication) is to think of the Lorenz energy diagram of Norquist et al. (1977) shown in Fig. 7.10. In their energy diagram, they were unable to measure the generation of AZ by the pattern of diabatic heating and the conversion of AZ to KZ by the weak Hadley-type circulation. Had these been measurable, Reed is confident that the conversion of AZ to KZ would be sufficient to account for the transformation of KZ to KE and that the generation of AZ would be of sufficient magnitude to keep the whole system going.

f. Convective energy, convergence, and convective activity in the A/B ship array

In the A/B ship array, Thompson et al. (1979) calculated the net energy released for an undilute parcel lifted upward from 1000 hPa in each of the wave categories. The net energy required for upward displacement is smallest in the region of least convective activity near the ridge and greatest around the trough. Using the previously mentioned simplifying assumptions, Thompson et al. computed a relation between the net energy available and the convective activity as represented by

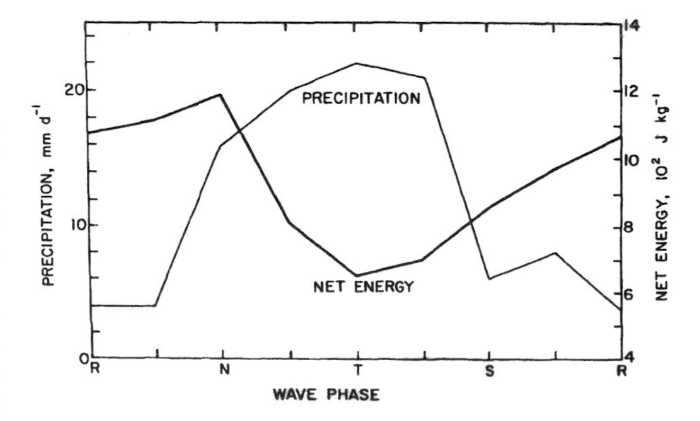

FIG. 7.11. Precipitation rate (thin line) and net energy in lifting undilute parcel from 1000 hPa to the level of zero buoyancy (heavy line) as functions of wave position. Reproduced from Thompson et al. (1979).

the amount of precipitation shown in Fig. 7.11, which illustrates that the available energy reaches a minimum near the trough where the convective activity is a maximum. After the trough passage, the energy rises until the arrival of the ridge. Of course, a realistic description of the energy available for convection must include the processes of entrainment and compensating subsidence in the environment of the convection. No matter how sophisticated the computation of available convective energy, Thompson et al. (1979), who used upper air data from the center of the GATE ship array (Fig. 7.12) in their calculations, conclude that the large-scale convergence field of the traveling AEWs primarily controls the degree of convective activity.

g. Mass, moisture, and heat budgets of the African easterly waves in the center of the A/B ship array

Thompson et al. (1979) computed mass, moisture, and heat budgets at the center of the B-scale ship array during Phase III of GATE (30 August–18 September 1974), but their results were based on the Quick-Look Data Set. Reed (1980) repeated their calculations with the Final Validated Data Set and obtained virtually identical results; his results are presented here. The B-scale ship array was in the eastern Atlantic intertropical convergence zone (ITCZ) during Phase III so the computations depict the waves moving through the ITCZ. The time series prepared for each quantity at the center of the ship array were determined with a least squares fit in time and a quadratic fit in space. Observations were recorded at 3-hourly intervals and each was assigned to one of the wave categories.

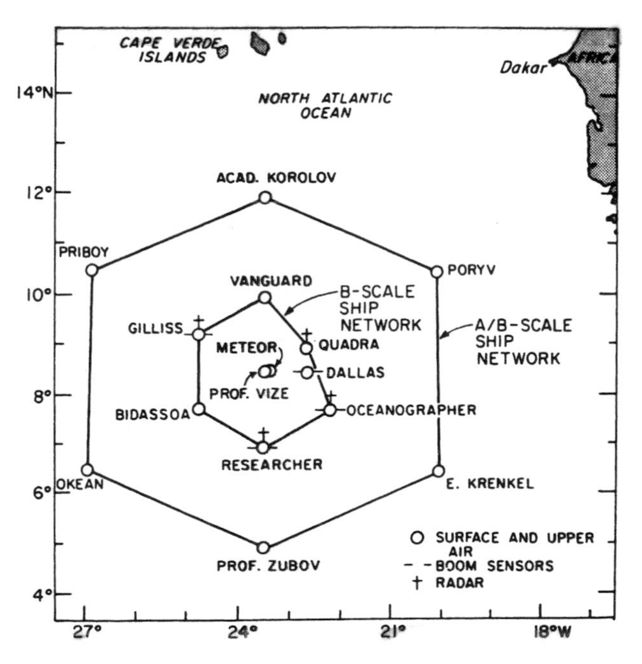

FIG. 7.12. The network of GATE ships in the A/B ship array. Reproduced from Thompson et al. (1979).

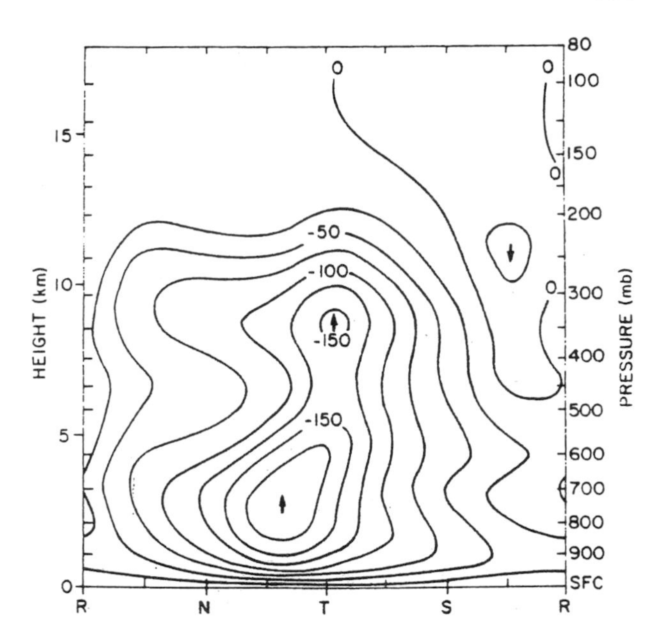

FIG. 7.13. Vertical cross section of vertical velocity where the units are 10^{-5} hPa s^{-1} (approximately hPa day^{-1}). R, N, T, and S are explained in Fig.7.5. Reproduced from Reed (1980).

1) MASS BUDGET

Reed (1980) presents the average vertical velocity ($\omega = dp/dt$) at the center of the ship array; he assumed ω to be zero at the surface and 100 hPa. The vertical velocity is upward everywhere except for a subsidence region in the upper troposphere near the ridge (Fig. 7.13). The strongest upward mass flux of ~200 hPa day^{-1} (~2 cm s^{-1}) occurs slightly ahead of the 700-hPa trough axis. The level of the maximum value of ω is much lower than Reed and Recker (1971) found for synoptic-scale waves in the equatorial Pacific, a difference possibly resulting from the presence of the midtropospheric jet in the eastern Atlantic.

2) MOISTURE BUDGET

In this and subsequent budgets, Reed (1980) adjusts the mean storage terms to zero, consistent with the negligible change in thermodynamic parameters that occurred during the 20 days that were studied. Very small adjustments were required to achieve this balance. In the mean budget, the average precipitation rate is 12.5 mm day^{-1}, which is nearly balanced by the sum of the moisture convergence (10.1 mm day^{-1}) and the surface evaporation (3.8 mm day^{-1}). Assuming small measurement errors, Reed determines that in the eastern Atlantic ITCZ region, almost three times more moisture is imported from the surrounding environment than is evaporated locally from the sea surface.

As the wave trough approaches, water vapor is stored and, as the ridge replaces it, water vapor is depleted. The fluctuations with wave category of the moisture convergence and precipitation are the largest in the budget. The precipitation is ~20 mm day^{-1} in and ahead of the trough and <5mm day^{-1} near the ridge. The evaporation is largely determined by the wind speed.

3) DRY STATIC ENERGY BUDGET

In the mean, condensation heating is approximately balanced by the sum of the net horizontal divergence of dry static energy and radiative cooling, which is about half the size of the horizontal outflow term. The surface flux is small and the residuals are negligible. The primary fluctuations in the wave budget are largely limited to the dry static energy outflow and the condensation heating terms.

4) TOTAL HEAT OR MOIST STATIC ENERGY BUDGET

In the mean part of the budget, three terms contribute positively. The first is the horizontal import of moist static energy (h), a small positive number that implies the h converged at low levels is slightly greater than that diverged at high levels. The two other terms that increase the moist static energy are evaporation from the ocean surface and a smaller transfer of sensible heat from the ocean to the atmosphere. Radiative cooling compensates for the three sources of h. An unexplained negative residual is needed to achieve the balance of h.

In the different parts of the wave, the balance of h is somewhat inconsistent. In general, the total heat budget shows that the largest values of atmospheric import and storage occur ahead of and near the trough while the largest values of heat supply from the ocean surface are only near the trough.

The pattern of vertical eddy flux of total heat is shown in Fig. 7.14 and its vertical convergence, the cumulus heating, is presented in Fig. 7.15. Thompson et al. (1979) explain their method for estimating the vertical profile of radiative cooling as a function of wave category and their procedure for adjusting the column integrals to the total heat flux from the surface. In the convectively active part of the wave (categories N–T) where rainfall has been shown to be large, the flux increases upward from near

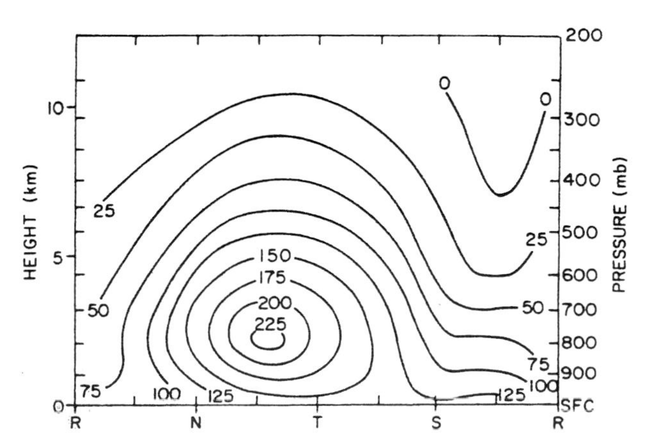

FIG. 7.14. Vertical eddy flux of total moist static energy as a function of wave category. Units are W m^{-2}. Reproduced from Reed (1980).

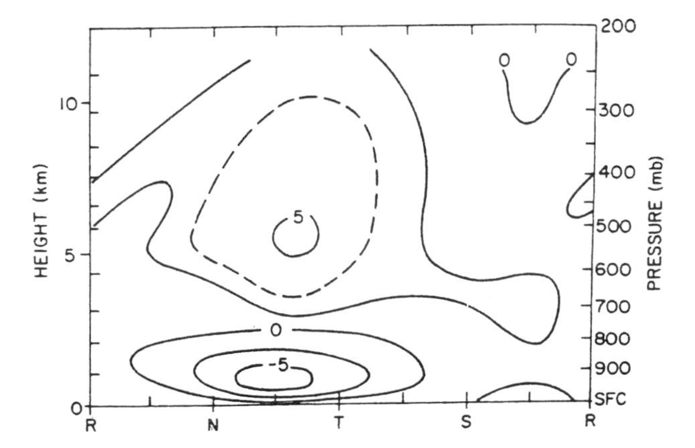

FIG. 7.15. Vertical distribution of cumulus heating as a function of wave category. The units are 10^{-2} W kg^{-1}, or approximately °C day^{-1}. Reproduced from Reed (1980).

the surface to 800 hPa and then decreases monotonically upward. Near and ahead of the ridge, in the region of generally suppressed convection, the flux decreases monotonically with increasing height from the surface. The associated flux convergence (Fig. 7.15) is mostly positive, indicating that turbulent and convective heat transports add energy to the large-scale flow in most parts of the wave. The maximum rate of energy input to the wave at 500 hPa is $\sim 5 \times 10^{-2}$ W kg^{-1} or nearly 5°C day^{-1} in the form of a rate of temperature change. The depth of the layer of cumulus detrainment is large compared with the deep convection of the U.S. plains.

The eddy flux of moist static energy and the vertical distribution of cumulus heating are presented in Figs. 7.14 and 7.15 as a function of the wave categories and pressure. The flux convergence is slightly positive, indicating that turbulent and convective heat transports deposit energy aloft in the large-scale flow in most wave categories. This process results in extensive heating that is maximum near 500 hPa. An energy loss or cooling of equal strength occurs at low levels in the convectively active part of the wave. In this region, the enhanced upward flux of moist static energy related to the convection results in an energy depletion that adds to the loss due to radiational cooling. The energy required to balance this cooling is supplied by the large scale, which transports high moist static energy air into the volume horizontally and removes lesser energy air aloft.

5) SURFACE ENERGY BUDGET

The mean budget shows that the net incoming radiative energy is used mainly for evaporation and, to a lesser extent, for sensible heat transfer to the atmosphere with a Bowen ratio of 0.21. Heat flux into the ocean is negligible. Reed and Lewis (1980), who calculated their results with data from an ocean buoy located in the C-scale ship array (Fig. 7.12) where

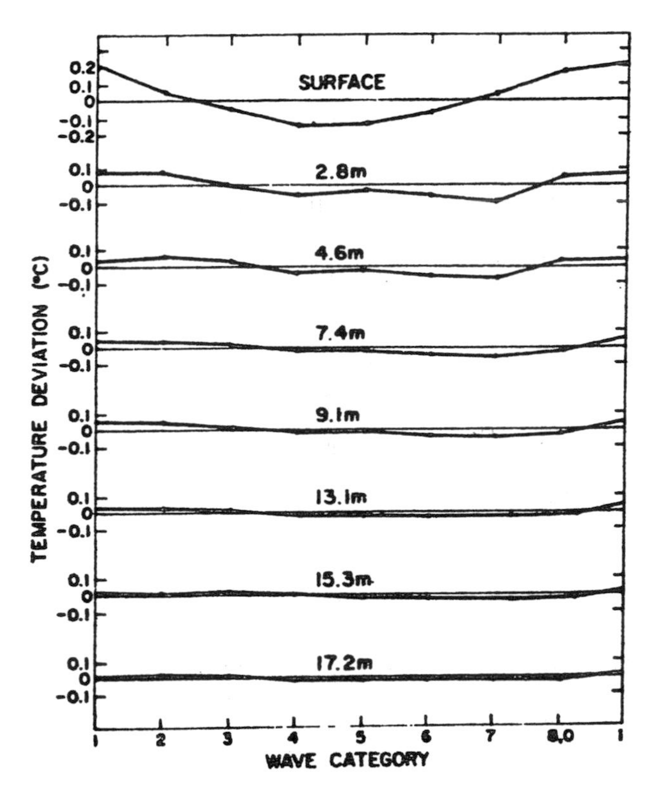

FIG. 7.16. Variations of ocean temperature (°C) plotted against wave category and depth (m). Reproduced from Reed and Lewis (1980).

the smallest-scale measurements were obtained by ships inside the B-scale ship array, reported a temperature variation in the upper-ocean layer that is similar to the variation of the surface heat flow. They found that the sea surface temperature is about 0.4°C cooler in the wave trough than in the ridge (Fig. 7.16). The temperature oscillation extends to a depth of ~20 m.

4. Diurnal variations of cirrus cloudiness generated by deep convection and rainfall in West Africa and the A/B ship array during Phases II and III of GATE

McGarry and Reed (1978) studied the diurnal variations of convective IR cloudiness and precipitation during GATE Phases II and III and the interphase break period between them, 28 July–19 September 1974. This study was another important aspect of Reed's research on the first GATE objective: the determination of relationships between the large and small scales — in this case, the occurrence of deep, moist convection in association with the diurnal cycle. They examined IR images from the *SMS-1* and subjectively estimated the percent coverage of cold convective clouds in square areas, 3° latitude and longitude on a side, in the region from the equator to 20°N and 8°E to 30°W at 0000, 0600, 1200, and 1800 UTC on each day. They confined their calculations to those areas of the IR images that appeared white (cold), which were likely the result of cirrus shields produced by deep convection. McGarry and Reed also analyzed hourly rainfall records from 33 land stations over West Africa and hourly or 3-hourly rainfall records from 13 ships in the A/B ship array for the same period.

McGarry and Reed estimated the percent of white-appearing cirrus cloud in the IR images that they referred to as convective cloudiness because the bright clouds were nearly always of convective origin. For each synoptic time, they determined the mean coverage of convective cloudiness in each square. The 54-day time series of cloud percentage estimates was harmonically analyzed to determine the phase (time of day) of the maximum coverage of convective cloudiness and the amplitude of the harmonic. The amplitude of each square was normalized by its daily mean. A normalized amplitude of 0.33 means that, in the absence of higher harmonics, the maximum cloud coverage at the phase of the harmonic would be 1.33 times the mean value and the minimum coverage would be 0.67 times the mean. A normalized amplitude of 0.33 is equivalent to a 2:1 modulation of cloud coverage from maximum to minimum during the diurnal cycle.

The hourly and 3-hourly rainfall amounts were harmonically analyzed in a similar manner to determine the phase and amplitude at each location. The total length of the records was only 54 days and rainfall has significant small spatial variations so the diurnal analyses varied widely from station to station, but averages of the diurnal values over large areas seem to give meaningful results. Thus, the rainfall amounts are grouped together for four land areas and all of the ships are combined in a single offshore position. For purposes of easy comparison of the 6-hourly cloudiness with the rainfall data, the cloudiness harmonic data were averaged for the same five areas as the rainfall (Figs. 7.17 and 7.18). The averaged diurnal oscillations of the cloudiness reveal maximum coverage near midnight over land with particularly large amplitudes in the latitudinal belt from 15° to 20°N (not shown). The GATE ship area, on the other hand, has a maximum of cloud coverage in the early afternoon. Figure 7.18 displays the results for the diurnal analyses of the rainfall data in which the northwest, northeast, and southeast African areas have a rainfall maximum just before midnight (the southwest area is probably unrepresentative because it includes only three stations, two of them coastal). The largest amplitude (0.67) occurs in the northeast area, the smallest (0.09) in the southeast area. The ship array had maximum rainfall in the early afternoon and the diurnal range was half of the daily mean amount.

McGarry and Reed (1978) concluded that the northern African area (15°–20°N) has thunderstorms in the late afternoon or early evening that frequently develop into squall lines. They speculated that the squalls require several hours to become organized, mature, and produce maximum rainfall before midnight and then maximum convective cloud cover occurs a couple of hours later. In the GATE ship array, they determined that the rainfall maximum for the entire ship array occurs about 1400 LST and that the diurnal range was half the daily mean. They did not offer an explanation for the early afternoon peak in oceanic rainfall, but remarked that it agreed with the timing of the large-scale convergence field.

5. Searching for African easterly waves in the analysis/forecast system of an operational, global model

During the early 1980s (R. Reed 2001, personal communication), with prospects of an upcoming sabbatical year from the University of Washington, Reed's interests focused on the soon-to-be-available 1978–79 First GARP Global Experiment (FGGE) dataset. Because of the expected easy access to software that would process and analyze the FGGE dataset at ECMWF, he planned to spend his sabbatical there, where he later was a visiting scientist from September 1985 to July 1986. The availability of the FGGE dataset was delayed and Reed looked for a way to redirect his efforts. Tony Hollingsworth, then the director of ECMWF's Data Division (including a diagnostics section), wanted to improve the ECMWF global model's initializations and forecasts over West Africa. With the encouragement of Hollingsworth and his colleagues, Reed decided to evaluate the capability of the ECMWF operational global model to represent AEWs.

Reed et al. (1988b) were apparently the first to conduct a broad review of a current operational global model in analyzing and forecasting AEWs. They evaluated the performance of the T106 version of the ECMWF analyses and forecasts of AEWs over western Africa and the tropical Atlantic during August and September 1985. The evaluation included a comprehensive synoptic study of the operational products. All pressure levels had significant data gaps from 35° to 10°E at longitudes to the east of those frequented by the waves, and east of the area where AEWs formed during GATE (Albignat and Reed 1980). Another region through which the waves propagate that was largely data void is the trade wind area from West Africa to the Lesser Antilles. This area had a few ship observations and satellite cloud-track winds, but no conventional upper-air observations. Reed developed a wave chronology, based on subjective determination of the 700-hPa wave trough axes on the objective streamline analyses, and with his ECMWF collaborators evaluated the forecast performance. They presented examples that illustrate the ability of the analysis system to represent AEWs. Some of the examples included regions that had widely scattered observations. They detected two main source regions for AEWs, one in the desert region west of the Ahaggar Mountains (about 22°N and the Greenwich meridian) and the other in the rain belt to the south near 10°N.

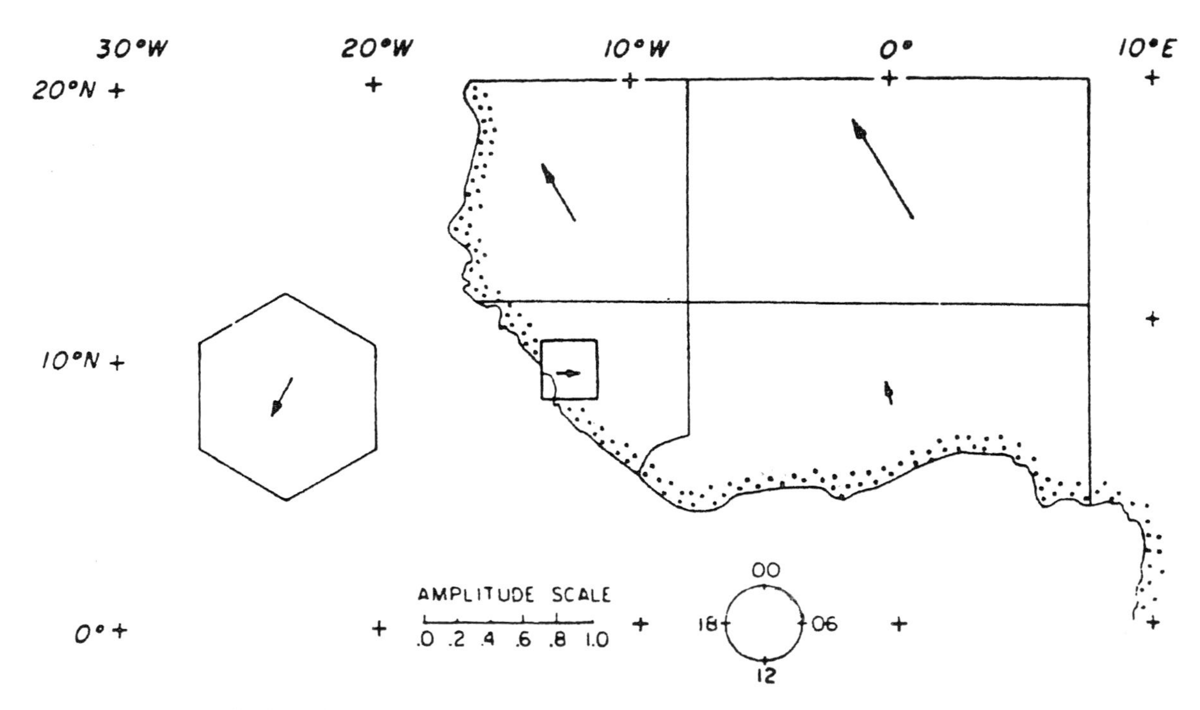

FIG. 7.17. Phase and normalized amplitude of the diurnal cycle of precipitation amount for Phases II and III of GATE based on rainfall amounts for five geographical areas. The square indicates that only three coastal or near-coastal stations were employed in the southwest area over West Africa. Time of maximum is indicated by direction of arrow according to 24-h clock. Amplitude scale is given. Reproduced from McGarry and Reed (1978).

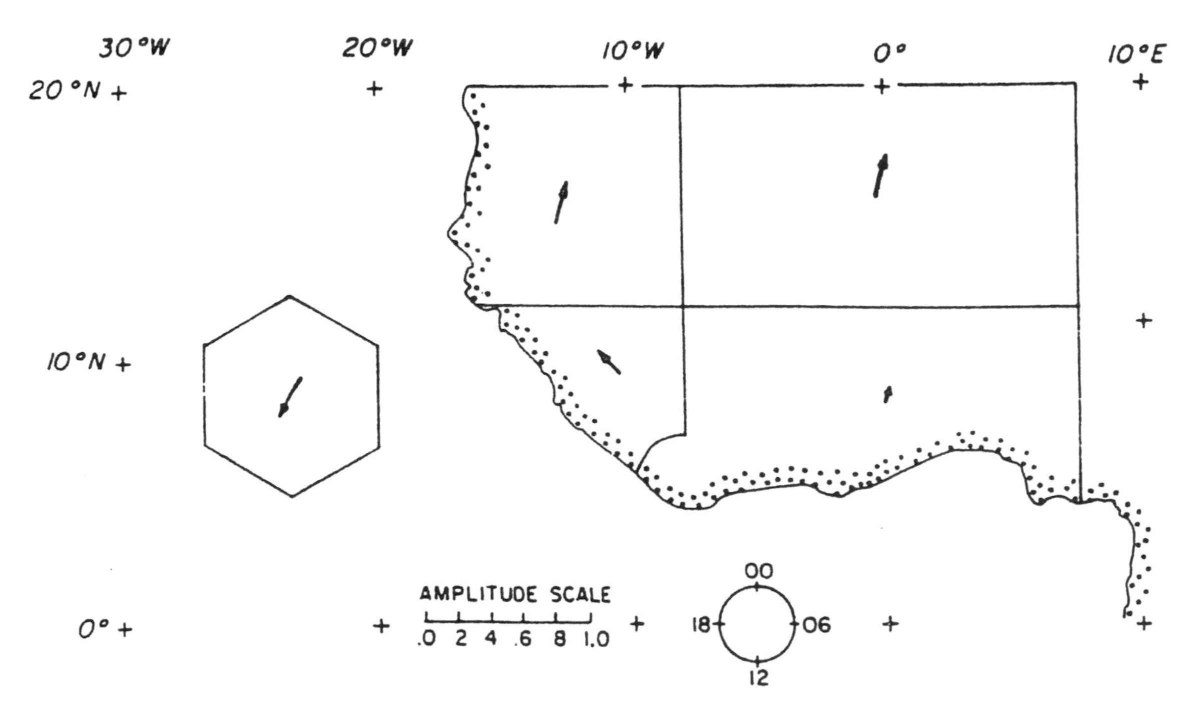

FIG. 7.18. Phase and normalized amplitude of diurnal cycle of convective cloudiness for Phases II and III of GATE based on 6-hourly *SMSH*-1 satellite photographs. See Fig. 7.17 for further explanations. Reproduced from McGarry and Reed (1978).

They followed 20 waves with similar characteristics to the observed AEWs summarized previously in this paper. The global model produced waves with wavelengths typically ~2000–3000 km, periods of ~3–5 days, and propagation speeds of ~8 m s^{-1}. They judged the 48-h forecasts of the 850-hPa AEW maximum vorticity to be encouraging.

Reed et al. (1988a) calculated spectral analyses of grided data from the ECMWF operational global model in the approximate area from 15°S to 25°N and 40°E to 50°W also for August–September 1985. They detected spectral peaks in the range of 3–5 day for the 850-hPa meridional wind. Cospectrum analyses indicated that the dominant waves propagated westward at ~8 m s^{-1} with an average wavelength of ~2500 km. Analyses of the vorticity spectra determined that waves in the 3–5-day band originated primarily in two areas: one near 22°N, 10°E and the other about 12°N, 10°E. These systems propagated westward over Africa, merged a short distance west of the West African coast, and continued moving westward across the Atlantic as a single system. Diagrams of covariance patterns of temperature, vertical velocity, and meridional and zonal winds corroborate the earlier observational findings of combined barotropic and baroclinic energy sources accounting for the formation and growth of the southern waves. Some evidence supports the wave-CISK (conditional instability of the second kind) process as a contributor to the growth of the waves that followed the southern tracks. The authors concluded that the ECMWF operational analysis and forecast system represented the wave origin and growth rather well (Lindzen 1974).

A few years later Pytharoulis and Thorncroft (1999) determined that the northern and southern low-level centers are parts of a single AEW mode that propagates simultaneously over the African continent.

6. Atlantic hurricane activity and Sahel droughts

In 1985, while R. Reed (2001, personal communication) was at ECMWF, his work on AEWs caught the attention of an oceanographer, Stefano Tibaldi, who was working at ECMWF on an Italian rotation. He was a member of a committee that advised the Pope on the selection of a topic for an annual scientific meeting held at the Vatican that focused on conditions adversely affecting human life and the discussion of possible methods to improve them. Tibaldi prevailed on the committee to schedule the upcoming meeting on the subject of Sahel (an approximately 10°-wide latitude band across Africa just to the south of the Sahara) droughts and their effects on the health and living conditions of the local inhabitants. Reed was invited to the ensuing conference held in September 1986.

At the Vatican meeting, Reed (1988) described the roles of the normal quasi-stationary, meteorological features of northern tropical Africa that affect rainfall in the Sahel and explored the possible impacts of the droughts on the transitory AEWs. Others (e.g., Kidson 1977; Lamb 1978) had previously discussed global-scale features that affected Sahel dry and wet conditions.

After reading the annual reports on Atlantic tropical systems written by forecasters at the National Hurricane

Center (NHC), Reed (1988) decided that the Sahel droughts do not appear to affect the number of AEWs or the favored location of the AEW tracks. He wondered whether Sahel drought conditions influence the extent of convective activity in AEWs in West Africa or the eastern Atlantic Ocean. If this were true, he felt that a Sahel drought might also affect the capability of AEWs to spawn hurricanes. Landsea (1993) and Avila et al. (2000) have estimated that in the long term (30 yr or so), ~65% of Atlantic hurricanes develop from AEWs.

Reed (1988) compared the annual NHC summaries of Atlantic disturbances with the standardized annual Sahel drought index (SDI) of Folland et al. (1986). The NHC disturbance counts were available from 1967 to 1985. This is an exclusively drier period than the long-term normal Sahel rainfall according to the SDI. In this period, the SDI indicates (not shown) that, in 5 of the 19 years, the dry anomaly exceeded any recorded earlier in the century. Based on the NHC reports, Reed determined that a greater number of the AEWs near the coast were more convectively active during the rainier years in the Sahel.

Reed (1988) plotted the number of hurricane hours per season in the Atlantic basin (including the Caribbean Sea and the Gulf of Mexico) relative to the reverse of the SDI from 1970 to 1985. Figure 7.19 reveals that when Sahel droughts are strong, the Atlantic hurricane activity, as represented by the annual hours of hurricanes, tends to be less than normal. [C.W. Landsea (2001, personal communication) believes that Reed (1988) was the first to postulate this kind of relationship in a formal publication.]

The years 1972, 1973, 1977, 1982, and 1983 in the sample were five of the six worst drought years according to the SDI, and these years had well-below-normal Atlantic hurricane activity. Reed pointed out that El Niños occurred in 1972/73, 1976/77, and 1982/83; Gray (1984) had shown previously that such years have fewer Atlantic hurricanes than average. In forecasting seasonal hurricane activity from 1 August to the end of the season, Gray et al. (1993) determined that June–July western Sahel rainfall explains ~25% of the hurricane variance in the remainder of the season in the Atlantic basin.

7. Concluding discussion

The international GATE atmospheric and oceanographic field program gathered unprecedented numbers of upper-air observations of excellent quality over tropical northern Africa and the eastern Atlantic Ocean. With this dataset, Reed and his collaborators analyzed AEWs and confirmed the findings of pre-GATE investigators, who used limited observations. These investigators had determined the basic wave structure over Africa, the general source region of the waves, rudimentary cloud and rainfall patterns relative to the waves, and the role of barotropic and the Charney–Stern instabilities of the midtropospheric easterly jet in the formation of AEWs. With the huge volume of GATE data, Reed and his collaborators verified much of the pre-GATE knowledge and far exceeded it. They put together detailed three-dimensional analyses that include much more than the basic meteorological measurements. With their composite analysis, they determined the average distribution of convective cloudiness, rainfall amounts, and reliable derived quantities like divergence, vorticity, and vertical velocity relative to wave features. Reed's wave chronology was a fundamental starting point for many other GATE scientists in their attempts to improve understanding of atmospheric processes of the boundary layer, deep convection, and radiation in the eastern Atlantic. Reed and collaborators were also able to estimate many of the terms in the Lorenz energy diagram. From these calculations, Reed and his students were the first to infer that latent heat release may be enhancing the growth of AEWs in West Africa.

In the GATE intensive ship array, Reed and his students computed detailed, wave-related variations of vertical velocity and vertical transports of mass, moisture, and heat by subgrid-scale motions. Furthermore, they provided their basic wave chronology for wave-related variations to other GATE scientists who calculated budgets of heat, moisture, vorticity, momentum, and surface energy.

Some questions remain about how well phase III and the preceding interphase period of GATE (a total of 28 days) represent other parts of the year and other years. GATE occurred during a drought year in Africa's Sahel region according to the SDI. Reed (1988) showed that the annual Atlantic hurricane activity is greater during wet years in the Sahel. It is also unclear whether the wave source region identified during the same 28 days would be identical in other years.

Nevertheless, Reed's design and use of the compositing procedure and his direction of his students and GATE collaborators led to significant advances in the understanding of AEWs. The combined total of his GATE research helped to achieve one of GATEs two primary objectives (Kuettner 1974): the estimation of the effects of smaller-scale tropical weather systems (convection in Reed's studies) on synoptic-scale circulations (AEWs). His research also was a major factor in making AEWs, excluding tropical cyclones, the best-documented synoptic-scale systems in the Tropics.

Acknowledgments. This paper was written in appreciation for the detailed insights about the structure and energetics of AEWs that I learned from Dick Reed. My friendship with Dick began in Dakar during GATE where our frequent end-of-the-workday discussions were nurtured by a beer and a few peanuts at the bar in the Diarama Hotel. I was fortunate to be appointed as the chief of the weather forecast team that guided the international GATE Mission Selection Team (MST) in its use of limited aircraft assets during the field program.

The primary goals of the GATE program were to observe the behavior of tropical oceanic deep convection and determine the control that large-scale systems may have on the organization of the deep convection. No one had ever had a reason to predict the development of deep convection in the eastern Atlantic location of the GATE A/B-scale ship array before the experiment so there were no guidelines to follow in preparing the forecasts. Reed assisted and encouraged me with the preparation of the daily weather briefings for the MST. We tried to develop subjective methods, based on tracking and extrapolating AEWs and an assumed pattern of deep convection and rainfall ahead of the midtropospheric trough of the AEWs. Later, in phase III, we began to account for the north-south position of the ITCZ in predicting the occurrence or nonoccurrence of deep convection in the A/B ship array that was the center of the GATE intensive observations. Proposed forecast ideas were, in part, based on our examination of some old satellite visible images of the region that I had borrowed from the basement archives of NHC and brought to Dakar. The old and current satellite images were helpful in preparing the daily convective ship array forecasts for the MST who in the afternoon selected the type of experiment to be held the next day and decided the aircraft to be involved (Burpee and Dugdale 1975). The subjective methods were subsequently, at least partially, corroborated by Reed et al. (1977), Thompson et al. (1979), and Chen and Ogura (1982).

Our friendship grew at national tropical meteorology and international meetings and with the writing of Burpee and Reed (1982). Recently, as each of us became ill, we exchange telephone and e-mail words of encouragement for each other's recovery and, later, I asked and Dick answered questions about the preparation of this manuscript.

My former colleagues at the Hurricane Research Division (HRD) encouraged and assisted me. Howie Friedman helped with transportation. Gary Barnes, on a 2-month visit to HRD from the University of Hawaii, suggested revisions to the first draft. Tony Hollingsworth of ECMWF and Chris Thorncroft of the State University of New York at Albany sent me copies of their papers. Neal Dorst of HRD helped with figure preparation. Use of the library at NOAA's Atlantic Oceanographic and Meteorological Laboratory (AOML), of which HRD is a division, allowed me to read many of the papers cited in the references. I appreciate the helpfulness of AOML librarians Linda Pikula and Maria Bello. Chris Labbe of AOML provided after-hours computer assistance at my residence. Chris Landsea of HRD answered several questions during the paper preparation. Sandy Taylor of HRD provided general assistance.

REFERENCES

Albignat, J. P., and R. J. Reed, 1980: The origin of African wave disturbances during phase III of GATE. *Mon. Wea. Rev.,* **108,** 1827–1839.

Arnold, J. E., 1966: Easterly wave activity over Africa and in the Atlantic with a note on the intertropical convergence zone during early July, 1961. Satellite Meteorology Research Paper 65, Dept. of Geophysical Sciences, University of Chicago, 23 pp. [Available from National Technical Information Service, 5284 Port Royal Rd., Springfield, VA 22601.]

Aspliden, C. I., Y. Tourre, and J. B. Sabine, 1976: Some climatological aspects of West African squall lines during GATE. *Mon. Wea. Rev.,* **104,** 1029–1035.

Atkinson, G. D., and J. C. Sadler, 1970: Mean cloudiness and gradient-level wind charts over the tropics. Vol. II. Tech. Rep. 215, Air Weather Service, U.S. Air Force, 50 pp. [Available from National Technical Information Service, 5284 Port Royal Rd., Springfield, VA 22601.]

Avila, L. A., R. J. Pasch, and J.-G. Jing, 2000: Atlantic tropical systems of 1996 and 1997: Years of contrasts. *Mon. Wea. Rev.,* **128,** 3695–3706.

Burpee, R. W., 1971: The origin and structure of easterly waves in the lower troposphere of North Africa. Ph.D. thesis, Massachusetts Institute of Technology, 100 pp.

——, 1972: The origin and structure of easterly waves in the lower troposphere of North Africa. *J. Atmos. Sci.,* **29,** 77–90.

——, 1974: Characteristics of North African easterly waves during the summers of 1968 and 1969. *J. Atmos. Sci.,* **31,** 1556–1570.

——, 1975: Some features of synoptic-scale waves based on compositing analysis of GATE data. *Mon. Wea. Rev.,* **103,** 921–925.

——, and G. Dugdale, 1975: GATE forecasting. *Report on the Field Phase of the GATE Scientific Programme*, GARP Publication Series, GATE Rep. 16, World Meteorological Organization, 12.1–12.17.

——, and R. J. Reed, 1982: Synoptic-scale motions. *The GARP Atlantic Tropical Experiment (GATE) Monograph*, GARP Publication Series, No. 25, World Meteorological Organization, 61–120.

Carlson, T. N., 1969a: Some remarks on African disturbances and their progress over the tropical Atlantic. *Mon. Wea. Rev.,* **97,** 716–726.

——, 1969b: Synoptic histories of three African disturbances that developed into Atlantic hurricanes. *Mon. Wea. Rev.,* **97,** 256–276.

——, and J. M. Prospero, 1972: The large-scale movement of Saharan air outbreaks over the northern equatorial Atlantic. *J. Appl. Meteor.,* **11,** 283–297.

Charney, J. G., and M. E. Stern, 1962: On the stability of internal baroclinic jets in a rotating atmosphere. *J. Atmos. Sci.,* **19,** 159–172.

Chen, Y.-L., and Y. Ogura, 1982: Modulation of convective activity by large-scale flow patterns observed in GATE. *J. Atmos. Sci.,* **39,** 1260–1279.

Erickson, C. O., 1963: An incipient hurricane near the West African coast. *Mon. Wea. Rev.,* **91,** 61–68.

Folland, C. K., T. N. Palmer, and D. E. Parker, 1986: Sahel rainfall and worldwide sea temperatures 1901–1985: Observational, modeling, and simulation studies. *Nature,* **320,** 602–607.

Frank, N. L., 1969: The "inverted V" cloud pattern—An easterly wave? *Mon. Wea. Rev.,* **97,** 307–314.

——, 1970: Atlantic tropical systems of 1969. *Mon. Wea. Rev.,* **98,** 307–314.

Gray, W. M., 1984: Atlantic seasonal frequency. Part I: El Niño and 30 mb QBO influences. *Mon. Wea. Rev.,* **112,** 1649–1668.

——, C. Landsea, P. W. Mielke Jr., and K. J. Berry, 1993: Predicting

Atlantic basin seasonal Atlantic tropical cyclone activity by 1 August. *Wea. Forecasting,* **8,** 73–86.

Hubert, H., 1939: Origine Africaine d'un cyclone tropical atlantique. *Ann. Phys. Globe France d'Outre-Mer,* **6,** 97–115. (Summarized by C. F. Brooks, 1940: Hubert on the African origin of the hurricane of 1938. *Trans. Amer. Geophys. Union,* **21,** 251–253.)

Kidson, J. W., 1977: African rainfall and its relation to upper air circulation. *Quart. J. Roy. Meteor. Soc.,* **103,** 441–456.

Kuettner, J. P., 1974: General description and central program of GATE. *Bull. Amer. Meteor. Soc.,* **55,** 712–719.

Kuo, H. L., 1949: Dynamic instability of two-dimensional nondivergent flow in a barotropic atmosphere. *J. Meteor.,* **6,** 105–122.

Lamb, P. J., 1978: Large-scale tropical Atlantic surface circulation patterns associated with subsaharan weather anomalies. *Tellus,* **30A,** 240–251.

Landsea, C. W., 1993: A climatology of intense (or major) Atlantic hurricanes. *Mon. Wea. Rev.,* **121,** 1703–1713.

Lawrence, M. B., and G. B. Clark, 1985: Atlantic hurricane season of 1984. *Mon. Wea. Rev.,* **113,** 1228–1237.

Lorenz, E. N., 1955: Available potential energy and the maintenance of the general circulation. *Tellus,* **7,** 157–167.

Mass, C., 1978: A numerical and observational study of African wave disturbances. Ph.D. thesis, University of Washington, 277 pp.

McGarry, M. M., and R. J. Reed, 1978: Diurnal variations in convective activity and precipitation in phases II and III of GATE. *Mon. Wea. Rev.,* **106,** 103–113.

Mouzna, N., 1984: Les characteristiques de propagation des ondes tropicales en Afrique de l'ouest observees pendant l'experience WAMEX. Ph.D. thesis, University of Clermont, 52 pp.

Norquist, D. C., E. E. Recker, and R. J. Reed, 1977: The energetics of African wave disturbances as observed during phase III of GATE. *Mon. Wea. Rev.,* **105,** 334–342.

Payne, S. W., and M. M. McGarry, 1977: The relationship of satellite infrared convective activity to easterly waves over West Africa and the adjacent ocean during phase III of GATE. *Mon. Wea. Rev.,* **105,** 414–420.

Piersig, W., 1936: Schwankungen von Luftdruck und Luftbewegung sowie ein Beitrag zum Wettergeschehen in Passatgebeit des ostlichen Nord-atlantischen Ozeans. *Arch. Deut. Seewarte,* **54** (6), 1–41. (Translation of Parts II and III, 1944: The cyclonic disturbances of the subtropical eastern North Atlantic. *Bull. Amer. Meteor. Soc.,* **25,** 2–17.)

Pytharoulis, I., and C. Thorncroft, 1999: The low-level structure of African easterly waves in 1995. *Mon. Wea. Rev.,* **127,** 2266–2280.

Reed, R. J., 1978: The structure and behavior of easterly waves over West Africa and the Atlantic. *Proc. Conf. on Meteorology over the Tropical Oceans*, Bracknell, United Kingdom, Royal Meteorological Society, 57–71.

——, 1980: Energetics and heat and moisture budgets of easterly waves. *Proc. Seminar on the Impact of GATE on Large-Scale Numerical Modeling of the Atmosphere and Ocean*, Woods Hole, MA, National Academy of Sciences, 31–38.

——, 1988: On understanding the meteorological causes of Sahelian drought. *Study Week on Persistent Meteo-Oceanographic Anomalies and Teleconnections September 23–27, 1986,* C. Chagas and G. Puppi, Eds., Pontificate Academiae Scientiarvm Scripta Varia, Vol. 69, 179–213.

——, and E. E. Recker, 1971: Structure and properties of synoptic-scale waves in the equatorial Pacific. *J. Atmos. Sci.,* **28,** 1117–1133.

——, and R. M. Lewis, 1980: Response of upper ocean temperatures to diurnal and synoptic-scale variations of meteorological parameters in the GATE B-scale area. *Deep-Sea Res.,* **26A** (suppl. 1), 99–114.

——, D. C. Norquist, and E. E. Recker, 1977: The structure and properties of African wave disturbances as observed during phase III of GATE. *Mon. Wea. Rev.,* **105,** 317–333.

——, A. Hollingsworth, W. A. Heckley, and F. Delsol, 1988a: An evaluation of the performance of the ECMWF operational system in analyzing and forecasting easterly wave disturbances over Africa and the tropical Atlantic. *Mon. Wea. Rev.,* **116,** 824–865.

——, E. Klinker, and A. Hollingsworth, 1988b: The structure and characteristics of African easterly wave disturbances as determined from the ECMWF operational analysis/forecasting system. *Meteor. Atmos. Phys.,* **38,** 22–33.

Regula, H., 1936: Druckschwankungen und Tornadoes an der Westkuste von Africa. *Ann. Hydrogr. Maritimen Meteor.,* **64,** 107–111. (Translation, 1943: Pressure changes and "tornadoes" (squalls) on the west coast of Africa. *Bull. Amer. Meteor. Soc.,* **24,** 311–317.)

Sadler, J. C., and L. K. Oda, 1978: The synoptic (A) scale circulations during the third phase of GATE, 20 August–23 September 1974. Dept. of Meteorology, University of Hawaii at Manoa, 41 pp.

Schove, D. J., 1946: A further contribution to the meteorology of Nigeria. *Quart. J. Roy. Meteor. Soc.,* **72,** 95–110.

Thompson, R. M., S. W. Payne, E. E. Recker, and R. J. Reed, 1979: Structure and properties of synoptic-scale wave disturbances in the intertropical convergence zone of the eastern Atlantic. *J. Atmos. Sci.,* **36,** 53–72.

Chapter 8

The Relevance of Numerical Weather Prediction for Forecasting Natural Hazards and for Monitoring the Global Environment

A. Hollingsworth, P. Viterbo, and A. J. Simmons

European Centre for Medium-Range Weather Forecasts, Reading, United Kingdom

"From the foregoing it may seem that the role of the forecaster has been greatly diminished by the development of numerical prediction and it is only a matter of time before he suffers the fate of the dinosaur. But this would be an unwarranted conclusion."— (Reed 1977)

1. Introduction

This volume is a tribute to the contributions of Prof. Richard Reed to the development of meteorological science, and to the development of the international collaborations that underpin that science.

a. Reverberations of Prof. Reed's 1985–86 visit to ECMWF

On this auspicious occasion, one is permitted to begin with a personal reminiscence. Professor Reed's extensive work on easterly waves in GATE and later is documented extensively in this volume (Perry 2003; Burpee 2003). A. Hollingsworth's first substantial task on the planning staff of the European Centre for Medium-Range Weather Forecasts (ECMWF) was a six-week mission in the summer of 1975 to the National Oceanic and Atmospheric Administration's (NOAA) Geophysical Fluid Dynamics Laboratory (GFDL) at Princeton University. His tasks were to familiarize himself with the GFDL model and to bring the model back to ECMWF. This generous assistance of Prof. J. Smagorinsky and Prof. K. Miyakoda was a tremendous help in developing and validating the first ECMWF model (Hollingsworth et al. 1980). Professor Miyakoda and his team were then deeply involved in global data assimilation of the Global Atmospheric Research Program (GARP) Atlantic Tropical Experiment (GATE) period (Miyakoda et al. 1976). Their work on the GATE analyses was Hollingsworth's first exposure to tropical meteorology, and imbued him with an enduring interest in the challenges of tropical data assimilation and tropical numerical weather prediction. In the early 1980s both GFDL and ECMWF took up the challenge of doing a global assimilation of the year-long First GARP Global Experiment (FGGE, later called the Global Weather Experiment or GWE) observational dataset. Reed's studies of easterly waves during GATE, published in the late 1970s and discussed elsewhere in this volume, provided invaluable synoptic and dynamical guidance in the Tropics for both assimilation teams. In 1985, ECMWF had the wonderful opportunity to host a year-long visit by Reed. The main purpose of his visit was to get an evaluation by the world's leading expert of the performance of the ECMWF assimilation system and the forecast model in analyzing and forecasting African easterly waves and Atlantic hurricanes.

As relief from his main task, Prof. Reed occasionally took time out to study the analyses and forecasts of explosive cyclogenesis and of polar lows. In particular A. Simmons continued to work with Reed on explosive developments, after his return to the University of Washington (Reed et al. 1988a; Reed and Simmons 1991).

Much of the scientific work done during that 1985 visit is discussed elsewhere in this volume (Burpee 2003). Here we record some later reverberations of the visit. Exposure to ECMWF's modern data analysis and forecast system was an acknowledged stimulus to Reed's later work. He foresaw how such systems could facilitate scientific insight into the dynamics and thermodynamics of weather systems. His long and fruitful collaboration with National Center for Atmospheric Research (NCAR) colleagues, particularly Dr. H.-Y. Kuo, was a direct outcome of that visit in 1985–86.

Prof. Reed's finding that low-level winds in the outer environment of a tropical cyclone are important in improving cyclone track forecasts was a valuable scientific outcome of his visit, a finding that had important repercussions. The production of low-level cloud-track winds from geostationary imagery poses difficult technical challenges, and Reed showed that the space agencies were doing a poor job in this area. ECMWF used Reed's results to motivate space agencies to invest in improving the quality of geostationary low-level cloud track winds near tropical cyclones. The same insight later persuaded space agencies to grant operational forecast centers early access to European Remote Sensing (ERS) satellite, National Aeronautics and Space Administration (NASA) Scatterometer (NSCAT), and NASA's Quick Scatterometer (QuikSCAT) data for the purposes of tropical cyclone location. In return, the operational centers supported the scatterometer missions with a wealth of results on calibration, validation, and long-term monitoring of the stability of the instruments.

An important recommendation stemmed from the visit:

> *"If higher yields of (low-level and mid-level) cloud-track winds cannot be achieved, it would be desirable to develop objective pattern recognition schemes for detecting and characterizing wave signatures. If the latter could be related to the fields of the basic variables—admittedly a task of enormous difficulty—greatly improved oceanic analyses would no doubt be achieved. Hurricane Gloria is an example of a system that could be located with almost pinpoint accuracy from the cloud pictures but for which few (low-level and mid-level) cloud-track winds were available, causing the disturbance to be badly misplaced at one stage in its history."* (Reed et al. 1988a, p. 866).

Since it would be several years before the space agencies could improve the low-level geostationary winds, and since the launch date of *ERS-1* was uncertain, work was undertaken to develop and evaluate a typhoon-bogus procedure as a stopgap (Andersson and Hollingsworth 1988). The procedure worked well for short-range forecasts, but caused problems for the medium-range forecasts if two or three tropical cyclones were interacting with each other in an ocean basin. For that reason, the typhoon-bogus procedure was not adopted for operational work at ECMWF.

Soon after completion, a presentation of ECMWF's work on the typhoon bogus to a Monterey audience stimulated interest at the Fleet Numerical Meteorology and Oceanography Center (FNMOC).

> "R. Hodur coded up the ECMWF algorithm to produce synthetic observations for the navy's Advanced Tropical Cyclone Model (ATCM). J. Goerss adapted Hodur's code for use in the Navy Operational Global Atmospheric Prediction System (NOGAPS) and operationally installed the synthetic observations ("the bogus") at FNMOC in June 1990, (Goerss et al. 1991). The tropical cyclone (TC) forecast skill of NOGAPS with the synthetic observations was superior to the other TC forecast aids in use at the Joint Typhoon Warning Center, (Goerss and Jeffries 1994). Later upgrades to the synthetic observation scheme are described in Goerss et al. (1998). For the past 12 years the U.S. Navy's NOGAPS system has run and continues to run with the bogus."(J. Goerss 2002, personal communication).

Other approaches to typhoon bogusing are discussed in a recent review of tropical cyclone forecasting (Chan 2000).

A further reverberation of Prof. Reed's visit was to stimulate interest at the Japan Meteorological Agency (JMA) in the capabilities of global models for typhoon track forecasting. During the 1980s JMA had invested much of its development effort for tropical cyclone forecasting in regional modeling. Prompted by Reed's papers, JMA offered to the World Climate Research Programme's (WCRP) Working Group on Numerical Experimentation (WGNE) to make detailed intercomparisons of TC track forecasts from a number of global forecast systems (including ECMWF's) in the western North Pacific. The intercomparisons showed that global models performed very well, perhaps because of their skill in forecasting the large-scale steering flow. In the early 1990s JMA increased their efforts in global modeling, leading to a dramatic improvement in the TC track forecasting skill of the JMA global model in the mid-1990s (Tsuyuki et al. 2002).

The Tsuyuki et al. (2002) summary of the WGNE intercomparisons of track forecast verifications in the northwest Pacific for 1991–2000 may be viewed in the context of similar verifications of the forecasts of the Joint Typhoon Warning Center (JWTC) in the period 1970–90, reported by Guard et al. (1992).

The tabulated comparison of the two reports in Table 8.1 shows a dramatic improvement in forecast skill for tropical cyclone tracks over a decade, coming probably from better data (especially satellite data), better models, and better assimilation systems. Both JMA and the U.K. Met Office use a typhoon bogus, while ECMWF does not. Persuasive clear evidence is available that a typhoon bogus improves short-range forecasts (0–24 h), but there is little persuasive evidence that a typhoon bogus consistently improves the forecasts at 48 h, 72 h, and at longer ranges.

The long-lived beneficial reverberations of Prof. Reed's activity in the course of one year, and in just

TABLE 8.1. Five-year mean tropical cyclone track errors (km) for JTWC (1986–90) and ECMWF (1996–2000) estimated, respectively, from Guard et al. (1992) and Tsuyuki et al. (2002). The ECMWF results for 1996–2000 are representative for the U.K. Met Office and JMA.

Forecast lead time (h)	1986–90	1996–2000	Error reduction (%)
24	200	170	15
48	410	230	44
72	620	380	39

one corner of his domain of interests, is typical of his lifelong contributions.

b. Forecasting natural hazards and monitoring the global environment

We now turn to the substance of this presentation, which has two objectives. The first objective is to summarize some of the progress made in numerical weather prediction since Reed's visit to ECMWF in 1985. The second objective is to indicate the potential of current data assimilation and weather forecast systems to address new issues of great practical importance to society.

Many natural hazards arise from weather events such as tropical cyclones or intense midlatitude storms, with associated floods, landslides, wind damage, heavy seas, and coastal damage. Because of the growing vulnerability of densely populated areas to natural hazards, society makes increased demands for reliable forecasts of natural hazards, and reliable forecasts of the consequences for the safety of life and property. As discussed in section 2, recent years have seen remarkable improvement in the quality and scope of numerical weather predictions. The accuracy of global numerical weather predictions of mean sea level pressure and 500-hPa height has improved by 1 day over the last decade in the Northern Hemisphere, and by 1 day over the last 3 yr in the Southern Hemisphere. Increasingly, ensemble weather predictions are used as the prime medium-range forecasting tool in many countries. The skill of deterministic forecasts and ensemble forecasts have benefited in largely equal measure from developments in data availability, data assimilation methods, and model physics, numerics, and resolution. New systems for forecasting severe weather are under development, based on the ensemble forecasts and on characterization of the climatology of extremes of the model used for the ensemble forecasts. Numerical weather prediction (NWP) will benefit further from forthcoming advanced satellite-sounding capabilities, effective assimilation of information on moist processes, more efficient numerics, and more accurate physical parameterizations.

Earth system scientists use a spectrum or hierarchy of earth system models to synthesize, or assimilate, the wide range of satellite and in situ observations of the earth system. The same models are used to make forecasts on a range of timescales: medium-range forecasts, seasonal-to-interannual forecasts (i.e., short-term climate forecasts), and decadal timescale forecasts. All of these forecasts can be used by the civil protection authorities to make advance preparations for coping with and mitigating the consequences of natural hazards. At one end of the spectrum of models are global general circulation models (GCMs) with coupled atmosphere–ocean models and with more or less simplified models for the land surface biosphere and hydrology, for ice, and for atmospheric chemistry. The GCMs are used for global data assimilation and to provide the global forecasts and simulations. At the other end of the spectrum of earth system models is a variety of specialized models such as chemical transport models, crop-yield models, forest/biosphere models, coastal zone models, regional atmospheric models, and hydrological models. Increasingly, the output of the GCMs is used to drive the specialized models so as to provide the best possible interpretation of the GCM results in terms of quantities that are of direct interest to end-users.

Efforts are under way in several countries to provide improved environmental forecasts of natural hazards, by coupling the specialized environmental models to weather prediction models. In such a context it is natural to explore the value of the ensemble weather forecasts in environmental forecasting. Transnational floods have caused much loss of life and property in Europe in recent years. The European Flood Forecast System (EFFS) project is a European Union (EU) funded project to assess the value of using deterministic and ensemble precipitation forecasts from both global and regional meteorological models to drive distributed hydrological models for flood-forecasting purposes for lead times of 3–10 days. The EFFS project will investigate if the existing level of meteorological forecast skill for precipitation can be translated into useful hydrological forecasts, whose skill should be easier to measure. As an introduction to such a study, we examine in section 3 a set of medium-range deterministic forecasts and ensemble forecasts for the precipitation in the Po valley on 14–15 October 2000. From the meteorological viewpoint the precipitation forecasts look encouraging. Ongoing work will assess the value of the precipitation forecasts for flood forecasting.

Similar approaches to environmental forecasting are being assessed in the seasonal forecast arena. Short-term climate fluctuations such as the El Niño–Southern Oscillation (ENSO) are frequently implicated in extended periods of tropical and subtropical drought or flood, with devastating consequences for human safety and for essential economic activities including agriculture and food and fiber production. Given the need to assess the uncertainties in all such forecasts, increasing use is made of ensembles of GCM forecasts that sample uncertainties arising from initial conditions and from the range of possible choices in GCM formulation. Such GCM ensembles can then be used to drive ensembles of specialized models, thus providing estimates of the uncertainties in the forecasts for the end-users. Ongoing studies will assess the value of ensemble seasonal forecasts for forecasting crop production and disease incidence.

Future developments in NWP are discussed in section 4. Forecast skill will benefit from advanced satellite sounding capabilities, effective assimilation of information on moist processes, more efficient numerics, and more accurate physical parameterizations. Section 5 discusses how the meteorological data assimilation systems can be extended to use a wide variety of satellite data to establish an effective global environmental monitoring of atmosphere ocean and land, while our conclusions are presented in section 6.

2. Recent improvements in operational forecasts

Figure 8.1 [from Simmons and Hollingsworth (2002), hereafter referred to as SH] presents root-mean-square

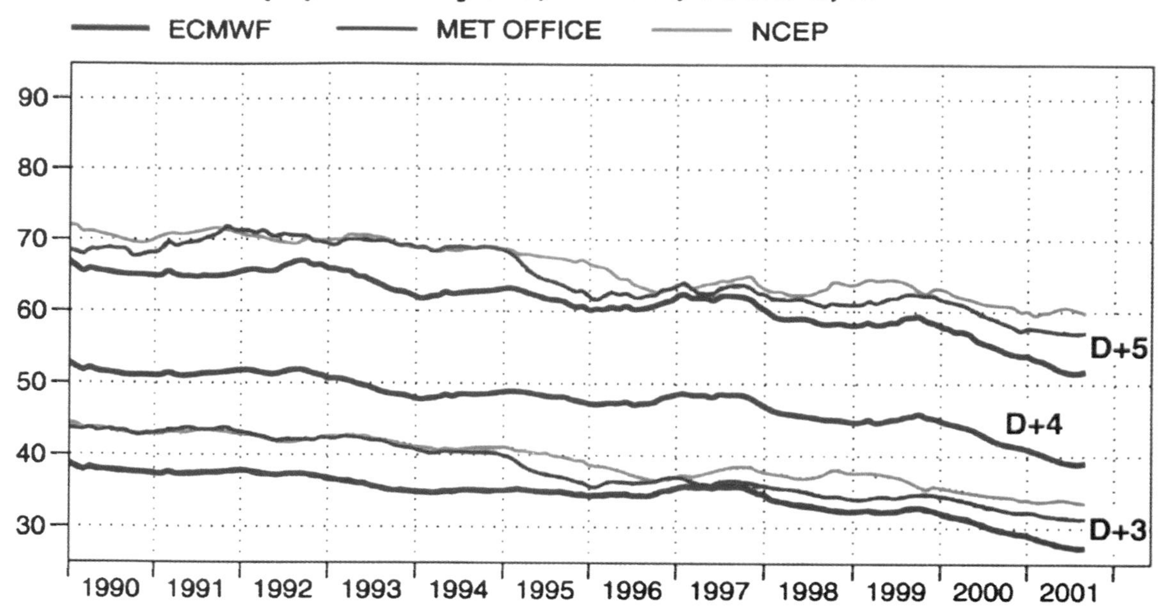

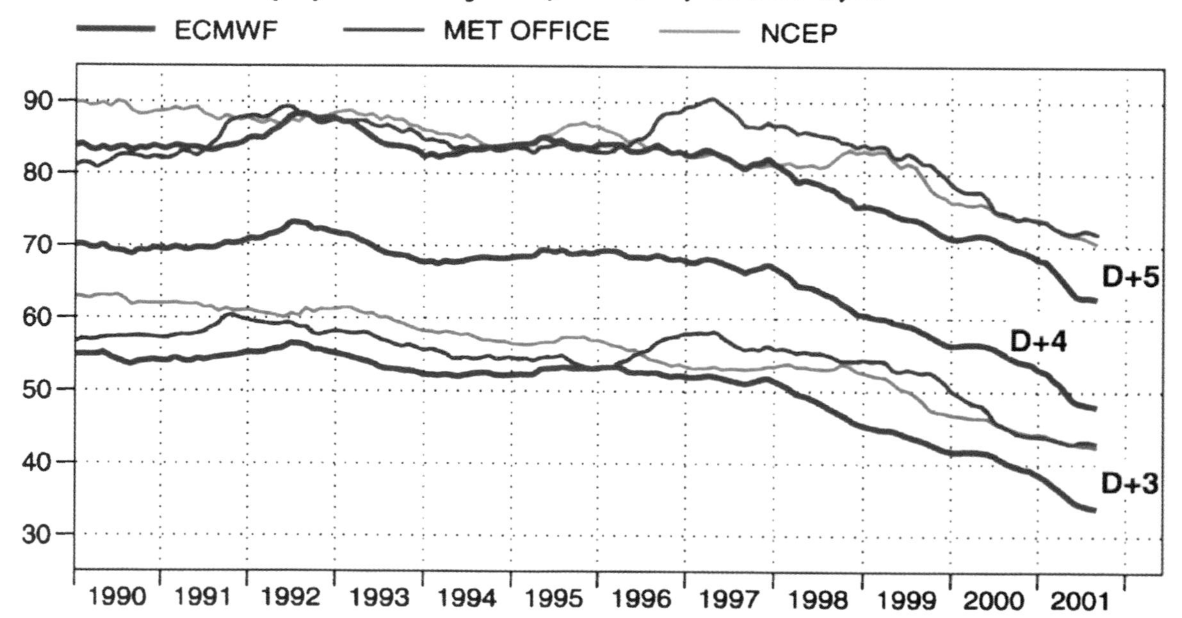

FIG. 8.1. Rms errors of 3- and 5-day forecasts of 500-hPa height for the extratropical (top) Northern and (bottom) Southern Hemispheres. Results from ECMWF, Met Office, and NCEP are plotted in the form of annual running means of all monthly data exchanged by the centers from Jan 1989 to Aug 2001. ECMWF 4-day forecast errors are also shown. Values plotted for a particular month are averages over that month and the 11 preceding months, so that the effect of a forecasting system change introduced in that month is seen from then onward.

(rms) errors of forecasts of 500 hPa for the extratropical Northern and Southern Hemispheres. Time series from 1990 onward are shown for 3- and 5-day forecasts from three global prediction systems, that of ECMWF and those of the Met Office and the National Centers for Environmental Prediction (NCEP), the two national centers closest to ECMWF's performance on these measures of forecast accuracy. ECMWF results are also presented for the 4-day range. The plots show annual running means derived from the verification statistics that forecasting centers exchange monthly under the auspices of the World Meteorological Organization. Each center's forecasts are verified by comparison with its own analyses. Results are presented for initial forecast times of 1200 UTC for ECMWF and the Met Office, and 0000 UTC for NCEP. The ECMWF forecasts are routinely produced with a cutoff time for data reception that is several hours later than is used by the other centers. Evidence from ECMWF forecasts produced with earlier cutoff times indicates that differences in the forecasting systems rather than in data reception are the primary cause of the differences in forecast accuracy illustrated here.

Figure 8.1 shows a general trend toward lower 500-mb forecast errors in both hemispheres; SH show similar results for mean sea level pressure. The improvement between 1990 and 2001 in ECMWF forecasts for the Northern Hemisphere amounts to around a 1-day extension of the forecast range at which a given level of error is reached. In other words, today's 4- and 5-day forecasts are, respectively, about as accurate on average as the 3- and 4-day forecasts of 10 years ago. The rate of improvement has recently been especially rapid in forecasts for the Southern Hemisphere, amounting to a 1-day gain in predictability in just 3 yr.

The starting point for the rapid recent improvement in ECMWF forecasts shown in Fig. 8.1 was the operational introduction of four-dimensional variational (4DVAR) data assimilation (Mahfouf and Rabier 2000, and references therein) in late November 1997. Subsequent data assimilation changes include improved utilization of surface (Järvinen et al. 1999) and radiosonde data, assimilation of raw microwave radiances from the Television and Infrared Observation Satellite (TIROS) Operational Vertical Sounder (TOVS) and the new Advanced TOVS (ATOVS) satellite-borne instruments (A. P. McNally 1999, personal communication), assimilation of retrievals of humidity (Gérard and Saunders 1999) and surface wind speed from the Special Sensor Microwave Imager (SSM/I) satellite-borne instrument, and general refinements and extensions of the 4DVAR analysis and use of raw radiances. The atmospheric forecast model has been coupled with an ocean wave model (Janssen et al. 2002) and improved in a number of other ways, including increased vertical resolution in the stratosphere (A. Untch and A. J. Simmons 1999, personal communication) and planetary boundary layer (Teixeira 1999), revisions to the representations of clouds and convection (Jakob and Klein 2000; Gregory et al. 2000), and new schemes for longwave radiation (Morcrette 2000) and for the land surface and sea ice (van den Hurk et al. 2000). Significant increases in the horizontal resolutions of the model and 4DVAR analysis were introduced in November 2000. Also noteworthy in Fig. 8.1 are the substantial recent improvements in the forecasts for the Southern Hemisphere produced by the Met Office and NCEP. Both of these centers have reported benefits from use of three-dimensional variational analysis and direct assimilation of TOVS and ATOVS radiances (Parrish and Derber 1992; English et al. 2000; Lorenc et al. 2000; McNally et al. 2000).

Recent improvements in short-range ECMWF forecasts can be linked very directly to the forecasting system changes summarized above. As detailed in SH, the ECMWF system undergoes constant development, with several major upgrades each year. Before implementation, the gain expected from each upgrade is assessed in preoperational trials, which may cover over 100 days in several seasons. Figure 8.2 shows the actual annual-mean rms errors of 1-day 500-hPa height forecasts for the past 5 yr, together with the errors that would have occurred had the changes introduced between November 1996 and November 2000 given exactly the same average forecast improvements in operational use as were measured in the preoperational trials. The agreement is remarkable and indicates that the overall recent improvement in short-range forecasts is indeed due overwhelmingly to changes to the forecasting system rather than to circulation regimes that were unusually easy to predict in the last year or two.

The extent of the reduction in 1-day forecast errors shown in Fig. 8.2 is also noteworthy. The error has been reduced by almost a third from 13.4 to 9.1 m over 4 yr for the Northern Hemisphere, and almost halved from 19.7 to 10.7 m over the same period for the Southern Hemisphere. As discussed in SH, the Northern Hemisphere results shown here are consistent with radiosonde verifications. Verification of forecasts by comparison with radiosonde observations provides a more independent validation than verification by comparison with a center's own analysis. The observations are, however, mostly located over land, and where sparsely distributed can give rise to difficulties in interpretation of verification statistics due to variations over time in the number of stations reporting. This inhibits the straightforward comparison of ECMWF and NCEP verifications against radiosondes over the Southern Hemisphere in particular, since the internationally exchanged verification statistics are for different forecast starting times, and the verification is quite sensitive to differences in radiosonde coverage between 0000 and 1200 UTC. For example, the 5-day Southern Hemisphere errors for the year to August 2001 are 72 m for both 0000 and 1200 UTC Met Office forecasts when verified against analyses, but are 53 and 57 m, respectively, for the 0000 and 1200 UTC forecasts when verified against radiosondes. Verification against radiosondes gives generally lower values than against analyses for this hemisphere because the observations are predominantly located away from the main band of variance over the Southern Ocean.

The levels of skill of Northern and Southern Hemisphere forecasts cannot be compared simply in terms of rms errors because of interhemispheric differences in the levels of variance of the fields. Comparison can, however, be made directly in terms of anomaly correlation coefficients, which are closely related to mean square errors normalized by corresponding variances (see, e.g., Simmons et al. 1995). Figure 8.3 presents anomaly correlations of 500-hPa height based on ECMWF's operational 3-, 5-, and 7-day forecasts from January 1980 to August 2001. Running annual means of the monthly mean skill scores archived routinely over the years are plotted for the two hemispheres. Figure 8.3 shows a higher overall rate of improvement in the forecasts for the Southern Hemisphere. In the early

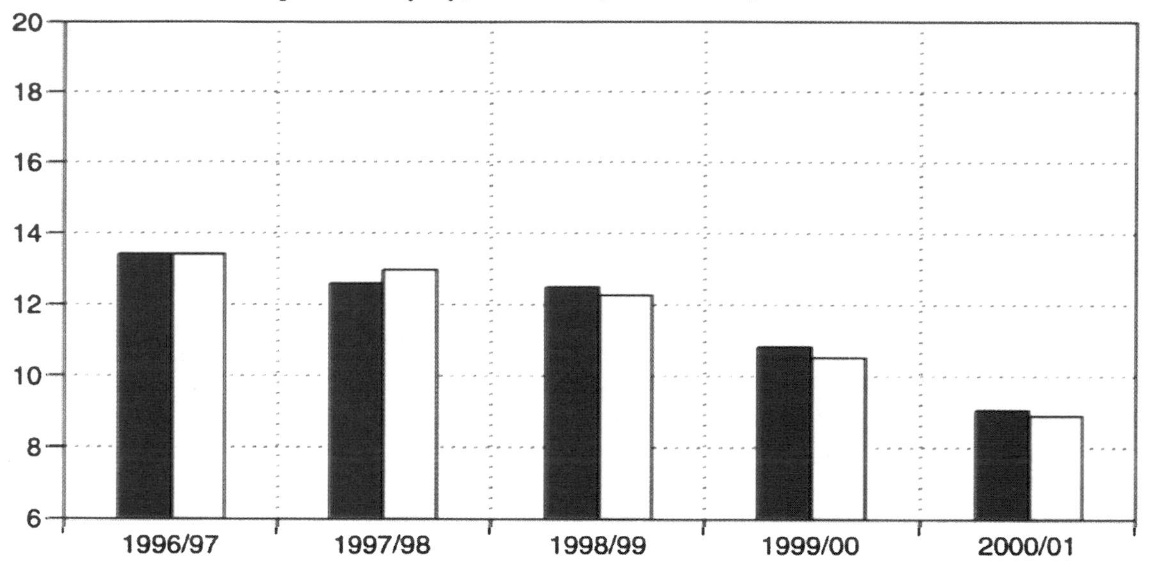

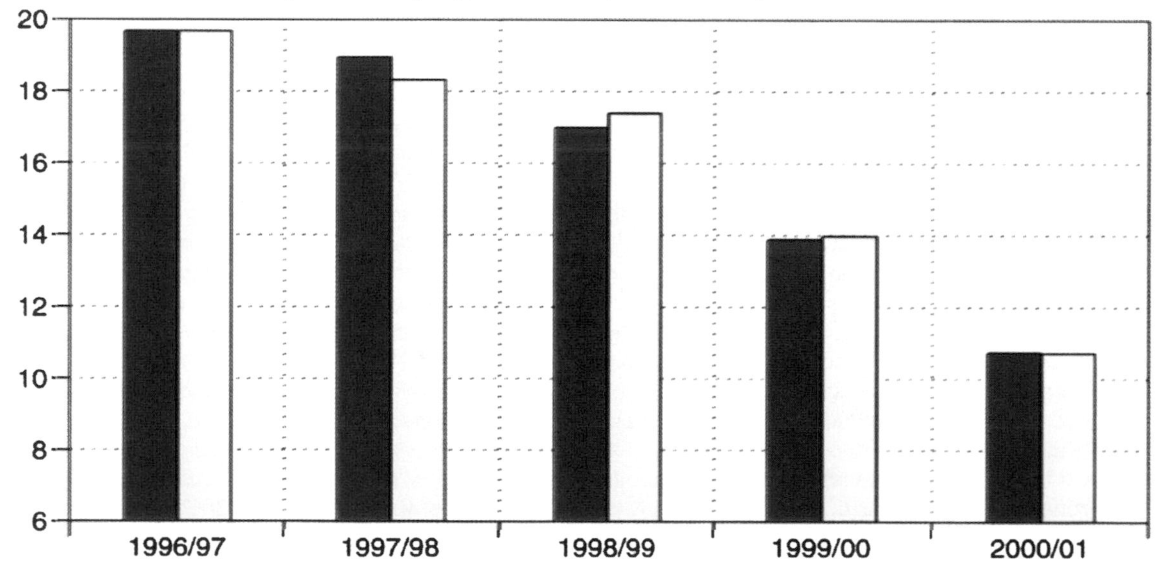

FIG. 8.2. Rms errors of ECMWF's 1-day forecasts of 500-hPa height (m) for the extratropical (left) Northern and (right) Southern Hemispheres. Twelve-month averages from Sep to Aug are plotted for the 5 yr up to Aug 2001. The red bars denote the actual annual-mean errors, and the yellow bars denote the errors that would have occurred had these operational forecasts been improved by exactly the average amounts as measured in the preoperational trials of the forecasting system changes introduced between Nov 1997 and Nov 2000.

1980s, the skill levels of the 3-, and 5-day forecasts for this hemisphere were only a little better than those of the 5-, and 7-day Northern Hemisphere forecasts. At the time this was not surprising in view of the sparsity of conventional ground-based and aircraft observations in the Southern Hemisphere (Bengtsson and Simmons 1983). Today, however, the skill at a particular forecast range in the Southern Hemisphere is only a little lower than that at the same range in the Northern Hemisphere.

There is little doubt that improvements in the availability, accuracy, and assimilation of satellite data have been major factors contributing to the relative improvement in forecast skill. In addition to the changes referred to earlier, information on marine winds has come also from scatterometers on the ERS satellites (Tomassini et al. 1998) while ERS altimeter data are used in the analysis of ocean wave heights (Janssen 1999). There have been evolutionary improvements in the wind estimates derived by tracking features in successive images from geostationary satellites (Tomassini et al. 1999; Rohn et al. 2001). Moreover, the newly assimilated raw ATOVS radiances are used more comprehensively over the oceans (where surface radiative properties are easier to characterize) than over land. All these developments would be expected to improve forecasts more in the Southern than in the Northern Hemisphere, both because satellite data provide a more important component of the observing system in the Southern Hemisphere and because of the greater extent of the oceans in that hemisphere.

Interannual variations in skill are also evident in Fig. 8.3, especially for the Northern Hemisphere at the 5- and 7-day time ranges. In particular, there is a pronounced minimum in the Northern Hemisphere scores arising from relatively poor performance over the year to August 1999. A corresponding maximum can be seen in the time series of rms errors shown in the left panels of Fig. 8.1. This is evident for the Met Office forecasts as well as for those of ECMWF.

3. Ensemble forecasts of high-impact weather

Weather forecasters categorize forecasts as short range (0–3 days ahead), medium range (3–10 days ahead), extended range (10–90 days ahead), and seasonal to interannual. A prime task of a medium-range forecast center is to provide early and quantitative warnings of high-impact weather. As shown above, there is a consensus view among meteorologists that there is skill to about 7 days in deterministic winter forecasts of the Northern Hemisphere pressure and wind fields; skill in forecasting these fields has improved by 1.5 to 2.0 days over the last 20 yr. There can be marked fluctuations in

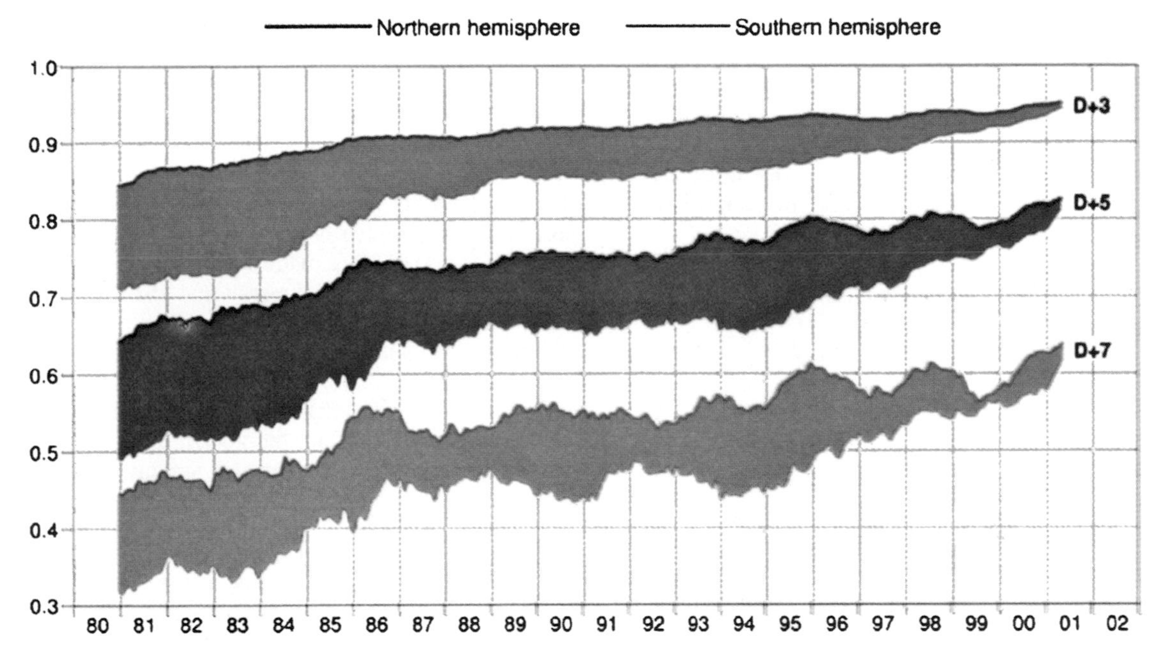

FIG. 8.3. Anomaly correlation coefficients of 3-, 5-, and 7-day ECMWF 500-hPa height forecasts for the extratropical Northern and Southern Hemispheres, plotted in the form of annual running means of archived monthly-mean scores for the period from Jan 1980 to Aug 2001. Values plotted for a particular month are averages over that month and the 11 preceding months. The shading shows the differences in scores between the two hemispheres at the forecast ranges indicated.

forecast skill, depending on the weather situation: some situations are easier to forecast than others.

Because of the many difficulties associated with precipitation verification [sampling difficulties in measuring a field that is finely structured in space and time, orographic effects, instrumental difficulties etc., see Cherubini et al. (2002)] there is less consensus among meteorologists on the range for which precipitation forecasts are likely to be useful. There is agreement that precipitation forecasts are better in winter than in summer because of the more dominant role of synoptic-scale dynamical processes in winter than in summer, and there is agreement that forecasts of area-averaged rainfall accumulations verify better than forecasts of point accumulations (e.g., Lanzinger 1996). However there is little material in the literature on which to formulate a statement of the form: "Based on generally agreed verification standards, midlatitude winter rainfall forecasts have improved by NN days over the last MM years." Here we illustrate an example where midlatitude precipitation forecasts were useful to 8 days or more.

Predictability studies have shown that high-impact weather is often associated with very energetic phenomena, which may have limited predictability, so that forecasts are best couched in probabilistic terms. Traditionally, numerical weather predictions have been issued in a deterministic form, using the highest affordable resolution in the forecast model. Such forecasts could be converted into probabilistic forecasts using the climatology of forecast error statistics. Starting in 1992, weather forecast institutes have begun to issue real-time forecasts based on ensembles forecasts (at lower resolution than the deterministic forecasts) to sample the uncertainties in the initial conditions and in model formulation (Palmer 1996; Molteni et al. 1996; Toth and Kalmay 1997; Buizza and Hollingsworth 2002). There is substantial evidence that in the later medium range (5–10 days ahead), and for many important weather quantities, probabilistic forecasts derived from the ensemble forecasts are more skillful than probabilistic forecasts derived from the deterministic forecasts (Richardson 2000).

a. The October 2000 flood in the Po valley

As noted by Lawrimore et al. (2001):

"Torrential mid-October rainfall led to floods and mudflows and contributed to numerous deaths in the Southern Alps in an area stretching from the Rhone valley in France to the Po valley in northern Italy. The town of Locarno, located in Switzerland near the Italian border, received more than 285 mm of rainfall from 11 to 16 October. This led to the overflow of Lago Maggiore and flooding throughout much of the surrounding area. A mudslide, brought down by heavy rainfall, also swept away much of the tiny town of Gondo, Switzerland. Nine-day precipitation totals were greater than 200 mm along much of the border region between Italy and Switzerland (Fig. 8.4). In the agriculturally rich Po valley heavy rainfall began on 14 October and continued for the next three days. More than 500 mm was recorded near Milan in the 3-day period from 15–17 October. Two-day rainfall totals in Turin, Italy, exceeded 115 mm. This extreme regional rainfall event brought the Po river to historic heights, causing it to break over its banks in some parts of Italy requiring the evacuation of thousands of people. Roads were closed, dozens of bridges were destroyed, and many rail services from Italy to France and Switzerland were suspended."

The heaviest precipitation started around Thursday, 12 October; it produced large accumulations in the 48 h from 1200 UTC on Saturday, 14 October to 1200 UTC on Monday, 16 October; there was little significant accumulation thereafter. Figure 8.5 illustrates the synoptic situation at 1200 UTC on Saturday, 14 October. The top panel shows a deep trough at 500 mb in the Western Mediterranean, with strong southerly flow across the central Mediterranean. The bottom panel shows the winds at 925 mb at the same time, together with the wet-bulb equivalent potential temperature for values above 316 K. The latter plot shows that the strong southerly flow across the central Mediterranean is extremely moist, and is likely to play a key role in contributing to the heavy rain over the mountains to the north. In the upper troposphere there is a deep intrusion of high potential vorticity air into the western Mediterranean, in association with the 500 mb trough. Such marked "PV banners" are regularly associated with heavy rain in the southern Alps and are a prime focus of study of the recent Mesoscale Alpine Experiment (Bougeault et al. 2001a,b).

Our verification data on rainfall accumulation is provided by the daily accumulation in successive ECMWF short-range 24 h forecasts from 1200 UTC on successive days. The forecasts were made with the 40-km (T511, 60 level) model that has been operational at ECMWF since 25 November 2000, and is described in the appendix. The quality of such estimates has been discussed by Cherubini et al. (2002) and by Rubel and Rudolf (2001). Figure 8.6 shows (top left) the sum of the precipitation in the two 24-h forecasts with the 40-km model from 1200 UTC on Saturday, 14 October and Sunday, 15 October. This sum is our proxy "truth" or best estimate of the total accumulated rainfall in southern Europe for the 48-h period from 1200 UTC on Saturday, 14 October. The bottom four panels of Fig. 8.6 show the rain accumulations for the same period in the forecasts from (i) 1200 UTC on Friday, 13 October (center left), (ii) 1200 UTC on Wednesday, 11 October (center right); (iii) 1200 UTC on Monday, 9 October (bottom left); and (iv) 1200 UTC on Saturday, 12 October (bottom right).

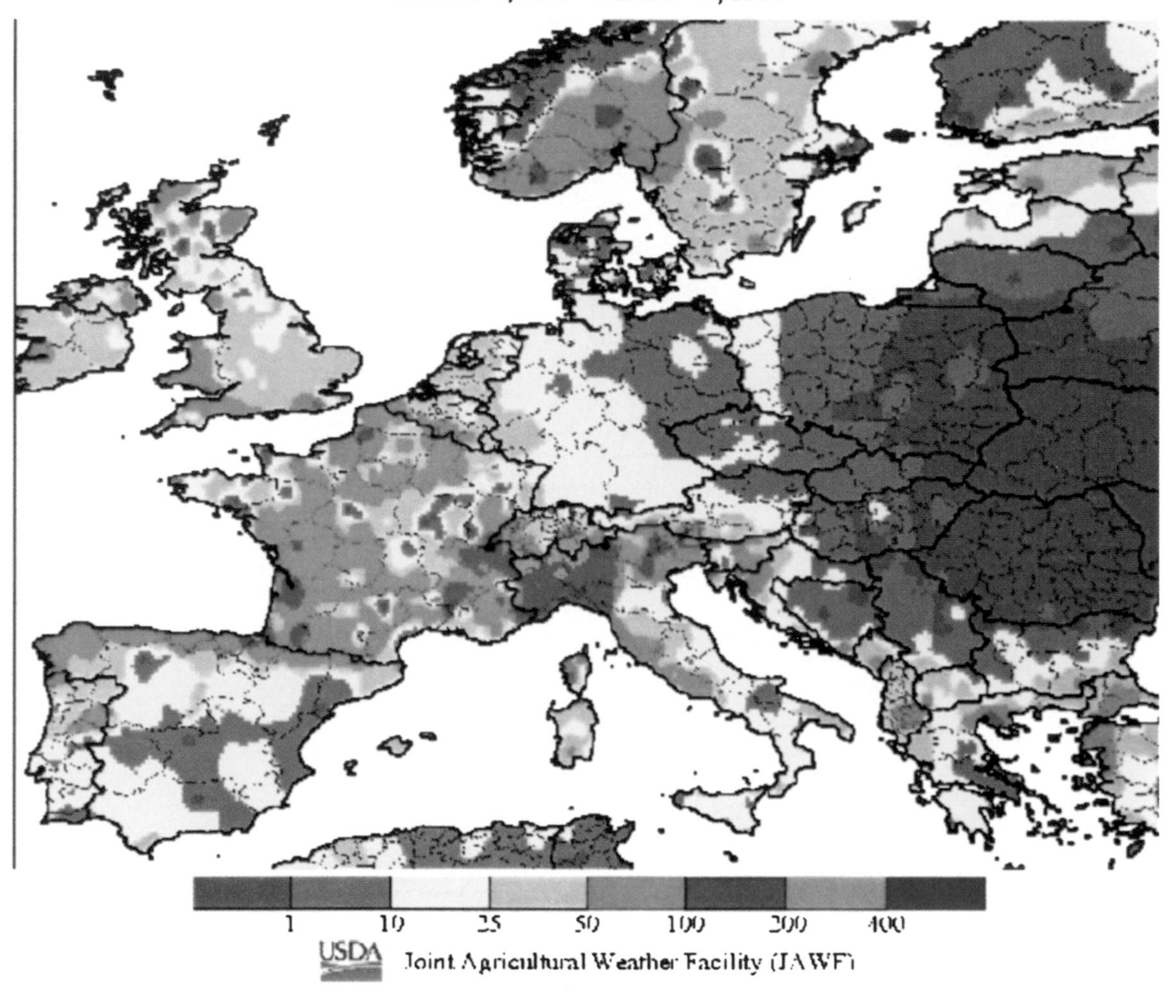

FIG. 8.4. Accumulated precipitation (mm) for the period 8–16 Oct 2000. [Courtesy of the U.S. Department of Agriculture; from Lawrimore et al. (2001).]

These forecasts may be thought of as 2-, 4-, 6- and 8-day forecasts, respectively. Judged synoptically, these deterministic forecasts provide clear warnings of heavy precipitation over the southern Alps during the 48-h period from 1200 UTC on Saturday, 14 October to 1200 UTC on Monday, 16 October.

b. Ensemble forecasts of integrated hyetographs for the Po valley

In establishing a successful interdisciplinary dialogue, it is helpful if one's presentations or displays communicate essential information in an effective manner. Displays based on integrated hyetographs have proven useful (K. Settler 2001, personal communication) in discussing ensemble forecasts with the hydrological community. A hyetograph is a plot of the rain rate at a given point. We define an integrated hyetograph as a plot showing the accumulation of rain (in mm) averaged over a particular area, and starting from zero at a particular time. For our purposes, the integrated hyetograph is set to zero at the beginning of each forecast, and the area average is taken over the Po catchment (including the Adige catchment). The catchment definition (resolution 30" of arc) is from the University of New Hampshire/Global Runoff Data Centre (UNH/GRDC) CD-ROM: Composite runoff fields V 1.0 (Fekete et al. 2000). The catchment area is 102, 183 km^2, corresponding to about 64 grid boxes of the T511 (40-km model), or 16 grid boxes of the T255 (80 km) model. The catchment is outlined in each frame of Fig. 8.5, and is defined following Fekete et al. (2000).

Figure 8.7 shows deterministic and ensemble ECMWF forecasts of the integrated hyetographs for the Po catchment for forecasts starting at 1200 UTC on 7, 9, 11, and 13 October 2000. Also shown in each frame is the proxy truth in green. This verification is produced by assuming

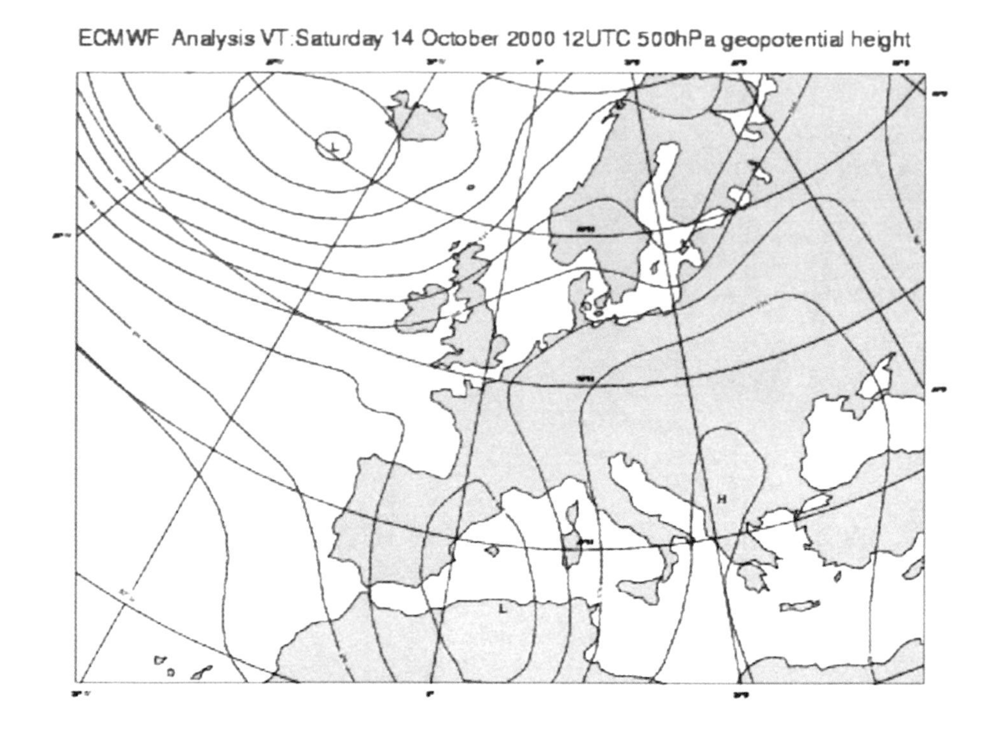

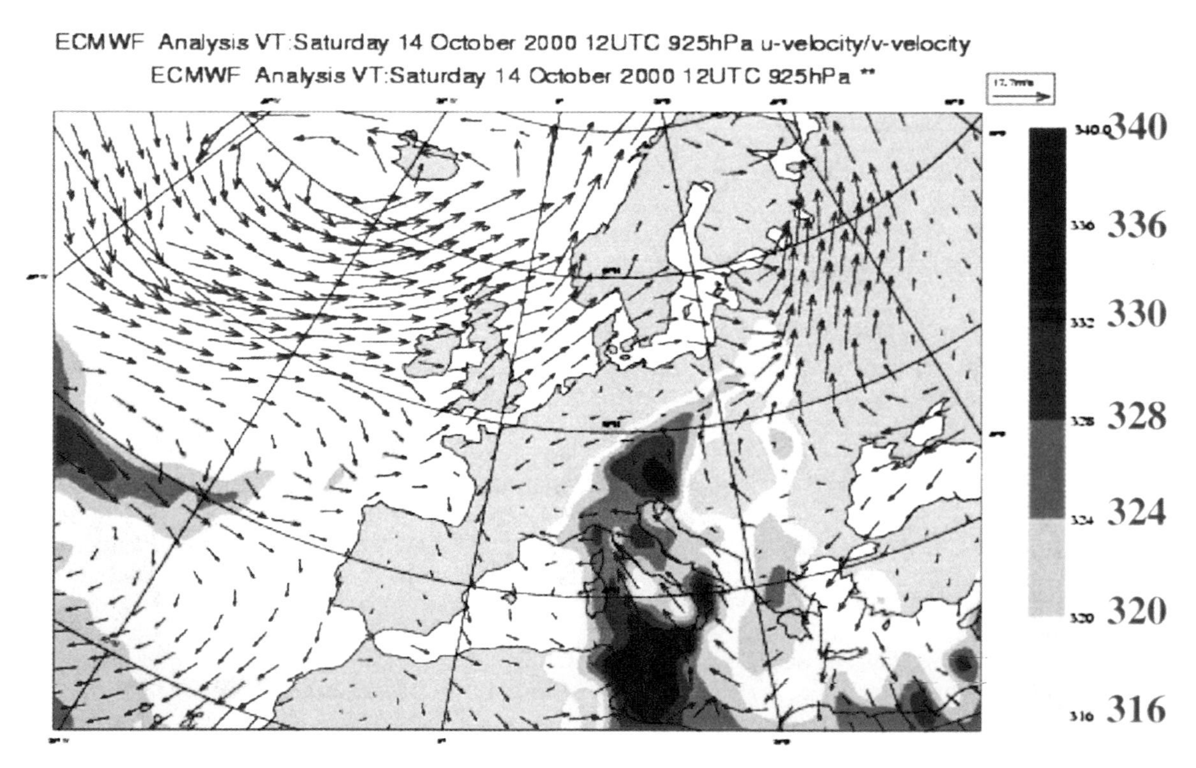

FIG. 8.5. (top) The 500-mb analysis at 1200 UTC on Friday, 14 Oct 2000: contour interval is 4 dam (bottom). Winds and wet-bulb equivalent potential temperature at 925 mb at 1200 UTC on Friday, 14 Oct 2000. The wet-bulb equivalent potential temperature is plotted only for values above 316 K.

20000114-15 + 0-24 H

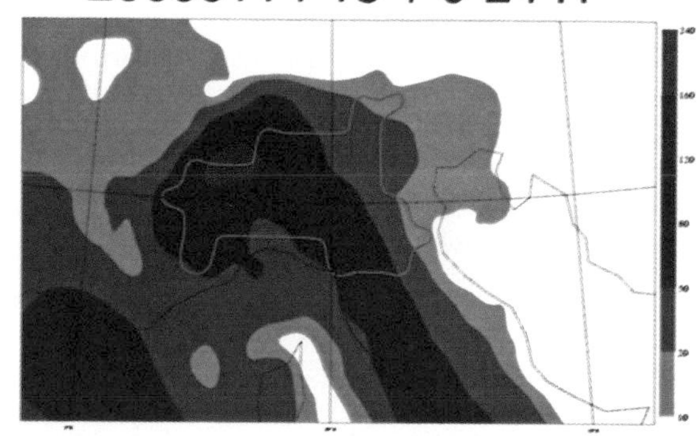

20000113 + 24-72 H

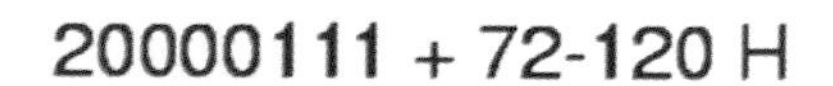

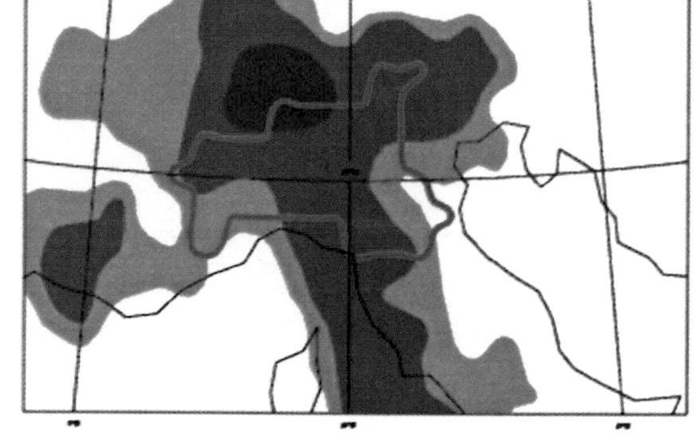

20000109 + 120-168 H

20000107 + 168-216 H

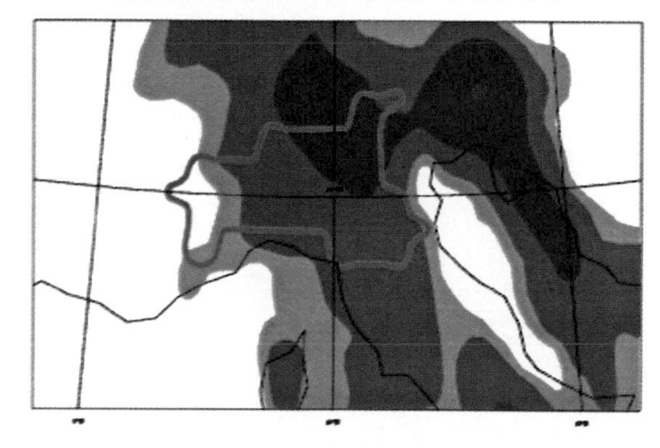

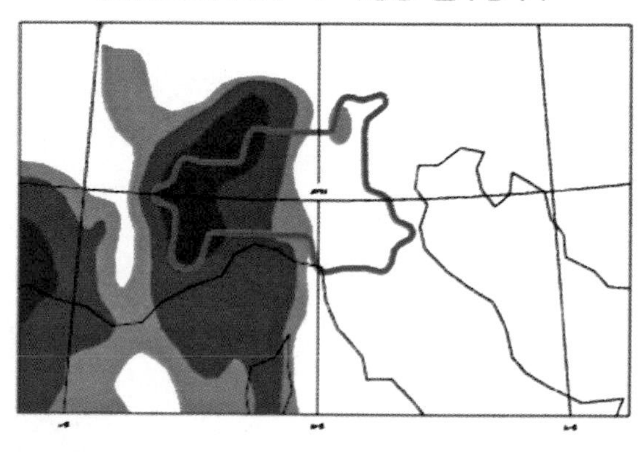

FIG. 8.6. (top) The accumulated precipitation in the two successive 24-h forecasts with the 40-km model from 1200 UTC on Saturday, 14 Oct and Sunday, 15 Oct. This accumulated precipitation is our verification estimate (i.e., proxy "truth") of the total accumulated rainfall in southern Europe for the 48-h period from 1200 UTC on Saturday, 14 Oct. The bottom four panels show the rain accumulations for the same 48-h period in the forecasts from (center left) 1200 UTC on Friday, 13 Oct (center right) 1200 UTC on Wednesday, 11 Oct, (bottom left), 1200 UTC on Monday, 9 Oct and (bottom right) 1200 UTC on Saturday, 12 Oct. These forecasts may be thought of as 2-, 4-, 6-, and 8-day forecasts, respectively.

that the successive T511 day-0 to day-1 forecast accumulations are truth (an assumption whose errors require quantification). The value of all integrated hyetographs is zero at time $t=0$, that is, at the start of each forecast. Hyetographs are shown in 24-h steps for the 40-km resolution deterministic forecast (blue), the 80-km resolution ensemble forecasts, the 80-km resolution unperturbed forecast, and the verification (green). Also shown in the plots are the 25%, 50% (median), and 75% percentiles of the ensemble accumulations on each day of the forecast. In reviewing the forecasts, we focus on three issues: (i) the quality of each deterministic forecast for the total precipitation in the Po basin from the start of the forecast (always at 1200 UTC) until 1200 UTC on Monday, 16 October; (ii) the success of the ensemble forecasts in encompassing the observed accumulation in the Po basin from the start of each forecast until 1200 UTC on Monday, 16 October; and (iii) the success of the prediction of the cessation of the rain forecast after 1200 UTC on Monday, 16 October. All three issues are of direct interest to water managers and to civil protection authorities.

1) FORECASTS FROM 1200 UTC SATURDAY 7 OCTOBER

Figure 8.7 (top left) shows the forecasts and verification from 1200 UTC on Saturday, 7 October. The profile of the observed rainfall shows a steady accumulation of about 60 mm between 1200 UTC on 10 and 14 October (i.e., between D+3 and D+7), and an accumulation of about 75 mm in the 2 days from 1200 UTC on 14 and 16 October (i.e., from D+7 to D+9). The observed 9-day accumulation to 1200 UTC on 16 October is about 135 mm. Both the 40- and 80-km deterministic runs forecast a 9-day accumulation of about 80 mm. The 40-km deterministic forecast did better at forecasting intensity than the 80-km unperturbed forecast from the same initial data. Turning to the ensemble forecast, we note that the ensemble forecast seems to be bimodal, with one cluster of forecasts showing a cessation of rain on Saturday, 14 October and another cluster showing the rain continuing over the weekend. We also note that the observed hyetograph lies well within the spread of the ensemble. For each day out to 14 October (D+7), the ensemble forecast gives a 40–50% chance of exceeding the 40-km forecast accumulation (and a 25% chance of exceeding the observed accumulation). The ensemble forecast also gives a 25% chance of a 9-day accumulation of more than 90 mm. Finally we note that the 40-km deterministic forecast shows a continuation of the rain over the weekend and a cessation of the rain on 16 October, as does one of the clusters of ensemble members mentioned above. Judged by the criteria set out above, we therefore conclude that the forecasts from 7 October were synoptically useful out to 9 or 10 days ahead.

2) FORECASTS FROM 1200 UTC MONDAY 9 OCTOBER

The top-right panel of Fig. 7.7 shows the forecasts and verification from 1200 UTC on Monday, 9 October. The

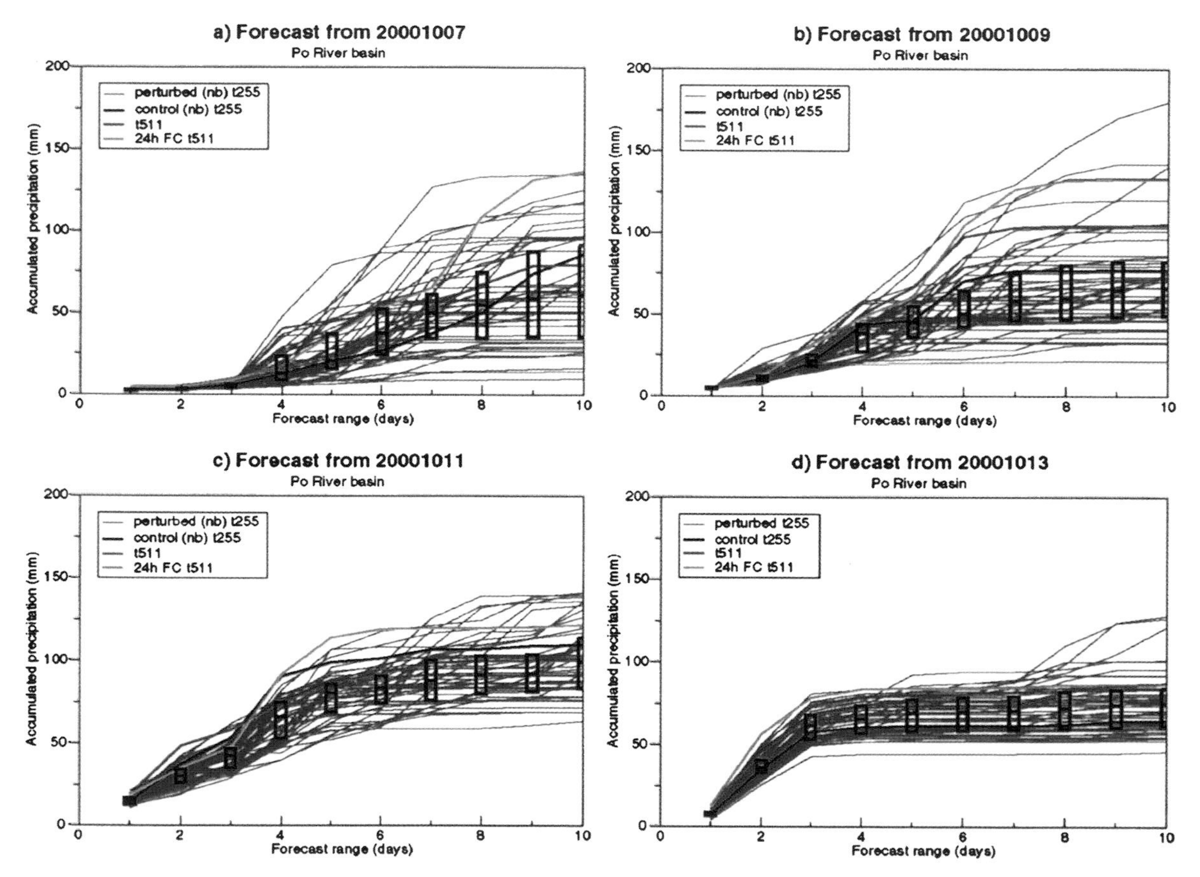

FIG. 8.7. Deterministic and ensemble ECMWF forecasts of the integrated hyetographs for the Po catchment for forecasts starting at 1200 UTC on 7, 9, 11, and 13 Oct 2000. Also shown in each panel is the proxy truth in green. This verification is produced by assuming that the successive T511 day-0 to day-1 forecast accumulations are truth (an assumption whose errors require quantification). The value of all integrated hyetographs is zero at time $t=0$, i.e., at the start of each forecast. Hyetographs are shown in 24-h steps for the 40-km resolution deterministic forecast (blue), the 80-km resolution ensemble forecasts, the 80-km resolution unperturbed forecast, and the verification (green). Also shown on the plots are the 25%, 50% (median), and 75% percentiles of the ensemble accumulations on each day of the forecast.

deterministic 40-km forecast gives a very good quantitative forecast for the accumulation out to D+6 (Sunday, 15 October) but stops the heavy rain a day too early. The control (unperturbed) 80-km Ensemble Prediction System (EPS) forecast from the same initial data shows a similar time evolution of the basin-wide accumulation, but with a smaller accumulation. The observed precipitation lies within the spread of the ensemble at all forecast ranges. The ensemble of forecast rain accumulations falls into two clusters, with a large cluster of ensemble members showing a cessation of rain at D+6 (Sunday, 15 October), and a smaller cluster showing the rain continuing until Monday and beyond. Out to D+5, the ensemble shows a 25% chance that the actual accumulation will exceed the deterministic forecast (and shows a 25% chance of exceeding the observed accumulation). The ensemble also forecasts a 25% chance that the 7-day accumulation (to noon on Monday, 16 October) will exceed 65–70 mm. The judgment must be that these were synoptically useful forecasts.

3) FORECASTS FROM 1200 UTC WEDNESDAY 11 OCTOBER

Figure 8.7 (lower left) shows the deterministic and ensemble forecasts from 1200 UTC on Wednesday, 11 October. The 40-km deterministic forecast is very consistent with the forecasts from the previous days, both in its good points (timing) and its bad points (20% underestimate of intensity from D+4 onward). At D+5 (1200 UTC on 16 October) the 40-km model accumulation is about 100-mm instead of the "observed" 125–130 mm. The observed rain accumulation lies on the upper fringe of the ensemble forecasts of rain accumulation. Compared to the earlier ensembles just discussed, this ensemble shows much more consistency, and smaller spread, in the forecasts of rain accumulation for the weekend. This result perhaps suggests that one may have more confidence in this forecast than in earlier forecasts. Most of the ensemble forecasts do well on the timing of the event, but underestimate the intensity of the event [the median accumulation forecast for 1200 UTC on 16 October (D+5) is about 75 mm]. After D+4, the ensemble forecasts a 25% chance of exceeding the accumulation in the unperturbed 80-km model. The similarity of the 40-km and the unperturbed 80-km forecasts adds to the confidence in the forecast that was engendered by the fact that the ensemble spread in the forecast for the weekend event is the lowest we have seen so far in the series. We also note that some of the ensemble members resumed substantial rain accumulations for a day or two after October 16-indicating a finite probability of such an eventuality.

4) FORECASTS FROM 1200 UTC FRIDAY 13 OCTOBER

The observed accumulation from 1200 UTC on Friday, 13 October (Fig. 8.7 bottom left) is very low in the first 24 h, and then shows a heavy accumulation of 75 mm between noon on Saturday (D+1) and noon on Monday (D+3). There is remarkable consistency between the 40-km deterministic forecast, the unperturbed 80-km EPS forecast, and the ensemble members in the timing of the heavy weekend precipitation, and on the cessation of the rain after D+3 (from noon on Monday), The 40-km model underestimates the accumulation by about 20% and the 80-km model underestimates it by about 35%. The verification again lies on the fringe of the ensemble. The ensemble forecasts show a 25% chance that the actual accumulation will exceed the forecast from the 40-km model. These have to be judged as successful forecasts.

c. Discussion

The "verifications" just shown of forecasts for basin-averaged rainfall accumulations indicate that the forecasts are quite consistent and show encouraging levels of skill even at 7–9 days ahead. It remains to be seen if the meteorological forecast skill can be translated into useful hydrological forecast skill in this case and in other cases. We understand (C. Schaer 2001, personal communication) that certain dam operators in the southern Alps had to make an important water-management decision on Wednesday, 11 October. If the rain continued at high intensity beyond Monday, 16 October, they faced the risk that their dams would be overtopped, with potentially severe consequences. Avoidance of such an eventuality would require the release of 30% of the active water in the reservoirs in the 36 h before the onset of the heaviest rain, as release of the water in an extended period of intense rain would exacerbate an already dangerous situation. Between Saturday, 7 October and Wednesday, 11 October, the consistency of the ECMWF forecasts for the cessation of the event on Monday, 16 October was therefore an important input to the water managers' decision not to make a precautionary release of the valuable stored water. Their decision was justified by subsequent events, and carried with it a considerable financial saving. We also understand (J. Ambuehl 2001, personal communication) that the ECMWF medium-range forecasts were very useful for planning civil protection in the southern Alps, where correct decisions were made to evacuate vulnerable communities to higher ground.

The rainfall forecasts show a number of systematic problems, such as erroneous distributions of heavy rain in mountainous areas, a marked sensitivity of forecast precipitation to the resolution of the forecast model, and a growing unrealiability of the forecasts beyond day 7. The spatial and temporal integrations involved in calculating and verifying integrated hyetographs for a basin mitigate these problems. Some of these issues will be difficult to resolve solely by developments of the meteorological forecast systems. For hydrological forecasting, a pragmatic approach may be worth investigating for example, downscaling of forecast rainfall accumulations to compensate for systematic model underestimates.

ECMWF is a partner in an EU-funded hydrometeorological research project the (European Flood Forecasting System) that is assessing the possibility of producing useful hydrological forecasts in the 3–10-day period using meteorological forecasts (both deterministic and ensemble) as forcing for distributed hydrological models. The project is studying major winter and summer flood events in Europe in recent years. Here we have commented on the ECMWF forecasts of accumulated rainfall for one of the EFFS case studies, namely the floods in the Po valley in mid-October 2000. The event is a good test of medium-range forecast skill.

4. Improved use of NWP for forecasting natural hazards

Reliable forecasts of natural hazards that threaten life and property are increasingly demanded by society because of the growing vulnerability of densely populated areas to natural hazards, and because of improved scientific and technical capability to provide better information. Indeed, European policymakers (Commission of the European Communities 2001) require the development by 2008 of a capability for operational forecasting of hazards such as floods, fires, and the consequences of oilspills at sea. Many natural hazards arise from weather events such as tropical cyclones or intense midlatitude storms, with their associated floods, landslides, wind damage, heavy seas, and coastal damage. Ensemble weather predictions are increasingly the prime medium-range forecasting tool in many countries. The skill of deterministic forecasts and ensemble forecasts have benefited in largely equal measure from developments in data availability, data assimilation methods, and model physics, numerics, and resolution.

a. Environmental prediction models

For both operational and scientific purposes, the output from global earth system data assimilations and forecasts are used to drive a variety of specialized Earth-system models or environmental prediction models including those listed in Table 8.2.

It is likely that the ensemble forecast methodology will prove as useful in environmental forecasts as it has proved to be in weather forecasting. Collaboration between the environmental and weather communities on optimization of the interfaces between the global earth system models and the specialized environmental models will be essential in translating assessments of severe weather risk into assessments of severe environmental risk. Ensemble methods for environmental forecasting are also being explored on the seasonal timescale. Short-term climate fluctuations such as ENSO are frequently implicated in extended periods of tropical and subtropical drought or flood, with their devastating consequences for human safety and for essential economic activities including food and fiber production. The ECMWF coupled atmosphere–ocean system has been used for real-time seasonal forecasting since 1996. ECMWF's tropical seasonal forecast products are published monthly on the Web, including forecasts for anomalies in rain, temperature, pressure, and sea surface temperature. The system gave good forecasts of the initiation, development, and decay of the 1997–98 ENSO event. Derived products such as forecasts of tropical cyclone frequency have also been successful. The seasonal forecast system runs at an atmospheric resolution of 200 km (shortest resolved half wavelength), while the meridional ocean resolution varies between 0.3° and 1.25°. A major effort is under way in the EU-funded Development of a European Multimodel Ensemble System for a Seasonal to Interannual Prediction (DEMETER) project to assess and document the capabilities of several European coupled systems in seasonal forecasting over the last 30–40 yr. An important element of the exercise will be the interfacing of the seasonal forecasts with a crop forecast model to assess the skill and utility of the ensuing crop forecasts.

b. High-resolution multimodel decadal scenarios

Climate change may affect the habitability of the continent by altering the frequency or severity of summer droughts and winter floods. Decision makers in the area of water management need credible assessments of the threats in the next 30 yr arising from changes in the habitability of the continent due to climate change. Ensembles of simulations at 500-km (T42) resolution are useful in assessing if there may be a threat, but not for assessing the regional variations of the threat. To meet the requirements of water managers, their political masters, and others concerned with contingency preparations for extreme weather arising from global change, one needs to provide 30–50-yr look-aheads at a resolution of at least 30–50 km. In order to sample uncertainty arising from model formulations and dynamical noise, the look-aheads should be multimodel ensemble simulations. Such simulations will require extremely efficient integration schemes, and extremely powerful computers. Operational weather services have implemented remarkably accurate and efficient semi-Lagrangian time integration

TABLE 8.2. Examples of the variety of existing applications where information from earth-system models provides boundary and /or initial conditions to drive specialized application models.

Category	Examples of existing applications
Atmosphere	Regional weather, chemical and aerosol transport, and trajectory models, and inverse models for carbon source attribution
Land	Hydrological, fire, crop, disease models
Ocean	Regional ocean, coastal zone, oil-spill, and storm surge models

TABLE 8.3. The successive scientific developments that were implemented between 1991 and 2002 in the semi-Lagrangian time-integration scheme at ECMWF, together with the gain in model efficiency from each step, and the appropriate bibliographical acknowledgment.

Algorithmic change	Gain in efficiency	Reference
Linear grid	×3.4 (=1.5**3)	Cote and Staniforth (1988)
Reduced Gaussian grid	×1.3	Hortal and Simmons (1991)
Three-time-level semi-Lagrangian scheme	×6	Ritchie et al. (1995)
Two-time-level semi-Lagrangian scheme	×2	Temperton et al. (2001); Hortal (2002)

schemes for high-resolution models. Over the last decade ECMWF and other forecast institutes have cumalatively gained a factor of up to 50 in model efficiency from the multiplicative factors shown in Table 8.3.

Then taking into account the fact that the time step is not strictly limited by strong stratospheric polar-night winds, nor by the thin layers in the well-resolved planetary boundary layer, it follows that the true gain in forecast model efficiency is closer to a factor of 150. In terms of Moore's law (which suggests a gain of a factor of 2 in computer power every 18 months), models that use these advanced techniques have a 10-yr advantage (about seven successive doublings) in computer productivity over models that do not use these techniques. These efficient schemes will no doubt be adopted for general circulation modeling in the near future, since remaining scientific issues such as exact mass conservation are being resolved.

To illustrate the status of current technology we note that ECMWF's Fujitsu VPP-5000 computer and the Japanese Earth Simulator computer both have processing elements (PEs) operating at almost 10 Gflops, with 100 PEs on the VPP-5000 and 5120 PEs on the Earth Simulator. The Earth Simulator thus has about 50 times the power of the VPP-5000. A version of ECMWF's model with 60-km resolution (T319) and 60 levels in the vertical (T319/L60) can deliver about 10 yr of integration in 1 day on the VPP-5000, so the Earth Simulator can deliver an ensemble of 50 such integrations (500 yr of integration time) of the TL319/L60 model in 1 day of elapsed time. The provision of a 50-member ensemble of 50-yr integrations of the TL319/L60 model would take 5 days on the Earth Simulator. If one makes the reasonable assumption that a 60-km ocean model costs no more than a 60-km atmospheric model, then these estimates apply to a coupled atmosphere-ocean GCM (AOGCM) with 30 levels in the atmosphere and 30 levels in the ocean.

Given the strong dependence of computing cost on resolution, a 50-yr look-ahead using a 30-km AOGCM (30 levels each in the atmosphere and ocean) to generate a (multimodel) 50-member ensemble would take about 40 days on the Earth Simulator. There is a demand from policy makers worldwide for answers to questions that can only be addressed if such simulation capabilities become available.

5. Prospects for global environmental monitoring

European policymakers (EU 2001) require the development by 2008 of a capability for global monitoring of specific aspects of the atmosphere ocean and land, as follows:

- global atmosphere monitoring delivering regular assessments of the state of the atmosphere with particular attention paid to aerosols, ozone, UV radiation, and specific pollutants;
- global ocean monitoring in support of seasonal weather predictions, global change research, commercial oceanography, and defense; and
- global monitoring to assess carbon fluxes and stocks in the biosphere.

Such capabilities will depend on earth observation from space. Some of the relevant satellite missions to be launched in middecade (2003–07) are listed in Table 8.4.

Given the current scientific status of modeling and data assimilation at the NWP centers, and given the new satellite capabilities coming onstream in the next five years, it is possible to plan the development of a global environmental monitoring system to meet policymakers' main requirements. Operational weather prediction centers such as ECMWF have strong research and development programs for operational forecasting, which includes de facto monitoring of the global hydrological and energy cycles. ECMWF, in partnership with the wider community, could extend its modeling and data assimilation framework to develop a global monitoring capability. In the initial phase, one envisages an experimental monitoring and assimilation system that could run daily, but perhaps with a delay of some weeks behind real time to facilitate a comprehensive data collection. The resolution is envisaged to be similar to that used for seasonal forecasting. The software would be built as an extension of the operational medium-range software, so that developments useful to medium-range work can be readily migrated into the operational environment. It is convenient to discuss the three global monitoring issues in sequence, beginning with the atmosphere, moving on to the ocean, and then closing the loop with the land issues.

a. Global monitoring of greenhouse gases and aerosols

1) ESTIMATION OF TOTAL-COLUMN AMOUNTS OF GREENHOUSE GASES

Feasibility studies (Chedin et al. 2003) indicate the possibility of mapping seasonal fluctuations in total

Table 8.4. Relevant satellite missions, 2003–07.

Expected launch	Mission	Objective
2003	Earth Observing System (EOS) *Aura*	Chemistry
2004	CloudSat	Clouds
	Cloud–Aerosol Lidar and Infrared Pathfinder Satellite Observations (CALIPSO)	Aerosol
	CryoSat	Sea ice freeboard
2005	Meteorological Operational (METOP-1) satellite	Operational meteorology including an advanced sounder and a scatterometer
	Geosynchronous Imaging Fourier Transform Spectrometer (GIFTS)	Advanced sounder in GEO Orbit
	Gravity Field and Steady-State Ocean Circulator (GOCE)	Geoid to T360
2006	Global Precipitation Mission (GPM)	Frequent sampling and complete Earth coverage of high-resolution precipitation measurements, for modeling weather and climate
	Soil Moisture and Ocean Salinity Mission (SMOS)	Global observations of soil moisture and ocean salinity, for modeling weather and climate
2007	Atmospheric Dynamics Mission (ADM-Aeolus)	Doppler Wind Lidar

column CO_2 using the AIRS and Advanced Microwave Sounding Unit (AMSU) instruments that were launched on NASA's AQUA mission, or the IASI and AMSU instruments to be launched on METOP-1 in 2005. Indeed, pilot studies with operational TOVS data have already demonstrated impressive capabilities (Chedin et al. 2002a,b). A new EU-funded project (COCO) aims to develop an operational system to estimate total column CO_2, and to validate the results, using inverse carbon modeling. Additional observations will be provided by the Scanning Imaging Absorption Spectrometer for Atmospheric Chartography (SCIAMACHY) instrument flying on Envisat since 2002. The COCO[1] project will use the data assimilation system to exploit the synergy between the AIRS infrared instrument (sensitive to both temperature and CO_2) and the AMSU microwave instrument (sensitive only to temperature) to back-out the CO_2 estimates. Because the assimilation system will provide consistent wind, temperature, and CO_2 estimates, it should be possible to estimate the export or import of carbon from the land to the ocean. Initially COCO will focus on CO_2 estimates over ocean where the surface radiative properties are well understood in the infrared and in the microwave. It will also be possible to estimate the CO_2 exchange between atmosphere and ocean. The technology being developed for the estimation of total column CO_2 should be readily adaptable to provide total-column estimates of other greenhouse gases such as CH_4, N_2O, and CO.

2) Modeling and assimilation of the ocean carbon cycle

One could establish an independent verification of the ocean uptake of CO_2, if one introduced a carbon cycle into the atmospheric and ocean models, as a basis for assimilating ocean color measurements, along with the ocean satellite data currently used directly or indirectly in the ocean data assimilation: scatterometer winds, SST, altimeter sea level heights, as well as altimeter wave data. By the end of 2002 five ocean color instruments were in orbit [the Sea-viewing Wide Field-of-view Sensor (SeaWiFS), the Moderate Resolution Imaging Spectrometer (MODIS) on board the Terra and Aqua satellites, the Moderate Resolution Imaging Spectrometer (MERIS) on board Envisat, and the Ocean Color and Temperature Scanner (OCTS) on board the Earth Observing Satellite (ADEOS)]. No plans are in place in the meteorological world for real-time use of this huge data source. To tie down the ocean CO_2 budget by assimilating ocean color, one needs to draw on the European science base for expertise in ocean-biology modeling, ocean-color modeling, and ocean assimilation, as well as for expertise in estimating primary production from ocean color measurements. For the research phase one envisages a development system running at the resolution of the seasonal forecast model. Since an advanced assimilation system can infer information on upper-ocean dynamics from ocean color measurements, there would be a direct benefit for the ocean dynamical assimilation and, thus, eventually for seasonal forecasts.

3) Modeling and assimilation of aerosol information

All the ocean color instruments provide a capability to estimate aerosol optical depth. The largest errors in the atmospheric clear-sky radiative calculations arise from uncertainties in aerosol. The introduction of an aerosol variable (say for desert aerosol) in the atmospheric model would enable one to assimilate a good deal of the aerosol information in the ocean color measurements. This would certainly improve the ocean color assimilation (by quantifying a part of the signal that would otherwise be treated as noise) as well as delivering a modest forecast benefit for the atmospheric model.

[1] COCO is an EU-funded project. Detailed information may be found online at http://www.bgc_jena.mpg.de/projects/Coco/index_coco.html.

4) MODELING AND ASSIMILATION OF INFORMATION ON THE LAND BIOSPHERE

Development of a capability to model the radiative properties of the land biosphere is a high priority for those concerned with estimating the stocks and fluxes of carbon related to the land reservoir. It is also a high priority for meteorologists because improved modeling of the land biosphere (i) will enable meteorologists to assimilate atmospheric sounding data over land and (ii) will lead to improved local weather forecasts. Until quite recently, our knowledge of the radiative properties of the land surface was so poor that almost all meteorological remote sounding data over land had to be discarded, because of our inability to model the radiative properties of the surface.

Satellite data on the land biosphere are available in the optical and near- infrared [Advanced Very High Resolution Radiometer (AVHRR), MODIS, MERIS, Spinning Enhanced Visible and Infrared Imager (SEVIRI)], in the thermal infrared [High Resolution Infrared Radiation Sounder (HIRS), AIRS, IASI], and in the microwave [Advanced Microwave Sounding Unit (AMSU), SSM/I, SSM/I Sounder (SSMI/S)]. The European Space Agency (ESA) instrument SMOS, which will be used to estimate soil moisture, is expected to fly in 2007. One needs to develop forward models that can simulate the satellite data, taking account of the dependence of the radiative properties of the land surface on the nature of the vegetation, on the nature of the soil, and on recent meteorological events (rain, drought, snow). Effective interpretation of the satellite data thus requires an effective assimilation system. Several research initiatives are getting under way. ECMWF is a partner in the EU-funded initiative ELDAS (European Land Data Assimilation System to Predict Floods and Droughts). ECMWF is also collaborating with EUMETSAT's Land-Satellite Applications Facility (Land-SAF), with International Satellite Land Surface Climatology Project (ISLSCP), and with the U.S. effort (GLDAS), Global Land Data Assimilation System) on land data assimilation.

Incorporation of a carbon cycle into the land, ocean, and atmosphere modules of an earth system model, and assimilation of the range of satellite data that is or will be available, will provide global fields consistent with our understanding of the main processes in the earth system. ISLSCP sponsors a worldwide network of about 100 land observatories making detailed boundary layer measurements of vertical fluxes of heat, moisture, momentum, and carbon. This station network would provide detailed validation and verification data on the key boundary layer exchange processes of the carbon cycle as they do at present for the water and energy cycles of earth system assimilations and models. The intercomparisons would thus provide a sound scientific basis for identifying shortcomings in the models and a global basis for assessing proposed model improvements. This scientific approach has been successful in other research areas. For example, the global ocean wave assimilation uses satellite altimeter data only, while the sparse in situ wave-buoy data provide independent and invaluable information for verification, diagnosis, and development.

An important limitation of current or imminent observations is the inability to sense fluctuations of CO_2 amounts in the planetary boundary layer. Proposals are under discussion to develop either active or passive satellite sensing capabilities to remedy the shortcoming. Experience with assimilation of AIRS, SCIAMACHY, and IASI data for greenhouse gas monitoring may be helpful in refining the requirements for new instruments or missions. When such missions eventually fly, their data will be immediately exploitable in the earth system assimilations.

b. Global monitoring of reactive gases and aerosols

Atmospheric chemical data assimilation has traditionally used chemical transport models, driven by gridded winds and temperatures from an independent source such as a meteorological data assimilation system. Atmospheric chemical data contains information on the advecting winds. This was one motivation for developing an interactive dynamical–chemical assimilation system for ozone in the framework of ECMWF's 4DVAR data assimilation system.

1) DYNAMICAL-CHEMICAL ASSIMILATION SYSTEM FOR OZONE

With support from ESA and EU [Framework IV project Studies of Ozone Distributions based on Assimilated Satellite Measurements (SODA)], ECMWF and Meteo-France have developed a capability to model the three-dimensional distribution of ozone, and to assimilate data on ozone from the following instruments presently in orbit or to be launched in the near future: Solar Backscatter Ultraviolet (SBUV), Total Ozone Mapping Spectrometer (TOMS), and HIRS on National Oceanic and Atmospheric Administration (NOAA) satellites; Global Ozone Monitoring Experiment (GOME) on ERS and METOP; SCIAMACHY on *Envisat*; and Ozone Monitoring Instrument (OMI) on *Aura*.

This meteorological/ozone assimilation system will be used to support the calibration and validation of real-time data from the Global Ozone Measuring by Occultation of Stars (GOMOS), Michelson Interferometer for Passive Microwave Sounding (MIPAS), and SCIAMACHY instruments on *Envisat*, and eventually for the assimilation of the products from these instruments. No plans are in place for use of data from NASA's chemistry mission *Aura*, due for launch in 2002–03.

The ozone assimilation system is also being used in a simpler 3DVAR context to provide a 23-yr assimilation

(1979–2002) of satellite data on ozone as part of the EU-supported ERA-40 reanalysis project. Useful satellite data on ozone first became available in 1979. The ERA-40 project will assimilate both SBUV and TOMS data on ozone for the period of record. The full reanalysis will cover the period 1958–2002. Prior to 1979, the ozone field in the model will evolve freely, unconstrained by observations.

The ozone assimilation system will become operational at ECMWF in the spring of 2002, around the time of the expected *Envisat* launch. The assimilation will use the real-time ozone data from GOME on *ERS-2*, from SBUV on the NOAA satellites, and from *Envisat* on completion of the validation process. As such, the system will provide valuable products for operations (e.g., UV_B forecasting) and for research. Ground-based Dobson spectrophotometer ozone data could be included in the system (either by real-time delivery of the data, or by a delayed mode run of the assimilation). Inclusion of the Dobson spectrophotometer data would make the assimilation fields a useful adjunct in the monitoring of the Montreal convention. A further scientific development of considerable interest is the direct variational assimilation of limb-sounding radiances from instruments such as MIPAS. This would provide a better net result than the use of MIPAS-retrieved profiles.

2) Coupled dynamical-chemical assimilat ion

Many chemical data assimilation activities use chemical transport models driven by specified winds. Unlike an interactive dynamical–chemical assimilation, they get no benefit from the wind information in the chemical data. The interactive dynamical–chemical assimilation approach could be extended from ozone alone to the entire NOx family of ozone precursors and perhaps also to the family of precursors of sulfate aerosol. This would offer substantial operational and scientific benefits by providing the most accurate possible wind and chemistry fields.

If a semi-Lagrangian model is used for the assimilation, then the marginal cost of the chemical advection computation is negligible. The additional costs are thus mainly memory and the costs of the chemical-interaction calculations.

3) Chemical weather forecasting

There is a large demand across Europe for accurate forecasts of air quality at times of environmental stress. Such forecasts are normally a national responsibility, and are usually made with specialized regional models that require meteorological and chemical boundary conditions, analogous to the requirements of regional meteorological forecast models.

The availability of a global assimilation and forecast system providing consistent global analyses and forecasts of the meteorological, chemical, and aerosol fields would offer substantial operational and scientific benefits.

c. Global monitoring and modeling of atmosphere, ocean, and land

The proposals just outlined involve comprehensive global monitoring of the atmosphere, the ocean, and the land at resolutions in the range 40–150 km corresponding to the range of models used for forecasting on medium-range, monthly, seasonal and decadal timescales. Success in the enterprise will require the participation from a wide spectrum of talent across the European science base. Success will also require improvement of the science of the atmospheric, ocean, and land models to the point where they can realistically simulate the observations. This immediately guarantees the use of the scientific advances in improved models for forecasts and simulations. It also guarantees that periodic reanalyses of the available data (e.g., for studies of low-frequency variability or trends) will benefit from the new science developed in the monitoring program.

6. Conclusions

Since 1985 there have been many developments in the technology of numerical weather prediction. We are at the threshold of a new era in observational capability. The modeling and data assimilation tools needed to exploit those new observational capabilities have been made ready. This contribution is a progress report to Prof. Reed on those developments. We salute Prof. Reed's achievements and we thank him for his friendship, his collaboration, and his sustained interest in our work.

Acknowledgments. Dick Reed's example has been a continuing for our work. Richard Johnson and Robert Houze have provided sustained encouragement. We appreciate the excellent work of the AMS technical staff, particularly Gretchen Needham.

APPENDIX

ECMWF's 2002 Operational Assimilation and Forecast System

ECMWF is an international organization supported by twenty-two European governments, and its primary function is the delivery of operational medium-range weather forecasts of increasingly high quality, over the range from 3 to 10 days and beyond; a complementary goal is to establish and deliver a reliable operational seasonal forecasting capability. ECMWF has pioneered

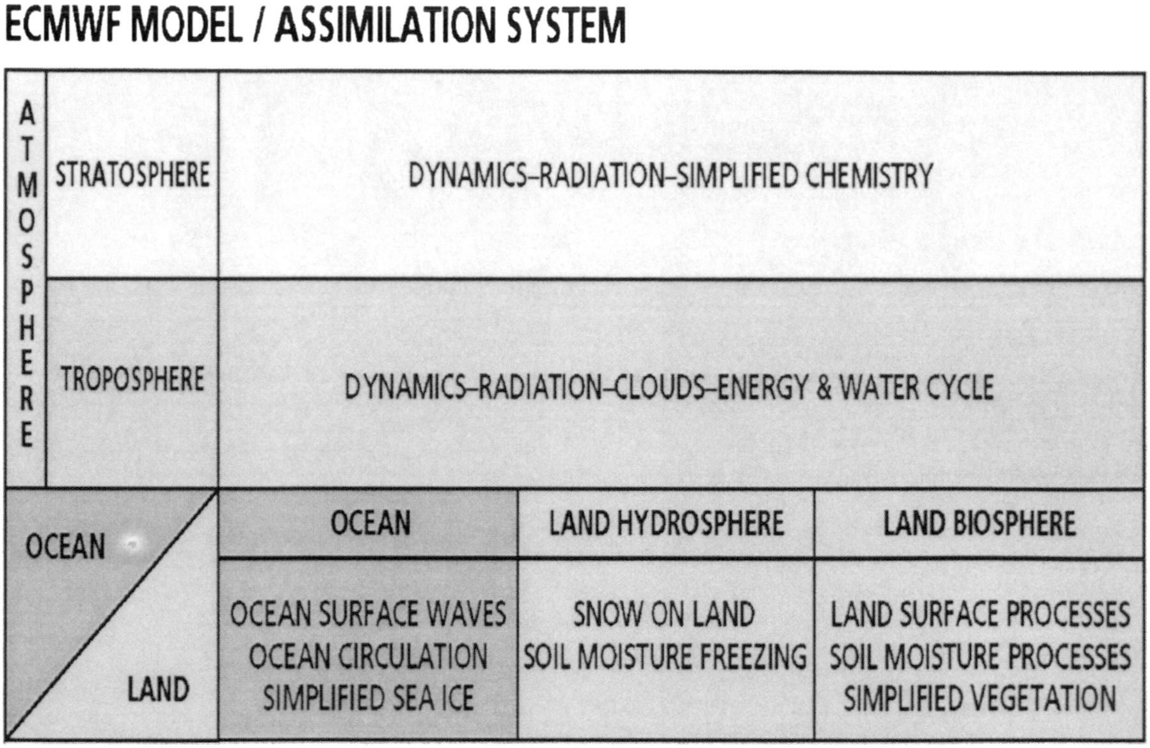

ECMWF MODEL / ASSIMILATION SYSTEM

ATMOSPHERE
STRATOSPHERE
DYNAMICS–RADIATION–SIMPLIFIED CHEMISTRY
TROPOSPHERE
DYNAMICS–RADIATION–CLOUDS–ENERGY & WATER CYCLE
OCEAN
LAND
OCEAN
LAND HYDROSPHERE
LAND BIOSPHERE
OCEAN SURFACE WAVES
OCEAN CIRCULATION
SIMPLIFIED SEA ICE
SNOW ON LAND
SOIL MOISTURE FREEZING
LAND SURFACE PROCESSES
SOIL MOISTURE PROCESSES
SIMPLIFIED VEGETATION

FIG. 8.A1. Cartoon outlining the structure of the ECMWF earth-system model.

numerous developments in the use of satellite data for determination of the state of the atmosphere's dynamics and composition, and of those aspects of the state of the land, ocean, and cryosphere that are relevant for medium- and extended-range forecasting.

ECMWF's earth system model (Fig. 8.A1) comprises the following coupled modules:

- *Atmospheric dynamics*—a global atmospheric 60-level general circulation model with a top at 65 km, and a horizontal resolution of 40 km
- *Ocean circulation*—a global ocean general circulation model and ocean ice processes
- *Ocean surface waves*—a global third-generation ocean surface wave dynamics model, directly coupled to the atmospheric model
- *Land*—a land biosphere module and a land surface, soil, hydrological, and snow model
- *Atmospheric composition*—a comprehensive model of the hydrological cycle, including all three phases of water, and dynamic ozone with parameterized chemistry

In preparation for the new generation of operational and research satellites currently coming onstream, ECMWF has developed a powerful four-dimensional variational assimilation system (4DVAR) as an effective method to use all observations. ECMWF has also developed scalable data ingest and processing techniques to enable the data assimilation process to cope with the vast volumes of new data.

To support its assimilation and forecasting activities, ECMWF has high-performance computing facilities and data handling facilities that are among the best in the world. ECMWF routinely assimilates a wide variety of data from operational and research satellites (both polar and geostationary) from instruments as diverse as infrared and microwave sounders, visible and microwave imagers, scatterometers, radar altimeters, and synthetic aperture

TABLE 8.A1. Satellite data used for operations and/or research in 2002.

Category of space mission	Products used
NOAA polar	Microwave and infrared radiances for temperature and humidity sounding
European, U.S., and Japan geostationary satellites	Cloud winds, water vapor radiances
Defense Meteorological Satellite Program (DMSP) series	SSM/I water vapor, wind speed
ERS series	C-band scatterometer ocean winds; altimeter—wave height, sea level height; synthetic aperture radar (SAR) wave spectra
Tropical Rainfall Measuring Mission (TRMM)	Rain profiler and microwave imager data on rain rate
QuikSCAT	SeaWinds—Ku-band scatterometer
SBUV/GOME/TOMS	Ozone profiles and total column amount

TABLE 8.A2. New satellite missions in 2002–03 used at ECMWF.

Mission	Mission instruments or products
EOS-Aqua	Advanced sounder (AIRS)
Envisat	Ozone (MIPAS, SCIAMACHY, GOMOS), waves (RA-2), spectra [Advanced Synthetic Aperture Radar (ASAR)]
ADEOS-II	SeaWinds—Ku-band scatterometer
SSMI/S	Microwave sounder and imager
JASON-1	Radar altimeter for sea level, waves
Meteosat Second Generation (MSG)	Winds, ozone, water vapor

radars. Table 8.A1 lists the satellite data currently used in operations (or recently used, in the case of ERS).

ECMWF has made extensive preparations for the calibration and validation (CAL/VAL) and eventual assimilation of the data from the missions listed in Table 8.A2, which were due for launch by the end of 2002. Use of the above data streams relies heavily on collaborations with mission teams at EUMETSAT, ESA, the Centre National d'Etudes Spatiales (CNES), the National Environmental Satellite, Data, and Information Service (NESDIS), NASA, the National Space Development Agency of Japan (NASDA), and with the wider scientific community through such groups as International Tiros Operational Vertical Sounder (ITOVS).

The analyses and deterministic 10day operational forecasts are global, at 40-km resolution in the horizontal, and cover the atmosphere to 65km. The ensemble 10-day forecasts are at a resolution of 80 km.

The ensemble prediction system samples uncertainties in the initial data through a set of specially formulated perturbations, and samples uncertainties in the model formulation through stochastic perturbations of the physical tendencies. A particular emphasis at ECMWF in the next few years will be the development of a medium-range severe weather forecast system, to complement existing forecast tools for quantitative forecasts of severe weather in the short range. The developments will comprise improvements to the existing forecast models, coupled with an extensive statistical processing of the ensemble forecasts to indicate the nature and severity of threatened natural hazards.

The seasonal forecast system issues forecasts monthly, and uses the ECMWF atmospheric model coupled to the Hamburg Ocean Primitive Equation (HOPE) ocean model (developed by the Max Planck Institute) through the OASIS coupler (developed at Centre European de Recherche et de Formation Avancée en Calcul Scientifique (CERFACS). The ocean assimilation system was developed by the Australian Bureau of Meteorology.

REFERENCES

Andersson, E., and A. Hollingsworth, 1988: Typhoon bogus observations in the ECMWF data assimilation system. ECMWF Tech. Memo. 148, 25 pp.

Bengtsson, L., and A. J. Simmons, 1983: Medium-range weather prediction—Operational experience at ECMWF. *Large-scale Dynamical Processes in the Atmosphere*, B. J. Hoskins and R. P. Pearce, Eds., Academic Press, 337–363.

Bougeault, P., E. Richard, and F. Roux, 2001a: L'experience MAP sur les phenomenes de mesoechelle dans les Alpes: premier bilan [The Mesoscale Alpine Programme (MAP) field experiment: First assessment]. *Meteorologie*, **8** (33), 16–33.

——, and Coauthors, 2001b: The MAP special observing period. *Bull. Amer. Meteor. Soc.*, **82**, 433–462.

Buizza, R., and A. Hollingsworth, 2002: Storm prediction over Europe using the ECMWF Ensemble Prediction System. *Meteor. Appl.*, **9**, 1–17.

Burpee, R. W., 2003: Characteristics of african easterly waves. *A Half-Century of Progress in Meteorology: A Tribute to Richard J. Reed*, Meteor. Monogr., No. 53, *Amer. Meteor. Soc.*, 91–108.

Chan, J., 2000: Understanding and forecasting of tropical cyclones: Progress and challenges. *Dealing with Natural Disasters: Achievements and New Challenges in Science Technology and Engineering*, The Royal Society, 106–134.

Chedin, A., A. Hollingsworth, N.A. Scott, S. Serrar, C. Crevoisier, and R. Armante, 2002a: Annual and seasonal variations of atmospheric CO_2, N_20, and CO concentrations retrieved from NOAA/TOVS satellite observations. *Geophys. Res. Lett.,* **29** (8), 1269, doi: 10.1029/2001GL014082.

——, S. Serrar, R. Armante, N. A. Scott, and A. Hollingsworth, 2002b: Signatures of annual and seasonal variations of CO_2 and other greenhouse gases from comparison between NOAA/TOVS observations and radiation model simulations. *J. Climate*, **15**, 95–116.

——, A. Hollingsworth, N. A. Scott, R. Saunders, M. Matricardi, J. Etcheto, C. Clerbaux and R. Armante, 2003: The feasibility of monitoring CO_2 from high resolution infrared sounders. *J. Geophys. Res.*, **108** (D2), doi:10.1029/2001JD001443.

Cherubini, T., A. Ghelli, and F. Lalaurette, 2002: Verification of precipitation forecasts over the Alpine region using a high density observing network. *Wea. Forecasting*, **17**, 238–249.

Cote, J., and A. Staniforth, 1988: A two-time-level semi-Lagrangian semi-implicit scheme for spectral models. *Mon. Wea. Rev.*, **116**, 2003–2012.

English, S. J., and Coauthors, 2000: A comparison of the impact of TOVS and ATOVS satellite sounding data on the accuracy of numerical weather forecasts. *Quart. J. Roy. Meteor. Soc.*, **126**, 2911–2931.

EU, 2001: GMES EC Action Plan (Initial Period 2001-2003), Final Communication from the EU Commission to the EU Council and the European Parliament, 23 October 2001. [Available online at http://www.gmes.info/library/index.php.]

Fekete, B. M., C. J. Vörösmarty, and W. Grabs, 2000: The UNH, GRDC, Global Composite Runoff Data Set (v1.0). The Global Runoff Data Centre (GRDC) and the Institute for the Study of Earth Ocean and Space, University of New Hampshire. [Available online at http://www.grdc.sr.unh.edu/.]

Gérard, E., and R. W. Saunders, 1999: Four-dimensional variational assimilation of Special Sensor Microwave/Imager total column water vapour in the ECMWF model. *Quart. J. Roy. Meteor. Soc.*, **125**, 3077–3101.

Goerss, J., and R. Jeffries, 1994: Assimilation of synthetic tropical cyclone observations into the Navy Operational Global Atmospheric Prediction System. *Wea. Forecasting*, **9**, 557–576.

——, L. Brody, and R. Jeffries, 1991: Assimilation of tropical cyclone observations into the Navy Operational Global Atmospheric Prediction System. Preprints, *Ninth. Conf. on Numerical Weather Prediction*, Denver, CO, Amer. Meteor. Soc., 638–641.

——, C. Velden, and J. Hawkins, 1998: The impact of multispectral *GOES-8* wind information on Atlantic tropical cyclone track forecasts in 1995, Part II: NOGAPS forecasts. *Mon. Wea. Rev.*, **126**, 1219–1227.

Gregory, D., J.-J. Morcrette, C. Jakob, A. C. M. Beljaars, and T. Stockdale, 2000: Revision of convection, radiation and cloud schemes in the ECMWF Integrated Forecasting System. *Quart. J. Roy. Meteor. Soc.*, **126**, 1685–1710.

Guard, C. P., L. E. Carr, F. H. Wells, R. A. Jeffries, N. D. Gural, and D. K. Edson, 1992: Joint Typhoon Warning Center and the challenges of multibasin tropical cyclone forecasting. *Wea. Forecasting*, **7**, 328–352.

Hollingsworth, A., K. Arpe, M. Tiedtke, M. Capaldo and H. Savijarvi, 1980: The performance of a medium-range forecast model in winter — Impact of physical parameterizations. *Mon. Wea. Rev.*, **108**, 1736–1773.

——, and A. J. Simmons, 1991: Use of reduced Gaussian grids in spectral models. *Mon. Wea. Rev.*, **119**, 1057–1074.

Hortal, M., 2002: The development and testing of a new two-time-level semi-Lagrangian scheme (SETTLS) in the ECMWF forecast model. *Quart. J. Roy. Meteor. Soc.*, **128**, 1671–1687.

Jakob, C. and S. A. Klein, 2000: A parametrization of the effects of cloud and precipitation overlap for use in general-circulation models. *Quart. J. Roy. Meteor. Soc.*, **126**, 2525–2544.

Janssen, P. A. E. M., 1999: Wave modelling and altimeter wave height data. *ECMWF Tech. Memo.* **269**, 35 pp.

——, J. D. Doyle, J. Bidlot, B. Hansen, L. Isaksen and P. Viterbo, 2002: Impact and feedback of ocean waves on the atmosphere. *Atmosphere-Ocean Interactions*, W. Perrie, Ed., Vol. 1, Advances in Fluid Mechanics, No. 33, WIT Press, 155–197.

Järvinen, H., E. Andersson, and F. Bouttier, 1999: Variational assimilation of time sequences of surface observations with serially correlated errors. *Tellus*, **51A**, 469–488.

Lanzinger, A., 1995: ECMWF forecasts of the floods of January 1995. ECMWF Tech. Rep. 77, xxx pp.

Lawrimore, J. H., and Coauthors, 2001: Climate assessment for 2000. *Bull. Amer. Meteor. Soc.*, **82** (6), S1–S55.

Lorenc, A. C., and Coauthors, 2000: The Met. Office global three-dimensional variational data assimilation scheme. *Quart. J. Roy. Meteor. Soc.*, **126**, 2991–3012.

Mahfouf, J.-F., and F. Rabier, 2000: The ECMWF operational implementation of four-dimensional variational assimilation. II: Experimental results with improved physics. *Quart. J. Roy. Meteor. Soc.*, **126**, 1171–1190.

McNally, A. P., J. C. Derber, W. Wu, and B. B. Katz, 2000: The use of TOVS level-1b radiances in the NCEP SSI analysis system. *Quart. J. Roy. Meteor. Soc.*, **126**, 689–724.

Miyakoda, K., L. Umscheid, D. H. Lee, J. Sirutis, R. Lusen, and F. Pratte, 1976: The Near-real-time, Global, Four-dimensional Analysis Experiment during the GATE period, Part I. *J. Atmos. Sci.*, **33** 561–591.

Molteni, F., R. Buizza, T. N. Palmer, and, T. Petroliagis, 1996: The new ECMWF Ensemble Prediction System: Methodology and validation. *Quart. J. Roy. Meteor. Soc.*, **122**, 73–119.

Morcrette, J-J., 2000: On the effects of the temporal and spatial sampling of radiation fields on the ECMWF forecasts and analyses. *Mon. Wea. Rev.*, **128**, 876–887.

Palmer, T. N., 1996: Predictability of the atmosphere and oceans: From days to decades. *Decadal Climate Variability: Dynamics and Predictability*, D. Anderson and J. Willebrand, Eds., NATO Advanced Study Institute Series I: Global Environmental Change, Vol. 44, Springer-Verlag.

Parrish, D. F., and J. C. Derber, 1992: The National Meteorological Center's spectral statistical-interpolation analysis system. *Mon. Wea. Rev.*, **120**, 1747–1763.

Perry, J. S., 2003: A life in the global atmosphere: Dick Reed and the world of international science. *A Half-Century of Progress in Meteorology: A Tribute to Richard J. Reed, Meteor. Monogr.*, No. 53, Amer. Meteor. Soc., 133–140.

Reed, R. J., 1977: Bjerknes Memorial Lecture. *Bull. Amer. Meteor. Soc.*, **58**, 390–399.

——, and A. J. Simmons, 1991: Numerical simulation of an explosively deepening cyclone over the North Atlantic that was unaffected by concurrent surface energy fluxes. *Wea. Forecasting*, **6**, 117–122.

——, A. Hollingsworth, W. A. Heckley, and F. Delsol, 1988a: An evaluation of the performance of the ECMWF operational system in analyzing and forecasting tropical easterly wave disturbances over Africa and the tropical Atlantic. *Mon. Wea. Rev.*, **116**, 824–865.

——, A. J. Simmons, M. D. Albright, and P. Unden, 1988b: The role of latent heat release in explosive cyclogenesis: Three examples based on ECMWF operational forecasts. *Wea. Forecasting*, **3**, 217–229.

Richardson, D. S., 2000: Skill and relative economic value of the ECMWF ensemble prediction system. *Quart. J. Roy. Meteor. Soc.*, **126**, 649–668.

Ritchie, H., C. Temperton, A. J. Simmons, M. Hortal, T. Davies, D. Dent, and M. Hamrud, 1995: Implementation of the semi-Lagrangian method in a high resolution version of the ECMWF forecast model. *Mon. Wea. Rev.*, **123**, 489–514.

Rohn, M., G. A. Kelly, and R. W. Saunders, 2001: Impact of a new cloud motion wind product from Meteosat on NWP analyses and forecasts. *Mon. Wea. Rev.*, **129**, 2392–2403.

Rubel, F., and B. Rudolf, 2001: Global daily precipitation estimates proved over the European Alps. *Meteor. Z.*, **10**, 408–418.

Simmons, A. J., and A. Hollingsworth, 2002: Some aspects of the improvement in skill of numerical weather prediction. *Quart. J. Roy. Meteor. Soc.*, **128**, 647–677.

——, R. Mureau, and T. Petroliagis, 1995: Error growth estimates of predictability from the ECMWF forecasting system. *Quart. J. Roy. Meteor. Soc.*, **121**, 1739–1771.

Teixeira, J., 1999: The impact of increased boundary layer resolution on the ECMWF forecast system. *ECMWF Tech. Memo.* 268, 55 pp.

Temperton, C., M. Hortal, and A. Simmons, 2001: A two-time-level semi-Lagrangian global spectral model, *Quart. J. Roy. Meteor. Soc.*, **127**, 111–127.

Tomassini, M., D. LeMeur, and R.W. Saunders, 1998: Near-surface satellite wind observations of hurricanes and their impact on ECMWF model analyses and forecasts. *Mon. Wea. Rev.*, **126**, 1274–1286.

——, G. A. Kelly, and R. W. Saunders, 1999: Use and impact of satellite atmospheric motion winds on ECMWF analyses and forecasts. *Mon. Wea. Rev.*, **127**, 971–986.

Toth, Z., and E. Kalnay, 1997: Ensemble forecasting at NCEP and the breeding method: *Mon. Wea. Rev.*, **125**, 3297–3319.

Tsuyuki, T., R. Sakai, and H. Mino, 2002: The WGNE intercomparison of typhoon track forecasts from operational global models for 1991–2000. *WMO Bull.*, No. 51, 253–256.

van den Hurk, B. J. J. M., P. Viterbo, A. C. M. Beljaars, and A. K. Betts, 2000: Offline validation of the ERA40 surface scheme. ECMWF Tech. Memo., **295**, 42 pp.

Chapter 9

A Life in the Global Atmosphere: Dick Reed and the World of International Science

JOHN S. PERRY[1]

Alexandria, Virginia

"The trend towards greater emphasis on 'big science' threatens to diminish the role of the individual in the scientific quest. In forcing new modes of operation, it is important to avoid organizational structures which will stifle individual responsibility and initiative." — (Reed 1971)

1. Introduction

By their very nature, the atmospheric sciences are the most international of the scientific disciplines. The winds and the substances they carry circulate freely across national borders. The storm that brings rain to London may bring snow to Stockholm. Even the climate of Siberia is moderated to some extent by the distant ocean. The volcanic plume of El Chichón traveled around the globe in a week, and pollutants emitted in Ohio land in Maine and Greenland. Viewed from space, the dominant features of our planet are not the static patches of land that we call nations, but rather its restlessly wandering clouds that know no fences. To study the atmosphere is necessarily to study the world as a whole, a world inhabited by many diverse nations but nonetheless one world.

This truism is particularly evident to anyone who ventures into the challenging arena of weather forecasting. In the dark ages before the computer — and indeed even today in much of the world — the first step in forecasting tomorrow's weather was to draw a pretty picture of the weather of today. Observations that magically appeared on a clanking teletype were laboriously plotted on large charts, mysterious multicolored lines were imaginatively drawn, and the forecaster bravely sought to divine from the resulting picture some notion of what sort of weather might lie ahead. Today, computers do much the same thing with far better algorithms, but the basic flow of data from a worldwide network to analysis and prediction mechanisms remains fundamentally unchanged. Indeed, research shows clearly that forecasting beyond even a few days requires observations covering the entire Earth. I believe that this existential reality impresses forecasters forcefully and lastingly with a frame of mind almost unique to our profession among the scientific disciplines: All forecasters know that without the cooperation and labors of unknown armies of distant and anonymous collaborators, they could accomplish nothing. Thus, to be a forecaster is to be an internationalist.

2. From Braintree to Seattle and beyond

As a member of what Tom Brokaw has termed the "greatest generation" (Brokaw 1998), Dick Reed inescapably became imbued with a sense of connection with the world beyond his native Braintree, Massachusetts. The Great Depression that dominated his generation's childhood was a global phenomenon, while the cancerous growth of fascism in Europe cast shadows even on Boston Common. These global-scale processes reached their flash point with the eruption of World War II, when the United States was catapulted out of its ocean-girdled isolation into the common cauldron of global conflict.

Seeking to contribute to America's part in this crusade, Dick left the security of an accounting program at Boston College and a part-time factory job to enlist in the U.S. Navy. There, he had a choice of numerous specialities. Always fascinated by weather, he unhesitatingly chose "aerology," which at the time meant mostly observing and forecasting, and under the pressure of war he mastered the black art of forecasting as it existed at that time, drawing some from books and much from the real world of data plotting, analyzing, and forecasting. His baptism in meteorology was characterized by restless curiosity, growing passion for the sheer esthetics of the

[1] John S. Perry, who received his M.S. and PhD degrees under Dick Reed's tutelage, is retired from his two careers in the U.S. Air Force and the National Research Council. He currently resides in Alexandria, Virginia, where he writes, edits, and consults on environmental issues.

atmosphere and the sinuous lines describing its structure and flow, and an intuitive understanding of its global interconnections. After the war, his energy, intelligence, and talent rapidly took him to advanced study at Dartmouth, the California Institute of Technology (Cal Tech), and then the Massachusetts Institute of Technology (MIT), where this passion connected with hard science. But undoubtedly, Dick's meteorological baptism in the fiery waters of operational forecasting in a conflict of global scale surely played a major role in shaping his future contributions to atmospheric sciences as an international endeavor.

Dissecting one thread from the complex fabric of a long, complex, and fruitful career is a difficult task. Like the atmosphere itself, practically everything an atmospheric scientist does has far-flung connections. However, in this brief snapshot through a narrow window I will focus primarily on two streams of activity in his diverse career — the Global Atmospheric Research Program and the American Meteorological Society (AMS).

After receiving his doctorate at MIT in 1949 — coincidentally the year in which I received my high school diploma — Dick stayed on for five years on the research staff. A chance encounter with Phil Church, and strong encouragement by MIT's Henry Houghton, led to his subsequent life-long career at the University of Washington. There, his passion for synoptic meteorology flowered again, as he replaced two experienced synopticians — coincidentally again the two who initiated me into the weather game. But despite the charms of the Pacific Northwest, Dick's wide-ranging interests and research took him far afield. I lack a complete dossier, but I recall well his accounts of visits to the Soviet Union — then a mysterious and feared domain — and to the more comfortable shores of Britain. Very characteristically, he found both the Russians and the Brits to be splendid fellows. Characteristically also, I recall his admiration of the British practice of assigning senior and distinguished researchers to regular forecasting shifts. To Dick, the connection between theory and the real world was always paramount. Certainly the students in his famous advanced synoptic meteorology course were never allowed to forget it! But the point of these rambling anecdotes is that the international linkages of the atmospheric sciences were always prominent in Dick's life and work.

3. The Global Atmospheric Research Program

These international linkages acquired special prominence in Dick's life in the 1960's with the emergence of the Global Atmospheric Research Program (GARP). A few words on the development of GARP are perhaps needed to set the stage.

GARP's history has been told many times (e.g., Malone 1968; Perry 1975; Perry and O'Neill 1979) by many veterans, but a brief precis may be in order. The roots of GARP may be traced back many centuries to the exchange of ships' logs among mariners. In the nineteenth century, formal international systems for data exchange were organized, and the International Meteorological Organization was formed. In 1882–83, a highly successful International Polar Year was carried out by a group of nations (a second was conducted in 1932–33). International scientific unions were developed in a number of disciplines, and in 1931 the International Council of Scientific Unions (ICSU) was established. Thus, by the outbreak of World War II, world science had developed a complex and strong international structure.

The war severely battered that structure, and in the postwar years scientists sought mechanisms to rebuild it. One of the first ventures was the highly successful International Geophysical Year (IGY) conducted in 1958–59 with the goal "to observe geophysical phenomena and to secure data from all parts of the world; to conduct this effort on a coordinated basis by fields, and in space and time, so that results could be collated in a meaningful manner." The aim was to allow scientists from around the world to take part in a series of coordinated observations of various geophysical phenomena employing rapidly advancing observational techniques, notably the then-new resources of space technology. Sixty-seven nations participated, and the program was designed by an ICSU Special Committee supported by national committees such as that organized by the U.S. National Academy of Sciences.

GARP was very much a child of the IGY. Its specific roots lay partly in the rapidly advancing technological opportunities opened by the development of successful meteorological satellites, particularly Tiros-1 launched on 1 April, 1960. At the same time the success of computer-based numerical weather prediction gave promise that complete global observations could be effectively utilized to markedly improve our capability to understand and predict the behavior of the atmosphere. These prospects – plus inspired prodding by Tom Malone, Jule Charney, Richard Goody, Verner Suomi, and other leading scientists — led President John F. Kennedy in 1961 to call upon the United Nations (UN) for international collaborative efforts in weather prediction.

Subsequent resolutions of the United Nations in 1961 and 1962 requested the World Meteorological Organization (WMO) and the ICSU to develop plans for international programs of research and development. With aid from the Ford Foundation, an international committee was established, and a planning office in Geneva, Switzerland, was established under Rolando García. A chain of meetings evolved the concept of two parallel and mutually supportive programs, the World Weather Watch (WWW) to develop world meteorological services, and GARP to develop the underlying scientific knowledge as a base for these services. The goals of the latter were formulated in the WMO/ICSU GARP Study Conference of 1967 (World Meteorological Organization 1967).

GARP emerged from the gestation process outlined above as an international program focused on improving our understanding of two complementary groups of processes: those determining the transient behavior of the atmosphere that could lead to better weather prediction, and the factors determining the statistical properties of the general circulation that could lead to an understanding of the physical basis of climate.

The strategy evolved in GARP to attack these problems centered upon the development of numerical models of the global atmosphere–ocean system. The underlying hypothesis for the quest to improve prediction was that the motions of the atmosphere could be observed and predicted by observations and calculations carried out on a feasible scale. Lorenz had already shown that atmospheric circulation systems were in principle unpredictable beyond about 10 days, however accurate the specification of the initial state. But this limit was vastly beyond the actual performance of weather forecasting at the time.

The gap between promise and performance could be closed, it was hoped, by improving prediction models. The indispensable prerequisite for this was to reduce the errors in the initial state to the point at which errors in prediction could be distinguished from errors in observation and analysis of the initial conditions. The goal of model building therefore required, first, the development of observing systems capable of defining the motions and state of the atmosphere with sufficient precision to permit meaningful prediction and verification and, second, the development of quantitative relationships between the explicitly treated variables and smaller-scale processes: a craft that became known as parameterization. For example, computer limitations constrain global atmospheric models to a spatial resolution of a few hundred kilometers, while in many regions the principal energetic systems are convective clouds on the scale of a kilometer. Specific efforts within GARP were devised in accord with this rationale; subprograms focused on individual problem areas such as tropical dynamics and radiation, while field experiments were planned as specific programs of observation and research. As recommended by a seminal report of the National Research Council's (NRC) Committee on Atmospheric Sciences (1966), the central focus of the program was to be a "global experiment" to produce a comprehensive worldwide dataset of unprecedented completeness and precision as a basis for prediction research.

An innovative and highly effective management structure was developed for GARP, largely through the efforts of Tom Malone and Bob White. The program as a whole was guided by a Joint Organizing Committee (JOC), an international group of scientists supported by a full-time Joint Planning Staff (JPS) in Geneva, both sponsored jointly by WMO and ICSU, and supported by the superb administrative and logistical facilities of the WMO Secretariat. Individual programs and experiments were shaped by international panels of scientists, and boards of government representatives arranged for coordinated support. This structure was backed up within many of the major participating nations by similar committees. U.S. participation, for example, was guided by the U.S. Committee for the Global Atmospheric Research Program (USC-GARP) of the National Research Council. Thus, a unique and highly effective partnership was forged between a broad base of fundamental science and the solidly established international network of operational weather services. More complete accounts of GARP's genesis have been given by Malone (1968), García (1969), O'Neill (1971), O'Neill and Sargeant (1971), Bolin (1971), and Reed himself (1969).

4. From professor to executive scientist

The USC-GARP was established in March 1968 at the invitation of the U.S. government, with Jule Charney as chairman and such figures as Joseph Smagorinsky, Verner Suomi, Richard Goody, Gordon MacDonald, Thomas Malone, Norman Phillips, and Herbert Riehl among the membership (Table 7.1). A network of liaison members assured coordination with and support from the government. In all, the committee's roster constituted a "who's who" of American meteorology of the day.

The committee's charge was daunting: 1) to develop scientific objectives, specify observational requirements, and evaluate technological feasibility; 2) to effect coordination between the government and the scientific community; 3) to advise on detailed project design, logistics, and field work; 4) to serve as the U.S. link with international activities; and 5) to study education and manpower needs.

As this apparatus took shape, it became clear to U.S. leaders that a full-time senior "point man" would be needed to move GARP from dream to reality by executing the committee's plans. Tom Malone recalls that it was through Bob Fleagle that Dick Reed's name first came to the fore. But for a program directed equally at theory and forecasting, he was an obvious choice. Moreover, he was ripe for a well-earned sabbatical. Dick himself was motivated not only by the scientific promise of the program, but also by its potential to rebuild links with other nations, notably Russia. At any rate, the deal was closed rapidly, and the July 1968 issue of the *Bulletin of the AMS* announced that Dick would take a one-year leave of absence from the University of Washington to serve in what Dick has termed a "strange post"—*executive scientist* for the committee. The committee's work was housed in the National Academy of Sciences in the offices of the Committee on Atmospheric Sciences, then managed by the astute and dedicated John Sievers as executive secretary.

The duties of "executive scientist" were described as "coordinating the GARP committee's many activities and...carrying out its directives and duties assumed on

TABLE 7.1. U.S. Committee for the Global Atmospheric Research Program.

Chairperson Jule G. Charney	Massachusetts Institute of Technology
Alfred W. Firor	National Center for Atmospheric Research
Robert G. Fleagle	University of Washington
Richard M. Goody	Harvard University
Douglas K. Lilly	National Center for Atmospheric Research
Gordon J. F. MacDonald	University of California, Santa Barbara
Thomas F. Malone	Travelers Insurance Company
Walter H. Munk	University of California, San Diego
Owen M. Phillips	The John Hopkins University
Herbert Riehl	Colorado State University
Vice Chairperson Joseph Smagorinsky	Environmental Science Services Admin.
Henry M. Stommel	Massachusetts Institute of Technology
Vice Chairperson Verner E. Suomi	University of Wisconsin-Madison
James A. Van Allen	University of Iowa
Executive Scientist Richard J. Reed	National Research Council
Executive Secretary John R. Sievers	National Research Council
Ex-officio members	
John R. Borchert	University of Minnesota
E. L. Goldwasser	National Accelerator Laboratories
Merle A. Tuve	Carnegie Institution of Washington
Invited participants	
George S. Benton	Environmental Science Services Admin.
Homer E. Newell	National Aeronautics and Space Admin.
Randal M. Robertson	National Science Foundation
John W. Townsend	Environmental Science Services Admin.
Robert M. White	Environmental Science Services Admin.

behalf of the atmospheric sciences." The precise chronology of his activities is shrouded in the mists of time, and the archives of the academy. However, I recall from my own years as one of Dick's several successors—who also included Douglas Sargeant, David Rodenhuis, and Mike Hall—the bulging files memorializing those early years, as well as my own busy round of meetings and reports as I tried to fill Dick's large shoes and follow his rapid footsteps. In practice, he served as the full-time channel between the mercurial genius of Jule Charney and the exceptional core group of USC-GARP leaders—notably cochairmen Smagorinsky and Suomi—more attuned to practical implementation. His work also crucially depended on establishing and maintaining effective liaison with the supporting federal agencies—particularly the National Oceanic and Atmospheric Administration (NOAA) and the National Aeronautics and Space Administration (NASA), primarily focused on providing the hardware, and National Science Foundation (NSF), which funded much of the university-based research.

Whatever Dick did, the results were clear. With strong planning offices in place in both Washington and Geneva (not to mention a vigorous and growing network in other countries), GARP planning rapidly accelerated with close coordination between the United States and international efforts. Six U.S. scientific working groups on broad topic areas were set up: predictability and data requirements, planetary boundary layer, internal atmospheric turbulence, structure of the tropical atmosphere, cumulus convection, and large-scale, long-period, air–ocean interactions.

Within a year, the USC-GARP developed a slim blue volume proposing an ambitious scientific framework for U.S. contributions to the program (U.S. Committee for GARP 1969). Formally, this seminal document was authored by the illustrious parent committee. However, a smaller band—Jule Charney, Doug Lilly, Bruce Lusignan, Verner Suomi, and of course Dick Reed—were the primary hands on the collective pen. To my eye, the conciseness, rationality, and lucidity of the document suggest that Dick had a larger role in its final shape than he modestly acknowledges.

Even at this distance in time, the scope and audacity of this document is breathtaking. It boldly lays down requirements for global observations with a scope, resolution, and accuracy far beyond the then-operational system.

(i) *Horizontal resolution:* 400 km in extratropical latitudes (not specified for Tropics).
(ii) *Vertical resolution:* eight data levels, with two at low levels (50 m and 1 km in midlatitudes), three in the middle and lower stratosphere (10, 50, and 100 mb), one at 200 mb, one at around 500 mb, and one at around 700 mb.
(iii) *Frequency:* once or twice daily.
(iv) *Variables and errors of measurement:* wind components ± 3 m s^{-1}, temperature $\pm 10°$C, pressure $\pm 0.3\%$ (3 mb at surface), vapor pressure ± 1 mb, and sea surface temperature $\pm 0.25°$C.

The document then proposes a composite observing system of polar and geostationary satellites, rawinsondes, balloons, buoys, aircraft, etc., to provide them. A number of subprograms were defined to lay the groundwork for model improvement and to prepare for a global observing experiment. Particular emphasis was placed on filling observational gaps in the Pacific, studying key processes in the Tropics, and improving numerical modeling and analysis methods.

This *Plan for U.S. Participation in the Global Atmospheric Research Program* (U.S. Committee for GARP 1969) was presented at a major conference organized at the National Center for Atmospheric Research in October 1969. Photographs taken at that event confirm my own recollections as an awestruck U.S. Air Force observer: It was a unique convocation of the elite of American meteorology, and the organizers took care to ensure that linkages with government and politics were not neglected.

The framework developed in the pressure cooker of those few months of 1968–69 guided the U.S.—and to a great extent the world's—efforts in GARP for the next decade or more. An ICSU–WMO Planning Conference held in Brussels, Belgium, early in 1970 had the benefit of the U.S. document, and developed a crucial international consensus on the shape of the program (International Council of Scientific Unions and World Meteorological Organization 1970). As elaborated by the international Joint Organizing Committee, the overall program included subprograms dealing with the monsoons, the polar regions, air–surface interactions, mountain effects, and climate dynamics. These were designed to support the overall goal of developing improved global models for weather and climate. However, it was clear at that point that GARP would initially focus on two core activities: preparation for the First GARP Global Experiment (FGGE), and execution of a tropical observing experiment to fill what was perceived as the greatest gap in understanding of atmospheric behavior.

5. The Tropics beckon

With GARP steaming ahead on a clear course, Dick returned to Seattle, and Douglas Sargeant took his place at the NRC. But Dick was not to remain aloof from GARP. It had long been recognized that atmospheric processes in the Tropics posed special problems for development of global models. Observational programs such as the Barbados Oceanographic and Meteorological Experiment (BOMEX) and the Line Islands Experiment had obtained intriguing clues, but there was a need for a comprehensive set of research data. In particular, the processes governing the formation of "cloud clusters," and the evolution of easterly waves and tropical cyclones presented many mysteries. Thus, a focused tropical observing and research effort in the Tropics was a major element in both the U.S. and international plans. Scarcely a year after leaving the academy, Dick found himself as chairman of an Ad Hoc Tropical Task Group of the USC-GARP, the task of which was to develop a revised and expanded version of the tropical experiment sketched in the committee's initial plan. By 1971, planning for a tropical experiment had largely crystalized, and his task group published its *Plan for U.S. Participation in the GARP Atlantic Tropical Experiment* (U.S. Committee for GARP 1971).

The GARP Atlantic Tropical Experiment (GATE) took place in June–September of 1974 in an experimental area in the tropical Atlantic Ocean between Africa and South America. The observational program involved 40 research ships, 12 research aircraft, and numerous buoys from 20 countries, directed by an International Project Office located in Senegal. The data collected were processed by the participating nations and disseminated throughout the world.

Not surprisingly, because of his strong interests in tropical processes, Dick Reed personally participated actively in this effort, serving in essence as the Joint Organizing Committee's on-site representative in Dakar, and contributed much to the subsequent research. His accomplishments in this arena are described in another chapter of this monograph (Burpee 2003). Suffice it to say that research using these data still goes on today. It is estimated that over a thousand papers have been published based on the data collected during this short period in 1974, and the scientific results of the program have been documented in a series of national and international reports.

6. The Global Weather Experiment

The other elements of the plans developed with Dick's leadership and support moved into fruition at varying paces. The crowning achievement of GARP, the First GARP Global Experiment—later more euphoniously rechristened by Bob White as the Global Weather Experiment (GWE)—took place in 1979, after a number of postponements due to slippages in satellite and in situ observing systems. The weather services of the 170 WMO member nations, many space agencies, and a drove of research organizations participated. The observing program was designed through extended series of observing system simulation experiments using the best models of the time. Geostationary and polar-orbiting satellites covered the earth, special observing equipment was deployed on commercial aircraft, reconnaissance aircraft deployed dropsondes to cover observational gaps over the oceans, and a fleet of buoys plied the seas. The resulting masses of data were processed and analyzed through a complex network of institutions and methods over the next decade, using highly innovative techniques of model-based data assimilation. The results of GWE laid the foundation of the global system of geostationary and polar-orbiting satellites that now comprise the space-based observing system of WMO's World Weather Watch.

The GWE dataset is regarded as the most comprehensive compilation of meteorological variables ever assembled, and has provided the basis for an immense body of research and development to support operational weather forecasting. This research rapidly bore fruit. By the mid-1980s, an international assessment concluded that improved data, plus increases in computer power, had "led to substantial improvements in the range and accuracy of weather forecasts." (Joint Scientific Committee 1985). Forecasts in midlatitudes of the Northern Hemisphere were deemed to be reliable out to 4–5 days ahead, with useful guidance up to 6–7 days — an advance of about 3 days in predictive skill over the pre-GWE era. The GARP planners had promised much — and had delivered!

7. Breaking barriers

But let us now backtrack to follow our hero in the affairs of our profession. In 1971, Dick was anointed as president-elect of the AMS, and — in the hallowed tradition of this august institution — assumed office in January 1972. In the same month, another President, named Richard Nixon, announced a historic breaking of the ice that had long entombed U.S. relations with China. Dick immediately saw an equally historic opportunity for meteorology, and wrote to Nixon (and Henry Kissinger) suggesting that the time was ripe for a renewal of scientific exchanges and cooperative scientific endeavors between the U.S. and Chinese meteorological communities. He cited the traditional collaboration of meteorologists worldwide in observational networks, operational services, and research, and proposed that the AMS could send a small delegation to follow in Nixon's footsteps.

While the official U.S. government response was not encouraging, the State Department, NOAA, and eventually the WMO did suggest that the AMS could profitably pursue the idea directly through its own channels. Thus, Dick wrote to the Chinese Academy of Sciences, citing the "people to people" contacts advocated in the talks between Nixon and Chairman Mao Tse-Tung, and expressing the desire to send a delegation to establish friendly relations and to achieve scientifically productive contact and exchange. Exploratory letters and contacts followed, and in 1973 positive official responses and an invitation were transmitted by the Chinese. Their openness to the proposal doubtless owed much to Ken Spengler's long cultivation of AMS international connections in the course of his many years of faithful participation in WMO Executive Committee meetings and Congresses.

By this time, several presidents of the AMS had been involved in the project. The final delegation consisted of Dick Reed, who had initiated the invitation; Will Kellogg, his successor; Dave Johnson, the then-current president; Dave Atlas, the president-elect; the indispensable Ken Spengler; and — at the kind invitation of the Chinese — their wives. At Chinese expense, the group visited meteorological organizations and facilities in Beijing, Canton, and Shanghai, and generally had a wonderful time. The visit was extensively documented in the *Bulletin of the AMS* in an article largely penned by Will Kellogg and illustrated by Thor Kellogg's excellent photographs (Kellogg et al. 1974). The following year, a corresponding Chinese delegation visited the United States. Parenthetically, I recall a reception at the Chinese mission in Washington in honor of the delegation, at which the visitors were clearly astonished that *all* the American guests employed chopsticks expertly! Through Dick's daring and imaginative initiative, the AMS became the first U.S. scientific society to reestablish working relationships with the Chinese scientific establishment.

8. From Braintree to the globe

Many more snapshots of Dick Reed's activities in the international environment could be cited. He has been no stranger to institutions such as the European Centre for Medium-Range Weather Forecasts, for example. These, together with the examples discussed at some length above, illustrate his consistent vision of meteorology as a global endeavor, and his conviction that the theory of atmospheric science and the practice of operational forecasting must move ahead hand in hand through concrete, practical actions.

The atmospheric sciences — and indeed world science as a whole — have advanced immeasurably since the days when Dick was phoning in weather forecasts to a Quincy, Massachusetts, radio station. Much of this progress is very clearly linked with the international efforts in which he played a seminal role. Within our own craft, weather forecasting has evolved from a seat-of-the-pants black art guided by arm-waving qualitative notions to a scientifically based quantitative technology. The GARP Global Weather Experiment for which Dick's efforts provided an indispensable foundation was a milestone in this development. Assessing the impacts of the GWE, B.J. Mason concluded that,

> The best models now regularly produce good 4–5 day forecasts. . .with useful indications of major developments for a further 1–2 days ahead. This represents an increase of 2–3 days in predictive skill over the last decade. . . — (Mason 1986).

Before GATE, the dynamics of the tropical atmosphere — and ocean — were little better understood than in Hadley's day. The observing tools developed in GATE provided quantitative fuel for the engine of research, and paved the way for today's operational observing and prediction systems for El Niño and seasonal climate.

GARP fostered new linkages among the disciplines of the earth sciences. Early planning, it is true, focused narrowly on atmospheric problems. However, GATE needed ships and the oceanographers owned them. Dragged in, at first reluctantly, oceanographers found to their surprise that the supposedly uninteresting patch of the Atlantic chosen for GATE held in fact much of surprising interest to their science, for example, oscillations in the equatorial undercurrent. Moreover, meteorologists found that they could not understand the development of tropical weather systems without knowing what was going on in the underlying ocean. Before GARP, it was unusual to find meteorologists and oceanographers in the same room; after GARP, it was often hard to tell them apart! The subsequent evolution of GARP into the arena of climate has entrained still other disciplines, so that it is no longer unusual to find meteorologists, oceanographers, chemists, and even economists arguing over a table of refreshing fluids.

GARP richly accomplished its covert purpose — to rebuild the international scientific linkages wounded by war. GARP planning entrained a multitude of nations, and a veritable United Nations assembled in Dakar for GATE. When I worked on the GWE in Geneva, my boss was a Swede and my coffee-break companions were a Finn, an Australian, and a couple of Russians. On a parallel track, Dick's daring initiative toward China dramatically broke the ice with the then-Communist world as a whole. In the 1960s, Chinese and Russian scientists were seen in American institutions only as rather exotic curiosities; by the end of the 1970s, scientific exchanges were commonplace and American academic institutions were thronged with scientific collaborators and students from China and Russia. This evolution had to start somewhere and be consistently fostered over time. Dick's initiative to China was the "start button" for the U.S. scientific community, and GARP was one of the primary engines of its progress.

Finally, GARP effected a sea change in what I will call the sociological technology of science. Today, the machinery of attacking some large problem in the earth sciences is as well understood as the organization of a PTA fundraiser. One puts together a few workshops to develop interesting ideas; these are presented to some suitable umbrella organization such as the Joint Scientific Committee for the World Climate Research Program for a blessing, that august body commissions a study conference with broad participation to develop a sound scientific rationale and a supportive constituency, a program is formally recognized, a steering committee is formed, some friendly institution hosts a little secretariat and a Web site, and the program moves merrily on to stimulate a host of activities that would not have occurred as soon, or as well, or perhaps at all without it.

All this is the result of a process that the sociologists term "social learning" — the process through which societies develop and institutionalize capabilities to perceive, evaluate, and respond to problems (Social Learning Group 2001). I believe that the social learning that occurred through GARP is arguably its most important single legacy. Early in his term as president of ICSU, Sir John Kendrew remarked that the distinguishing characteristic of twentieth century science was its willingness to tackle large, complex, ill-posed — but important — problems that awkwardly fall in the untilled landscape between the fortresses of our academic disciplines. Building on the foundation to which Dick Reed contributed so significantly, GARP made major and lasting contributions to our global society's ability to deal with such issues. In my view, the remarkable effectiveness with which science has addressed messy problems such as ozone depletion, El Niño–Southern Oscillation (ENSO) prediction, and climate change derives directly from the arsenal of programmatic tools developed in GARP.

In conclusion, the "greatest generation" that includes Dick Reed and so many other pioneering leaders of our profession guided our science and our world with imagination, wisdom, and courage from the despair of the Great Depression through the horrors of world war, and then into a better world of scientific progress, global comity in attacking global problems, widely distributed prosperity, and growing freedom. They enriched our science, advanced our capabilities to deal with the global environment within which we live, and taught us how to focus and organize our work across political and disciplinary boundaries. Dick and his cohort need no granite memorials — their monuments surround and pervade our lives; but they do deserve our enduring respect and gratitude.

Acknowledgments. The author is grateful to Diane Rabson, UCAR, for making available tapes of an oral history interview of Dick Reed very ably conducted by Earl Droessler.

REFERENCES

Bolin, B., 1971: *The Global Atmospheric Research Programme*, ICSU/WMO, Geneva, Switzerland, 28 pp.

Brokaw, T., 1998: *The Greatest Generation*. Random House, 432 pp.

Burpee, R. W., 2003: Characteristics of African easterly waves. *A Half-Century of Progress in Meteorology: A Tribute to Richard J. Reed, Meteor. Monogr.*, No. 53, Amer. Meteor. Soc., 91–108.

Committee on Atmospheric Sciences, 1966: *The Feasibility of a Global Observation and Analysis Experiment*. National Research Council, 172 pp.

García, R. V., 1969: An introduction to GARP. GARP Publications Series, No. 1, World Meteorological Organization, Geneva, Switzerland, 22 pp.

International Council of Scientific Unions and World Meteorological Organization, 1970: Report on the Planning Conference on GARP, Brussels, March 1970. GARP Special Rep. 1, Geneva, Switzerland, 42 pp.

Joint Scientific Committee, 1985: International conference on the results of the global weather experiment and their implications for the World Weather Watch. GARP Publications Series, No. 26, World Meteorological Organization, Geneva, Switzerland.

Kellogg, W. W., D. Atlas, D. S. Johnson, R. J. Reed, and K. C. Spengler, 1974: Visit to the People's Republic of China: A report from the AMS Delegation. *Bull. Amer. Meteor. Soc.*, **55**, 1291–1330.

Malone, T. F., 1968: New dimensions of international cooperation in weather analysis and prediction. *Bull. Amer. Meteor. Soc.*, **49**, 1134–1140.

Mason, B. J., 1986: Recent improvements in numerical weather prediction. *Proc. Int. Conf. on the Results of the Global Weather Experiment and Their Implications for the World Weather Watch*, Geneva, Switzerland. Global Atmospheric Research Programme and WMO–ICSU Joint Scientific Committee, GARP Publications Series 26, WMO Tech. Doc, 107.

O'Neill, T. H. R., 1971: GARP chronology. *Bull. Amer. Meteor. Soc.,* **52,** 879–885.

——, and D. H. Sargeant, 1971: GARP—The evolving fabric of an international program. *Bull. Amer. Meteor, Soc.,* **52,** 1082–1088.

Perry, J. S., 1975: The Global Atmospheric Research Program. *Rev. Geophys. Space Phys.,* **13,** 661–795.

——, and T. H. R. O'Neill, 1979: The Global Atmospheric Research Program. *Rev. Geophys. Space Phys.,* **17,** 1753–1762.

Reed, R. J., 1969: A guide to GARP. *Bull. Amer. Meteor. Soc.,* **50,** 136–141.

——, 1971: The effects of GARP and other future large programs on education and research in the atmospheric sciences. *Bull. Amer. Meteor. Soc.,* **52,** 458–462.

Social Learning Group, 2001: *Learning to Manage Global Environmental Risks.* vols. 1 and 2, The MIT Press, 583 pp.

U.S. Committee for GARP, 1969: *Plan for U.S. Participation in the Global Atmospheric Research Program.* National Research Council, 172 pp.

——, 1971: *Plan for U.S. Participation in the GARP Atlantic Tropical Experiment.* National Research Council, 25 pp.

World Meteorological Organization, 1967: *Report of the study conference on the Global Atmospheric Research Program.* WMO, jointly organized with the International Union of Geodesy and Geophysics, Geneva, Switzerland.

(left) Fred Sanders, Master of Ceremonies;
(right) Mike Wallace, Banquet Speaker

Dick Reed

Dick Reed, Dick Johnson, and Joan Reed